MITTELBACH
P9-CFU-658

Management Systems

A Global Perspective

STEVEN CAVALERI
Central Connecticut State University

KRZYSZTOF OBLOJ
University of Warsaw

Wadsworth Publishing Company
BELMONT, CALIFORNIA
A DIVISION OF WADSWORTH, INC.

ASSISTANT EDITOR: **Marnie Pommett**
PRODUCTION: **Helen Walden**
DESIGN: **Eve Mendelsohn Lehmann**
MANUFACTURING: **Lisa Flanagan**
COMPOSITION AND ART: **Graphic Composition**
PRINTING AND BINDING: **Arcata/Martinsburg**

Printed in the United States of America
2 3 4 5 6 7 8 9—97 96 95 94 93

This book is printed on recycled acid-free paper

Library of Congress Cataloging-in-Publication Data
Cavaleri, Steven.
Management systems: a global perspective / Steven Cavaleri, Krzysztof Obloj.
p. cm.
Includes bibliographical references and index.
ISBN 0-534-92511-1
1. Management. 2. System theory. I. Obłój, Krzysztof.
II. Title.
HD31.C375 1993
658—dc20 92-33146
CIP

Dedicated with love to our wives,
Linda and Zofia

Preface

If there is a single theme for this book it is that the systems approach has value for all managers, not just those with technical backgrounds or those with mathematics and computer skills. The purpose of this book is to introduce the major systems concepts and applications in a way that will create pragmatic benefits for the greatest number of students and managers.

The impetus for writing this book arose from two main factors. First, we shared the belief that systems theory is best understood when presented together with timely examples of how those theories are actually used in organizations. Second, we believe that both systems theories and the world have changed significantly since the early work on systems approaches was done in the 1950s. There is a need to illustrate the various ways existing systems theories for managing have been refined, and to examine newer approaches such as Checkland's 'Soft' Systems Methodology, Hannan and Freeman's population ecology model, Senge and Diehl's "Microworlds," and Prigogine's perspective on dissipative structures.

The business environment has changed in fundamental ways over the past decade. Innovations in communication and other technologies have made the global business environment more relevant for a greater number of organizations. The changing landscape of economic and political relations around the globe are more closely tied than ever to business decisions. The complexity of business has grown dramatically as a consequence of these changes. We believe that the systems approach offers extraordinary power to managers as a tool for understanding the complexity arising from the interaction among these many forces and factors.

While we have not attempted to include comprehensive coverage of all systems approaches to managing, we have examined those concepts we think have the greatest potential value to those interested in both theory and practice. We believe that readers will appreciate the variety of perspectives and diversity of examples included in the book.

This book was written for use in courses (most often titled Management Systems) that adopt a systemic approach to managing. The book is also appropriate for Organization Theory courses. The systems approach can provide a unifying theme to the course, and may shed a new light on many of the topics customarily taught in organization theory. This text is also useful as a source book for a seminar or special topics course.

Management Systems: A Global Perspective can be used at both the undergraduate or graduate level, depending on the goals of the instructor. The comprehensive cases at the end of the book provide an additional challenge to students in those settings where more rigorous analysis and discussion are appropriate.

The book is organized into four parts:

Part I Introduction
Part II Systems Thinking: Basic Approaches
Part III Systemic Tools
Part IV Integrative Approaches and the Future

Part I is the starting point for most instructors in the management systems course, after which the instructor may wish to reverse the order of parts II and III or combine parts II and IV before concluding with Part III. Part III includes coverage of traditional managerial functions and may be used as an effective follow-up to Part I. Instructors who are very familiar with sociotechnical systems or system dynamics may want to cover them immediately after Part II to strengthen the connection between basic systems approaches and these applications. Additional ideas for using the book are included in the instructor's manual, which includes test questions, chapter summaries, exercises, case notes, and transparency masters.

Three main pedagogical features distinguish this book from others:

1. It integrates a diverse yet well-established body of theory into a single model, the Management Systems Model. This model allows students to visualize how the various systemic threads of managing are woven into a single, whole concept.
2. The Management Systems Model combines theoretical and practical dimensions. The fundamental roots of systems theory, external-internal organizational alignment, and the balance between the social and technical dimensions of organizations, are integrated with the systemic managerial tools.
3. A global orientation is woven throughout the text as well as in a chapter dedicated to this topic (Chapter 2, The Global Systemic Framework). The global theme is reinforced by global cases, some addressing contemporary issues in eastern Europe.

The book includes *System Notes* on topics of interest, and *System Profiles,* which describe contributions of people influential in shaping the field. The comprehensive cases are based on firms whose influence is felt around the globe: General Electric, International Business Machines, and Toyota. Each case was written specifically for this text, and focuses on the applications of one of the systemic managerial tools in practice. In addition, each chapter includes chapter objectives, end-of-chapter discussion questions, and a list of key terms and concepts.

~ ACKNOWLEDGMENTS

Many people contributed to this project in diverse but important ways. We wish to thank several former students for their contributions to this book: Daniel Bach, Hibeh Bakr Dilzer, Glen Colley, Luke Giroux, and Lisa Mattei. We also offer special thanks to Robert Rarus of the U.S. Department of Labor for his editorial comments and insights into ways of improving the comprehensive cases.

Krzysztof Obloj wishes to thank the following people for their cooperation, research, and critique, which have made this book possible: Andrzej K. Kozminski, Professor of Management at The Warsaw University, a friend and mentor throughout; Pat Joynt of the Norwegian School of Management; and Donald Cushman of SUNY—Albany.

Steven Cavaleri expresses his gratitude to Dale S. Beach, Professor Emeritus at Rensselaer Polytechnic Institute, for his encouragement and for serving as a role model; and Peter Senge and John Sterman of the MIT System Dynamics Group, for profoundly shaking his ideas concerning systems.

We also thank the following reviewers:

David Fearon
Central Connecticut State University

William Murin
University of Wisconsin—Parkside

Conrad Kasperson
Franklin and Marshall College

Jim Stoner
Fordham University

Brian Kleiner
California State University—Fullerton

We also express our gratitude to the many family members and friends who have offered their encouragement and support over the years since this project began.

Finally, such a project could not become reality without the inspiration and follow-through from the people at PWS-KENT Publishing. In particular, we wish to gratefully acknowledge Rolf Janke for his enthusiastic support and clear vision of what he believed this book could be. We also wish to thank an unsung hero who has contributed to this book in so many ways with her professional dedication and work ethic: Marnie Pommett, assistant editor, has been a beacon in the night as we traversed the sometimes turbulent seas of publishing a book.

Contents

~ PART ~

I Introduction

THE SYSTEMS PERSPECTIVE of managing includes several major approaches, yet all of them are based on two fundamental elements: developing an understanding of the concept of a system and its properties, and mastering systems thinking. Because that first element is essential to understanding the systems perspective, the first part of this book is devoted to defining the concept of a system. The emphasis in this part is placed on how managers can use this knowledge to take action. Chapter 1 provides a foundation of the concepts, terms, and basic philosophies underlying systems and systems thinking. By understanding the basic properties of systems, managers may gain insight into the inner workings of organizations. The lessons they may learn about how systems actually work can reinforce their ability to think systemically. Systems thinking is an active process of applying systems concepts in problem solving, planning, and otherwise doing the work of managing. A knowledge of systems concepts addresses the *what* of managing; systems thinking is concerned with the how. This book will examine both of these dimensions of the systems perspective in detail.

Managers are increasingly aware of the changing nature of the global business environment. Each manager may have his or her own view of how these changes will impact work responsibilities, but in order to understand fully how his or her piece fits into the overall puzzle, it is necessary to gain a broader view of the situation. Chapter 2 examines the fundamental systemic forces that are transforming the global business environment of the 1990s: new economic patterns, high-speed communications, changing structures in trade blocs, and the globalization of competition. The essence of this book is to help the reader to understand and appreciate the help that the systems perspective can provide for managing in a global environment.

These changes in the global business environment represent one example of how systems can be transformed over time by the interaction of major forces. In general, systemic change forms recognizable patterns. By recognizing these patterns managers may identify many of the common root causes of changes in a system's behavior. Chapter 3 explores how systems change and how these changes are relevant to the process of managing. The primary mechanisms that create systemic change, feedback and time delays, are examined in this chapter. By appreciating how these mechanisms work, managers may more thoughtfully experiment with ways to improve organizational performance. The systems perspective can offer managers insight into how a system works, but it provides no guarantee of success. In highly complex systems, even the most thoughtful actions can create unintended results.

The problems of managing complex organizations are not new. For centuries philosophers and scientists have been thinking about ways to solve the problems of humankind. A continuous thread of systemic thinking in these disciplines can be traced back to the earliest recorded philosophers. Great thinkers such as Aristotle and Plato, in Western thought, and Lao-tze, in the Eastern tradition, have considered these issues and provided many lessons for the modern manager. While people may not normally associate philosophy with managing, the connections are clear and unmistakable. Chapter 4 traces systems thinking from its roots in the earliest philosophies to modern views. Philosophy deals with how people think about the issues that are of concern to them. From this point of view, what could be more important to managers than understanding the perspective from which they see their work and give meaning to their experience?

▲

1 Systems Thinking

A system is the simplest . . . experience we humans can have. A system must always have insideness and outsideness. Recognition of a system begins with initial discovery of either self or otherness.

Buckminster Fuller (1992), p. 124

1. To understand how rational-analytical thinking and systems thinking can complement each other.
2. To define reframing and explain its benefits for managers.
3. To comprehend the role of causality in distinguishing rational-analytical thinking from systems thinking.
4. To appreciate the role of mental models in systems thinking.
5. To understand the concept of a system and how it serves as the basis for systems thinking.
6. To explain the basic functions of human systems.
7. To explain what is meant by system boundaries and system hierarchy.
8. To learn how to define a system.
9. To recognize the differences between static and dynamic systems.
10. To describe the primary role of systemic structure in relation to system behavior.

SECTION I
THE SYSTEMS VIEW OF THE WORLD

THE SYSTEMS PERSPECTIVE

Charles Dickens's novel *A Tale of Two Cities* begins, "It was the best of times, it was the worst of times." This phrase may well describe the feelings of a growing number of managers about the present. Shifts in the global economic balance of power have proven beneficial to some organizations and disastrous for others. In North America and Europe there is a growing realization that many of the concepts that have served as the standard for management practices in the past are no longer effective in meeting the managerial challenges of the 1990s. On the other hand, there is increasing cause for optimism due to advances in the use of innovative management approaches such as total quality management (TQM), organizational learning, lean manufacturing systems, and systems thinking. These approaches are being fine-tuned by world-class businesses, consultants, and academics at an ever-growing pace. Consequently, the range of alternative perspectives for managing is being stretched as never before.

One effect of these changes is that managers are faced with an increasing number of choices regarding ways to manage. Managers may ask themselves the following questions: Should I change the way I manage? If so, how fundamentally do I want to change my current style? Which management practices will I adopt? These questions may be exceedingly difficult for these managers to answer. Even in the best of times, it is not easy to forsake the tried and true for newer, less familiar ways of managing. Changing the way one manages is not as simple as wearing a new hat; it requires the investment of time in learning new concepts and techniques. The difficulty inherent in change is compounded by the fact that large numbers of organizations have down-sized and now operate with fewer managers than before. At the same time, organizations striving to remain competitive in global markets are taking stringent cost-cutting measures. The net effect of growing global competition is that organizations of the future must accomplish more with fewer people. The managers of these global businesses are likely to find more opportunities and rewards than ever before for managing in novel ways.

Virtually every industry has been rocked by the transition from national to global industries. Global industries are characterized by a few organizations controlling a large share of the market. Much like the process of natural selection in wildlife populations, only the fittest will survive. In the past—in the time of regional and national economies—being fit meant being efficient. In the global economy, fitness is a multidimensional idea. The organizations that will survive must create high-quality products and services, design innovative new products and processes, be more efficient, and learn—all at an ever-increasing speed. Managers who identify ways to help organizations become increasingly fit at a faster pace will be making a substantial contribution to the survival of those organizations.

One way to achieve those four goals of fitness is for managers to visualize organizations from a fundamentally different perspective by mastering systems thinking. The systems perspective differs from the many other schools of thought that have developed over the years in the field of management. These other perspectives—the scientific, administrative, process, human relations, and contingency approaches to management—all are action oriented; they do not explicitly consider managerial thought processes. Systems thinking, however, is based on the assumption that a manager's actions cannot be separated from the way he or she thinks.

By focusing on the mind of managers in relation to action, systems thinking can address situations on a more rudimentary level than can be achieved with those approaches biased toward considering action alone. Attempts to apply theories without thought are lame. Actions taken without a conceptual base are superficial. On the other hand, systems thinking without action is futile. Systems thinking connects thoughts and actions by linking them in a circle of causality. Thoughts must precede action, and through the mechanisms of reflection and discussion, actions come to influence future thinking. The result is portrayed graphically in Figure 1-1.

Systems thinking is based on a philosophical foundation different from that underlying most other management approaches. To effectively use systems thinking means to integrate an additional set of assumptions about how things work into the way one manages. Through this integrative process, one's capacity for managing in novel ways can become more robust, and one's thinking can become more balanced.

SYSTEMS THINKING

During the twentieth century, managers have become highly competent at using a specific style of thinking to address the problems found in organizations. Evolved in Europe during the eighteenth century, this style is known as **rational-analytical thinking.** Rational-analytical thinking is, specifically, a problem-solving approach and, generally, a way of viewing the world. Its logic suggests that complex situations can best be understood by breaking them apart into simple components. The underlying assumption is that by breaking apart the elements of a system into ever smaller, simpler parts, complexity will be reduced to levels where the causes of problems will become evident. Then problems will be easier to understand and solutions easier to develop.

The value of this style of thinking was first recognized by scientists in the eighteenth century. Later it was adopted by social scientists and managers because it could easily be learned and could be used quickly. By now rational-analytical thinking has become deeply embedded in the thinking style of people in Western cultures. Over the past century, traditional American and

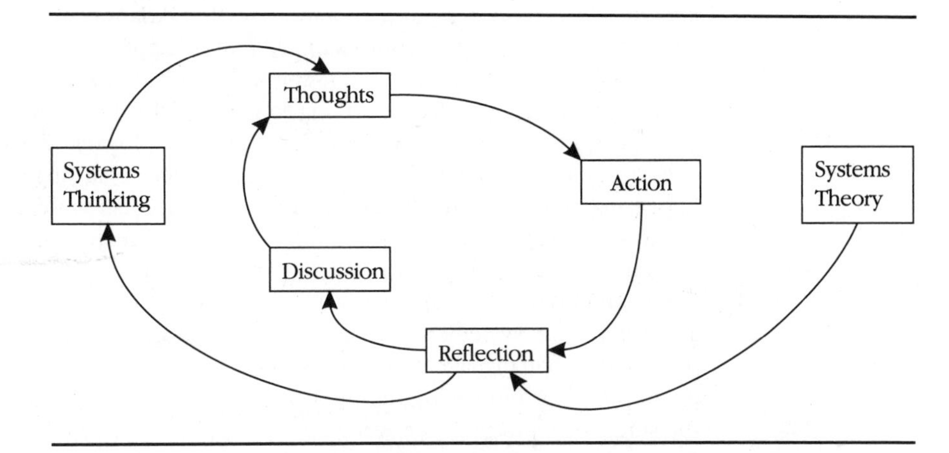

FIGURE 1-1 ***Cycle of Purposeful Action in Organizations***

European management approaches have tended to emphasize this. Despite its many advantages, this style of problem solving also has numerous limitations (these will be discussed in detail in Chapter 4). Among the most serious limitations is that it is not very useful in explaining how complex systems work. Systems thinking has considerably more power for this type of use (Ackoff, 1991).

There are many advantages to being able to think from both a systems perspective and an analytical view. Systems thinking is a complement to, not a replacement for, rational-analytical thinking. **Systems thinking** is a way of understanding the causal relationships between the parts of a system. It is based on a fundamental set of notions, which proposes that systems behavior is the result of the way a system is structured and circular patterns of cause-and-effect relations, often referred to as "vicious cycles" or "virtuous cycles." When both systems thinking and rational-analytical thinking are used together, they offer the potential to gain incomparable breadth of insight into how and why complex systems work as they do. When these ways of thinking are used in tandem, managers are increasingly likely to develop a more balanced perspective. This is one of the contributions of systems thinking to managers—to offer the broadest possible panorama of a system while still retaining the ability to examine details in a microscopic fashion. In effect, it means enabling people to see *both* the forest *and* the trees. The purpose of this book is to explore systems thinking as the mirror image of rational-analytical thinking. Furthermore, it will also explore the merits and limitations of the various systems approaches.

~ APPLYING THE SYSTEMS PERSPECTIVE

The use of systems approaches for managing dates back to the development of military strategies by British and American researchers during World War II. Today, systems thinking is being applied by managers throughout such well-known organizations as Apple Computer, Bell Atlantic, Federal Express, and Ford Motor Company (Hendrick, 1992). Such approaches show promise for changing the way managers intervene in organizations. Growing numbers of managers are using systems thinking to find novel ways to manage in an effort to build on the creativity and intelligence of employees. These new ways of thinking often challenge the conventional wisdom regarding how things work in organizations and which management styles are most effective.

Systems thinking focuses on identifying the relationships between the parts of a system. When managers use systems thinking, they gain insight into how changing these relationships may affect the behavior and performance of a system. Since systems thinking is founded on a different set of assumptions than traditional management, its use may enable managers to visualize new possibilities for improving the performance of a system. Systems thinking sets the stage for managing in novel ways by providing a way to reframe situations. **Reframing** is changing the way one thinks by replacing one set of mental references with another to create new meaning in a situation. When managers change their thinking in such a way that they can see a system in a new way and make sense of what had previously been contradictory, they have reframed it (Quinn, 1988). A manager's creative potential is closely related to his or her ability to reframe situations. A prime example of reframing that resulted in numerous breakthroughs is Chaparral Steel. Chaparral Steel has become a model of innovation by building the mental capabilities of its employees.

Chaparral Steel: An American Mentofacturer

Mentofacturing is a word to replace manufacturing in the English lexicon. Whereas manufacturing means "made by hand," mentofacturing means made by the mind. It describes the transformation currently underway in the leading, technologically sophisticated firms in America. Mentofacturing companies will look more like software companies, technology driven and operated by technicians and thinkers. Most importantly, they will be learning organizations which emphasize cognitive skill development.

Chapparal Steel Company, a state-of-the-art steel company, produces medium-sized steel products for sale to construction, automotive, railroad, mobile home, defense, appliance, and other firms. Chaparral was incorporated in July 1973 with the mission of becoming the international low-cost producer of high-quality construction and industrial grade steel products. It began operations in 1975 as a steel mini-mill with annual capacity of 22,000 tons. With a major plant expansion in 1982 and continuous design and application of mentofacturing techniques, the firm's current annual capacity is 1.5 million tons.

For the traditional integrated operation, several days are required to produce finished steel products. In Chaparral's mentofacturing process, products are finished in a few hours. Achieving success in implementing Chaparral's business strategy through mentofacturing requires a fit among primary design variables, including structure and operating characteristics, management, and workforce attributes and human resource policies and practices. [See Figure 1-2.]

Chaparral Steel has a unique culture which has its roots in the values and beliefs of the founding team. The five central values and beliefs of the founders were: 1. trust in people to be responsible, 2. take risks for achievement and success, 3. challenge to grow in knowledge and expertise, 4. be open to learn and to teach, 5. make work fun and pleasurable.

Chaparral Steel was built from the ground up. There was no existing culture or technology which had to be changed or modified. The founders did not accept the constraints and assumptions of conventional steel industry wisdom. They viewed the acceptance of industry constraints as part of the industry's bureaucratic pathology. Instead, they were guided by their imaginations, vision, and creativity.

Source: From G. Forward, D. Beach, D. Gray, and J. Quick. *The Executive,* Ada, Ohio (August 1991), pp. 32–44.

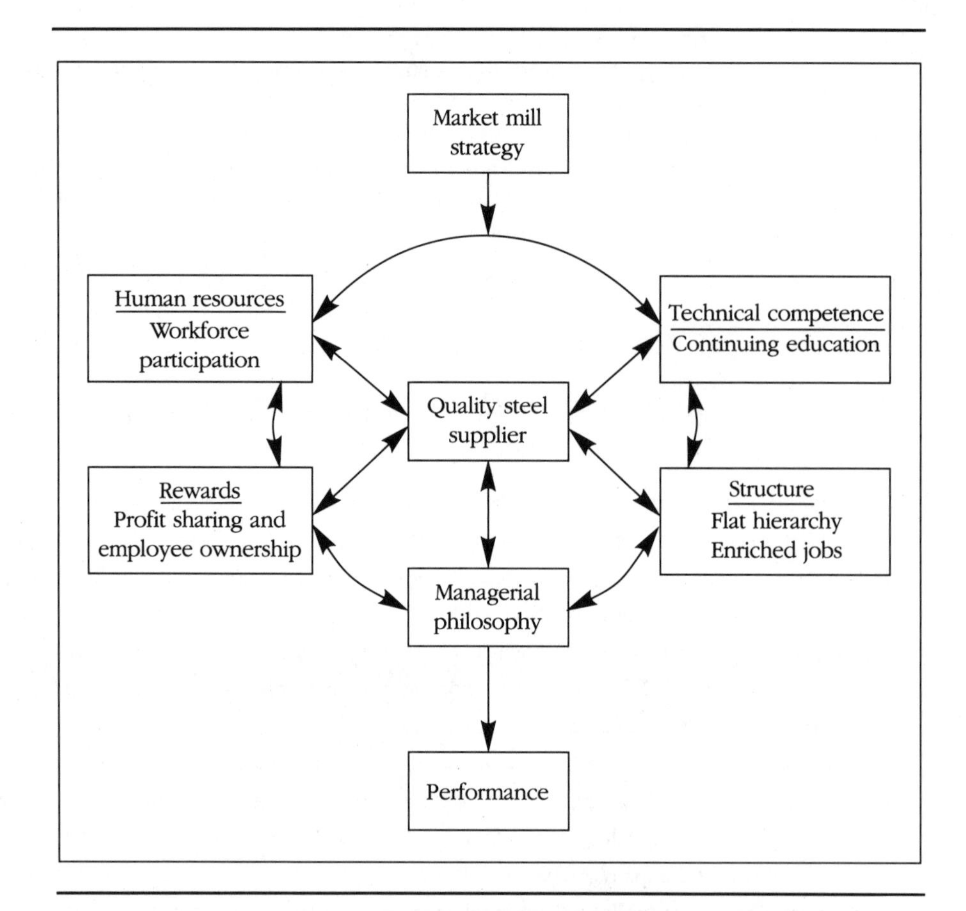

FIGURE 1-2 ***Factors Chaparral Blends to Remain Competitive***
Source: From G. Forward, D. Beach, D. Gray, and J. Quick. *The Executive,* Ada, Ohio (August 1991), pp. 32–44.

Mentofacturing at Chaparral Steel is based on the idea of gaining competitive advantage through challenging the assumptions that long were the legacy of its industry and developing unique ways of thinking about the world. As the minicase suggests, how people think and learn may be the ultimate competitive resource. Ray Stata, Chairman of Analog Devices, amplifies this point, "I would argue that the rate at which individuals and organizations learn may become the only sustainable competitive advantage, especially in knowledge-intensive industries" (Stata, 1989). Throughout this book, you will read many variations of this basic theme, that organizations can gain competitive advantage by creating new patterns of thinking. In particular, this book will explore the potential of systems theory and systems thinking as human technologies for managing organizations. This chapter will examine the relationship between systems theory, systems thinking, and managing. It will survey the various properties of systems and consider their implications for managing.

SECTION 2
MANAGING FROM A SYSTEMS PERSPECTIVE

SYSTEMS THINKING AND MANAGING

Over the past century, the systems approach has become associated with computer science, engineering, mathematics, and production or operations management (Churchman, 1968). On the surface, there is little evidence to suggest that systems approaches are any more relevant to today's manager than they were fifty years ago. From another vantage point, however, systems thinking may be seen as a thought process rather than as a technique for solving problems. In a broad sense, systems thinking can serve as a new perceptual lens that provides an alternative set of meanings to one's experience. In this context, systems thinking can be a sophisticated managerial tool. Managers can use it to put many of the subtleties of organizational reality in a new light. These subtleties, if ignored, can confound managers and give them the impression that organizations are disorderly arenas of action. Systems thinking is also oriented toward building the capacities of managers and is a tool for organizational renewal.

Becoming a master of systems thinking is, like any developmental process, a learning process. It requires opportunities for practice and reflection that help one develop the capacity to skillfully apply systems concepts (Kim, 1990). Effective systems thinking depends on integrating many of the key tenets of systems theory with one's own thought patterns. Effective systems thinking requires changing not only what one thinks, but also how one thinks. In this regard, systems thinking might be considered as being more comparable to becoming an ace airplane pilot, virtuoso jazz musician, or martial art expert. This book is not intended to serve as a guide to becoming a systems thinker. Rather, it serves as a platform to catapult readers out of nonsystemic patterns of thinking. It is particularly useful for managers to learn about systems thinking early in their career, because there are relatively fewer nonsystemic dimensions already built in to their way of thinking. Hopefully, the book will also serve as a catalyst to looking at the management process as if you were seeing it for the first time.

THE BASIS OF SYSTEMS THINKING

According to many systems theorists, human actions are the product of one's core set of basic beliefs and assumptions (Argyris and Schon, 1978; Churchman, 1971; Richmond, 1990; Senge, 1990). These core beliefs function as a person's internal set of explanations for how the world operates. People develop a general mental map that is capable of explaining why virtually anything happens. In essence, this is a personal set of theories of how things work. Whether or not the explanation is valid, most people seem to prefer to explain why things happen, rather than to adopt the discomforting alternative that they have no answer. Senge (1990) suggests that such mental representations of how the world operates are filled with inconsistencies and nonsystemic beliefs. The consequence is poor problem solving. Unaware of the presence of limitations, people take actions that have limited value in achieving the desired ends.

Among the most important of these explanations are those involving cause-and-effect relationships. Theories that explain causality generally focus on the direction or pattern of causality. That is, they provide a simple explanation of what caused what. One of the most common nonsystemic beliefs is the belief in the principle of **linear causality.** People who oversubscribe to this convention believe that cause-and-effect relationships happen through a one-way, straight line of sequential influence. Conversely, systems thinking is generally based on the premise that causality usually works in patterns of loops called **causal loops.** Figure 1-3 shows such a causal loop, in which the production of new products affects the desire for new products, which affects R&D on new products, which in turn affects the production of new products. (Causal loops, important dimensions of both cybernetics and system dynamics, will be discussed in greater depth in several later chapters.) Certainly, there are many systems that are dominated by linear causality, such as machines or accounting systems. In general, such systems are not major obstacles to organizational performance and tend to be generally well understood. Indeed, these are not the complex systems that pose the most formidable challenges to managers. Rather, managers are more often baffled by systems with less evident causal loops such as inventory management issues, stock markets, or national economies. Such explanations of causality are part of a larger mental framework, one's mental models.

MENTAL MODELS

Growing evidence indicates that the way managers think about situations is directly related to their effectiveness in achieving improvements in performance (Senge, 1990). In order to understand how a manager's thought patterns

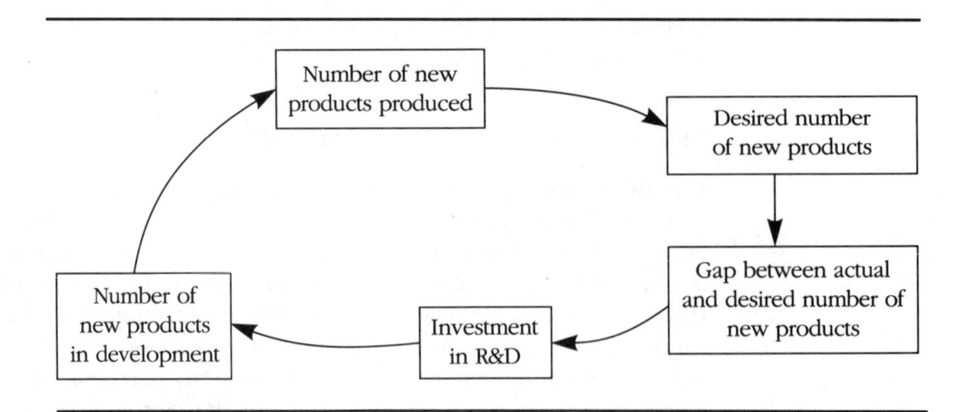

FIGURE 1-3 ***Causal Loop for New Product Development***

may influence his or her performance, it is useful to examine the source of these patterns. One school of thought suggests that the actions of managers are a consequence of their personal mental model (Senge and Sterman, 1991).

A **mental model** is a simplified representation of reality, an internal network of organized thought patterns that are integrated together based on a person's beliefs. A mental model is composed of beliefs that are compatible, although not always consistent with each other. Mental models are comprehensive cognitive maps that organize our beliefs. They are the force that generates the theories of action that people use to explain why things happen as they do. Mental models selectively direct people's attention to what they believe is important. Mental models also orchestrate how people see the relationships among different parts of their world. For example, if an investor believes that fluctuations in the performance of the London stock exchange are caused by government economic policy, his or her investment strategy will depend on observing changes in economic policy. On the other hand, an investor who believes that stock market fluctuations are generated by the structure of the global economic system will track long-term trends in the performance of the market.

Inappropriate mental models can be disastrous. In the 1991 Gulf War, Iraqi military leaders made a number of assumptions based on the belief that American and allied forces would primarily attack their coastal defenses. In fact, the allied forces attacked the Iraqi army from the flank and the rear, inflicting major losses.

Mental models are self-perpetuating and self-reinforcing mechanisms. In helping to frame one's perceptions of the world, they reinforce the beliefs that shaped the original perception (see Figure 1-4). Such a self-reinforcing loop can convince people that what they believe is actually the truth, rather than just a way of thinking about the world. Mental models can also dramatically influence a manager's behavior. First, they narrow his or her focus to those issues that are defined as significant in the mental model; second, they limit the range of possible outcomes that he or she can conceive of to those which are compatible with the mental model. Together mental models and theories of action can act as self-reinforcing forces that ignore any information that runs to the contrary. Argyris (1990) refers to this circular, self-referential pattern as **defensive reasoning**.

For example, an organization's management team may share the belief that most problems are caused by the business environment. Consequently, they devote the majority of their attention to attempting to control or react to their environment. In 1992, the CEOs of the Big Three American automobile makers traveled with President Bush to Japan trying to persuade the Japanese to remove trade restrictions. This effort was essentially externalizing the issue, rather than asking the more fundamental questions related to survival, such as, "How can we manufacture the highest quality automobiles in the world so that people in other countries will want them?" This thought pattern can provide an implicit justification for ignoring important internal factors that are necessary for survival.

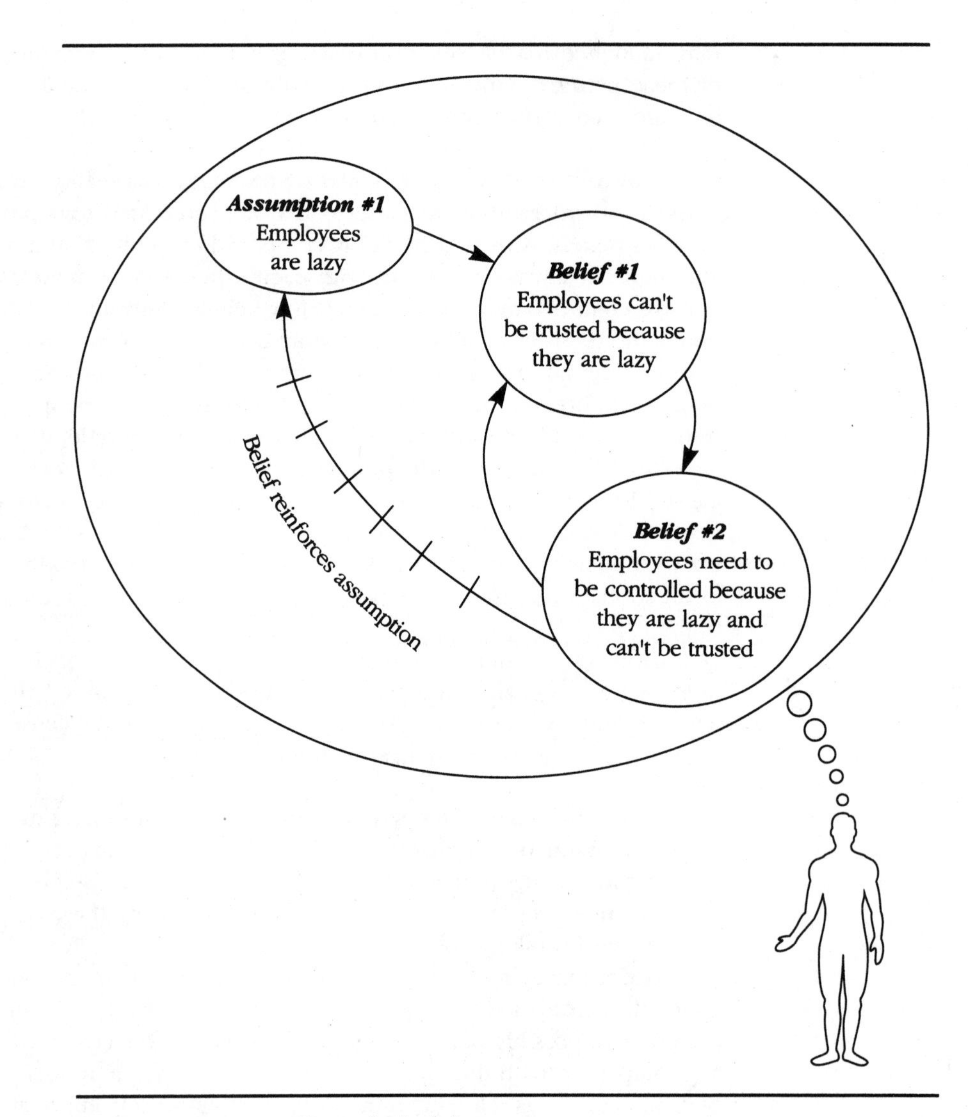

FIGURE 1-4 ***Mental Model***

When managers and organizations continually learn, they have abundant opportunities for gaining insight into the ways that their mental models may contain inconsistencies. By engaging each other in a shared examination of their mental models, managers may find ways that they are limiting their performance by subscribing to unrealistic beliefs (Senge, 1990). Clearly, the revision of one's mental models is a continuous learning process that is most fruitful when it occurs in concert with other people similarly engaged.

Mental models are the structural foundation that supports the process of systems thinking. Mental models are the power source of thinking; they are the generators that supply the electricity for systems thinking. Mental models are not the cause of systems thinking, but, rather the co-creators of meaning. Meaning is the fuel for managerial creativity and innovation. For managers, theory and action are linked by thought.

One of the most commonly cited limitations of nonsystemic mental models is that they lead people to picture systems as being unchanging. That is, there is a tendency to perceive the world as a series of fixed, unrelated events. This is in contrast to systems thinking, which sees things as a continuous flow, over time, of interrelated causes and effects. To understand systems theory, it is useful first to understand the simple concept of a system. Appreciating the concept of system is not the same as systems thinking. The concept of a system depicts a set of parts and relations at a point in time, rather than as a continuous flow over time. The concept does, however, provide a simple, concrete starting point for managers to develop comprehension of the more active concept, systems thinking.

SYSTEMS DEFINED

What is a system? In everyday life the term *system* is used broadly with a variety of meanings. A secretary may use a filing system to organize information. An investor may have a system for picking the most attractive stocks in which to invest funds. From these examples it is clear that a system is an entity that is composed of interrelated parts and that can, in some way, be separated from all that is outside of it.

Put another way, a system is a group of elements that act in concert and with some degree of coherent unity. But that does not tell us all about systems.

Every system is comprised of elements, and most systems are parts, themselves, of larger systems. For example, a manufacturing division is an element of a corporation. That organization, in turn, is an element of an industry. The human body is an element system composed of a number of interdependent subsystems—the respiratory system, nervous system, digestive system, skeletal system, and so on. A **system,** then, is a grouping of component parts that individually establish relationships with each other and that interact with their environment both as individuals and as a collective.

SYSTEMS AS IDEAS, SYSTEMS AS OBJECTS

During the past twenty years, use of the word *system* has increased in popularity. Phrases such as *global economic system, system of communication, organizational system, technological system, ecological system,* and *values system* have become part of everyday life. Nevertheless, the meaning of the word *system* still confuses many people because the word is used at different levels of abstraction. The result is confusion over whether the term *system* refers to something concrete that can be directly experienced or is simply a conceptual tool used to enhance one's ability to make sense of the world.

Systems theorist James Grier Miller (1978) proposed that systems can be divided into three types: (1) conceptual systems (for example, language); (2)

concrete systems (such as machines); and (3) abstract systems (including culture). For managers, systems are the objects of their attention and the basis of a way of thinking about the process of management. Systems thinking, then, is a way of seeing the world that is based on principles derived from the study of systems both as entities *and* as concepts. Since the prime concern of this book is with systems thinking, the focus will be on the relationship between the idea of a system and systems thinking. As such, our concern is with recognizing common patterns of behavior exhibited by systems and understanding the fundamental causes for this behavior.

Some of these patterns are recurrent and typical of certain systems. For example, the periodic swing of the business cycle from boom to bust is a signature behavior that can be found in most unregulated markets. This balancing type of behavior is created by the interplay of various elements within the framework of how markets deal with supply and demand. Another familiar pattern is seen in sales figures of new products, which usually start off slowly, grow rapidly, reach a plateau, and then decline.

By recognizing these recurrent patterns, it becomes possible to understand the logic of how systems operate and to identify common themes that link the behavior of systems in general. Pattern recognition is important for managers because patterns represent the essence of how a system behaves. Recurrent patterns are likely to be more valid or closer to the truth than any short-term oriented ways of understanding a system. Patterns can reveal the themes that are the signature or trademark of how a system will behave. Following a short-term perspective does not give the system time to reveal its true nature. Although love at first sight and whirlwind romances are surely exciting, agreeing to marry someone after two weeks does not allow two people time to know if they can accept each other's patterns—leaving the cap off the toothpaste, for instance. Pattern recognition gives managers powerful insight into what actually makes a system tick, without clouding their view with the complexity of recent events that are likely to color their memories.

In practice, a coherent set of explanations of how systems operate has already been developed. Known as **systems theory**, it serves as the conceptual foundation for the practice of systems thinking. By understanding systems theory, managers have the opportunity to gain insights that can assist them in revising their own mental models. The effect of learning systems theory is similar to the benefits felt by an athlete undergoing training. Most athletes engage in training to build their capacity to excel at their sport, not to be expert at a particular form of exercise. By learning more about systems theory, managers can incorporate relevant ideas into their mental models and fine-tune their systems thinking.

HUMAN SYSTEMS

Not all systems are living systems, yet most of the systems of greatest concern to managers involve living systems.

Systems, such as organizations, are often defined in terms of biological attributes such as growth, development, and survival. This is because organizations behave in many ways that are dynamic, lifelike, and shaped by their human components. Traditional management approaches have been criticized for adopting a mechanical perspective in explaining organizational behavior rather than a more organic view (Capra, 1992). Since people are the most critical dimension of the identity of living systems, we will refer to those living systems that involve people as **human systems.** Human systems have four main characteristics that differentiate them from other systems:

1. They are concrete.
2. They interact with their environment in significant ways.
3. They possess key attributes characteristic of life in general (Tracy, 1989).
4. Their identity is defined primarily in terms of their human element.

Among the most important examples of human systems that are of interest to managers are organizations, markets, economies, and societies.

In general, human systems engage in a number of important functions in an effort to perpetuate their own existence. These functions are adaptation, regulation, renewal, communication, and transformation. In order for systems to promote their own survival it is essential that these five basic functions are integrated. History is filled with examples of human systems that have not been able to fulfill these functions sufficiently to survive. Recent examples include failed airlines and financial institutions in the United States and the government of the Soviet Union. Let us examine the five functions in greater detail.

1. *Adaptation.* Systems are able to maintain their relationship with a dynamic environment by evolving in ways that better serve the system. Successful adaptive behavior depends on the system's ability to learn from past experience and discern which changes are most necessary. Adaptive responses may involve local alterations rather than system-wide change. Organizations usually need to adapt to factors such as changing technologies and changing patterns of customer demand (McKelvey, 1982). This response often takes place via formal analysis by managers of threats and opportunities and the formulation of a strategic response (Hannan and Freeman, 1977).
2. *Regulation.* All purposeful human systems require a mechanism to control their own behavior. This is necessary for the system to remain on track toward achieving its primary goals and purposes. Regulation is also instrumental in helping systems to adapt by restricting some maladaptive behaviors and not restricting those that support further adaptation. For example, the use of total quality management techniques helps control those forces in organizations that tend to produce products or services that deviate from the level of quality that customers expect.
3. *Communication.* Systems must link their various elements into a network that effectively transmits information. The signals that are received must be as nearly as possible identical with those that are transmitted. Unnecessary

noise or signal distortion should be minimized (Shannon and Weaver, 1949). In organizations, the values and assumptions of top management need to be presented clearly in a form that is consistent with the message. There must also be feedback mechanisms in place to notify the source whether the message was received and understood.

4. *Transformation.* The process of adaptation requires a recognition of the need for change and a realization of the type of changes necessary for survival. Normally these changes are gradual and occur through a process of evolution. However, in some cases, the demands for change placed on a system are overwhelming and necessitate a fundamental reshaping of the system. Even the basic identity of a system may be called into question. This type of change calls for a large-scale, rapid response known as *transformation.* Transformation is radical and often disordered, but necessary for survival (Liefer, 1989). Transformation tends to be system-wide rather than local. For example, in the late 1970s Lee Iacocca led Chrysler Corporation through a major transformation in order to ensure the corporation's survival.
5. *Renewal.* All human systems have natural tendencies toward decay and disorganization; these are known as forces of entropy. These pressures can be offset by countervailing initiatives that foster growth and development such as innovation, creativity, continuous improvement in quality, and organizational learning. Through various self-renewal processes, organizations may facilitate further adaptation as well as regenerate themselves to remain energetic and vital (Morgan, 1986). As a system becomes more internally sophisticated, it will have an enhanced capacity to respond effectively to a wider variety of environmental pressures. For example, organizations where continuous learning and improvement take place enjoy an increase in the availability of ways to gain competitive advantage in the market.

When all five functions are working together effectively, they enable human systems to move continuously toward attaining their purpose. One of the most important characteristics of human systems is that they are purposeful (Miller, 1978). **Purposeful systems** are able to regulate their behavior on the basis of a set of core values (Ackoff and Emery, 1972). In purposeful systems, such as organizations, systems thinking can have great impact. Systems thinking, in tandem with organizational learning, can serve as valuable tools to reframe and refine positions on important issues.

SYSTEM BOUNDARIES

Systems are limited in the space or realm over which they have influence. The extent of a system's influence can be visualized as a line of connected points that form a set of limits. Outside these limits lies a general zone in which the system has negligible ability to control other systems. This set of points represents a system's **boundary.** Most people are familiar with the physical boundaries shown on maps. In regard to organizations, boundaries are usually fuzzy, and are used as a tool for understanding and defining the scope of an organization's interests. The idea of a boundary is also a useful way to help managers visualize the extent of the relationship between an organization and its environment. Both of these dimensions of boundaries are relevant to the creation of the European Community, described in the global systems spotlight that follows:

GLOBAL SYSTEMS SPOTLIGHT

A System Is Born: Boundaries in a New System

In 1957 six European countries (Belgium, France, West Germany, Italy, The Netherlands, and Luxembourg) signed the Treaty of Rome aiming for the development of a common market in which goods, services, capital, and people could move without problems. For almost twenty years afterwards, however, Europe remained fragmented: common functions were underdeveloped, loosely coupled structures lacked a hierarchy, relations lacked cohesive and synergetic patterns, and common boundaries stayed ill-defined. Despite this fragmentation, more and more countries of Europe joined the European Community (EC) in the 1970s and 1980s. Meanwhile, the world economic and political environment was changing. Japan and the Far East countries became powerful economic competitors. The United States and Canadian economic systems slowly integrated and many commonalities helped them to develop systemic ties. Facing a new world pattern in 1985, the EC developed and implemented a master plan for the transformation of Europe from an aggregate of political and economic entities into a system. The framework of the system was political; the political and administrative hierarchy was developed with a European Parliament in Strasbourg and management centers in Brussels. Relations among European Community members were redefined, aiming at the development of common rules, laws, and standards and the elimination of frontier barriers and control. All conflicting issues were to be resolved and defined by the European Parliament and its regulatory agencies and courts. In this way the structure, hierarchy, and common functions of the new European system were developed.

The development of the EC offers an excellent example of a system's birth and growth. First, it shows that relations among European countries had to evolve into a coherent and sustainable pattern before any synergistic effect at the level of the total system could be seen and felt. The integration of Europe into a single system is expected to cause a surge in productivity, production growth, and savings and investment and to create many positive cultural, educational, and political effects. Second, it indicates how changes in the environment (such as Japan's dominance in many industrial sectors) can influence a system's development. Third, the situation in 1992 highlighted many issues related to a system's definition of boundaries. While in the 1980s it was understood who could and who could not be a member of the EC—where Europe starts and ends—the fall of communism in Eastern Europe; the resurrection of market economies in Poland, Hungary, and Czechoslovakia; and the transformation of former Soviet Union republics into independent states raises the question anew. In other words, the European Community is attempting to establish its own boundaries while at the same time striving to overcome the influence of very concrete political and economic boundaries that have existed for centuries. We believe that boundary definition will remain the most important systemic issue facing the European Community after 1992.

When adopted for the purpose of defining systems, boundaries are an abstract but necessary conceptual device. (See Figure 1-5.) Managers constantly make choices about how they will define a particular system. It is easy for managers to be unaware of how they have defined a system because most management approaches assume that the system has already been defined and the prime need is to understand the details of its parts. Further, even when

FIGURE 1-5 ***Boundary of a System***
Source: THE FAR SIDE copyright 1991 by Universal Press Syndicate. Reprinted with permission. All rights reserved.

systems are defined, managers do so on the basis of unquestioned assumptions that have evolved as artifacts of tradition. In such cases, it is likely that the way a system is defined will contribute little to building an understanding of what is important about a system. But such an understanding offers real benefits. From the time of Christopher Columbus's venture to the New World to the present, innovators have broken through many of the self-imposed limitations created by unrealistic definitions of boundaries.

SYSTEM HIERARCHY

Systems are composed of other smaller, less complex systems known as **subsystems.** Subsystems are also purposeful entities and contribute in varying degrees to achieving the purposes of the whole system. In most human systems, subsystems are stratified into levels on the basis of their importance to the overall system. Subsystems can also be broken down into ever-smaller, interrelated subsystems.

In organizations, a system's hierarchy is often assumed to be the same as the organization structure. The two are not equivalent, however. Organization structure cannot provide an accurate picture of how a system works. An x-ray will reveal if a person has a broken bone and is a useful diagnostic tool for examining skeletal structure, but it is useless in attempting to understand how the human respiratory system is related to the nervous system. Our use of the term *hierarchy* is not intended to imply an arrangement of power and authority, as is customary in management theory. In this book, hierarchy is viewed as a characteristic of a system's form, not of its structure. **Hierarchy** means, simply, that all systems can be seen as subsystems of larger systems and as being composed of smaller, simpler subsystems. A system is linked as part of a more extensive network of systems and subsystems.

In fact, all systems are subsystems; they only become systems when they become the focus of someone's attention. When a system becomes the object of a manager's attention, he or she is, in essence, selecting a slice of a larger system. By defining a system, the effect is to focus perception at a specific setting that gauges the scope and depth of one's attention. The system that people select to concentrate on is known as the *system-in-focus* (Beer, 1985). There are a plethora of potential systems that can be defined in the world, yet people often behave as if the systems they have defined are the "real" system. Figure 1-6 illustrates the many different levels that can be the focus of a person's perceptual lens. The process of defining a system is an important part of the management process, with major implications for determining the way managers will act to improve an organization.

DEFINING A SYSTEM: A MANAGERIAL PROCESS

The process of defining a system is often ignored by managers because nonsystemic managerial approaches focus on the individual parts of a system and not on the relations between parts, or the system as a whole. From a systems

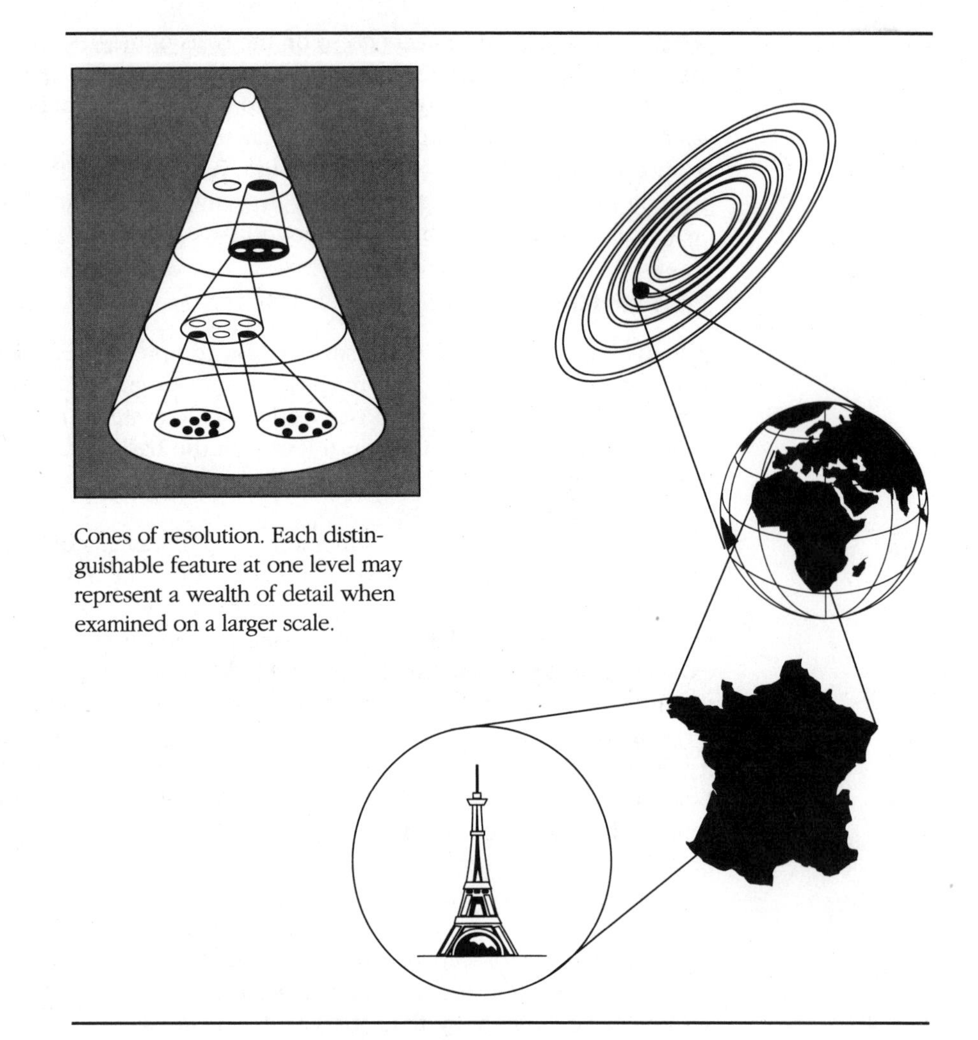

Cones of resolution. Each distinguishable feature at one level may represent a wealth of detail when examined on a larger scale.

FIGURE 1-6 ***A Conceptual Systems Hierarchy***

Source: Adapted from Stafford Beer, *Management Science: The Business Use of Operations Research.* Copyright © Aldus Books Limited, London, 1967.

perspective, however, system definition is a core activity in management. It causes managers to continually reclarify what are the most important issues, both to them and to the organization. The process of system definition adopts a dynamic perspective for managing. It suggests that the perspective one takes on a set of problems or issues will continually change based on what has been learned. In essence, the way a situation is perceived should be continually altered because the managers have changed; they have learned. Consequently, the process of system definition is critical to ensure that managers are addressing today's issues, not yesterday's. By adopting this framework, managers are able to continually think both about pragmatic concerns and conceptual issues. They are empowered to learn while they are managing and to build their capacity for future efforts to manage.

While systems need not exist in a physical form (language, for example), they clearly exist as something distinct and separate from the person observing them. Systems exist as a part of a person's experience and as an object of his or her thought process. The realization that something is indeed a system is governed by the heightened awareness that some parts of the field of experience, which appear to be independent, in fact exist together in some interrelated fashion.

Simultaneous with the recognition of the existence of a specific system is the mental act of separating the system from all things that are not part of the system. Simply recognizing or defining something as a system triggers a perceptual shift that causes the system to jump out of its surroundings. Merely concentrating one's attention automatically brings the object of attention into the foreground. All elements that are part of the surroundings fade into a fuzzy blur. Gestalt psychologists such as Fritz Perls (Perls, Hefferline, and Goodman, 1951) have elegantly described the visual process of separating a figure from a field of ground, or background. By recognizing a pattern, a figure comes to the foreground of a person's awareness while all else fades into the background. If the ground receives the focus, the figure fades from view. For example, in Figure 1-7 do you see the young woman or the old woman? By recognizing one of the images, you are forced to choose one over the other. A system and its environment have just been defined.

The thought process that helps make the existence of a system apparent to someone is known as **defining a system.** By defining a system, a person mentally declares that he or she believes that a designated set of parts exist together as a larger whole that can somehow be distinguished from all else. A system is given definition when someone acknowledges which parts are members of the whole and determines the scope of the system in terms of the number of interrelationships that constitute it.

Stafford Beer (1966) pointed out that relatively complex systems do not have an obvious form that makes their status as a system clearly evident. Rather, a perceptual process takes place; a system is recognized by someone, who by recognizing it, in effect defines it. In this regard, defining a system refers to the process of identifying a phenomenon as being a system and establishing the criteria that help to clarify the form of this system. The number of ways to define a system is virtually limitless. The way the system is defined depends on who is doing the defining and which criteria are used.

The way a system is defined largely determines the scope and direction of any subsequent efforts to improve the system. Ackoff's (1978) Fish Tale, recounting a study of the problems in the British fishing and food processing industry, illustrates several issues inherent in the process of defining a system.

FIGURE I-7 ***Figure and Ground***

Ackoff's Fish Tale

After World War II, as affluence increased in the British Isles, its inhabitants became less committed to discomfort: central heating and refrigeration became increasingly common. With the increased use of refrigerators, freezing compartments were found in many homes. Therefore, the frozen-food business became more attractive.

A large food company decided to develop a line of frozen fish, an important source of protein in the British diet. The company was already a completely integrated fish producer and marketer. At one end, it had its own fishing fleet, and at the other it had its own chain of fish markets. It installed freezing and packaging equipment at its dockside plants to which its fleet brought their catches. Accompanied by a vigorous advertising campaign, it introduced frozen fish to the British consumer.

The initial trial rate was very high, but sales dropped off sharply within a short time. It was clear that there were few second triers. The company put its market researchers to work to find out why. They learned from interviews of women who had tried their products that the taste of the fish was flat, not nearly as good as fresh fish.

After independently confirming the flatness by the use of taste panels, the company asked its food chemists to find the cause of the loss of flavor. The chemists put the blame on chemical changes that took place in dead fish, even though they were stored in ice on the trawlers. These changes, combined with the freezing process, resulted in the loss of flavor. The chemists recommended that the fish be frozen on board ship or that they be kept alive until they were brought to the freezing plant on shore.

The company then had its engineers carry out a comparison of the costs of these alternatives. They found it less expensive to keep the fish alive by converting the holds into pools in which the fish could be dumped from the nets in which they were caught. This was done. A new advertising campaign was launched. Again the initial number of trials was high, but the subsequent drop in sales was dramatic.

Another market survey was initiated. It revealed

that the taste of the frozen fish was still flat. The food chemists were called in again. This time they found that the density of the fish in the holds was so great that they did not move about. This inactivity, the chemists said, produced the chemical changes responsible for the loss of flavor. They advised that the fish be kept active.

Engineers were called in again to find out how to make the fish move around in very densely occupied water. They set up tanks in a laboratory, filled them with water and fish, and experimented with various ways of disturbing the water to make the fish move about. Everything they tried failed. The fish remained inactive no matter how, or how much, the water was disturbed.

One day the laboratory was visited (for an unrelated reason) by an expert on the natural history of fish. He saw the tanks and asked what the engineers were trying to do to the fish in them. He listened to their explanation and patiently watched their efforts without comment. When the engineers had finished and he was about to leave he asked: "Why don't you try putting a predator in there with them?"

They did and it worked: the fish moved to avoid being consumed before being frozen. Of course some failed and were lost, but this was a small price to pay for the tasty frozen fish. The market for them subsequently thrived.

Source: R. Ackoff, *The Art of Problem Solving,* John Wiley & Sons, 1978, pp. 50–53. Reprinted by permission of the publisher.

The visitor's outsider perspective helped him to define the system from a different angle. He defined it in biological terms rather than from a technological point of view. Ackoff's story demonstrates the potential value of systems thinking as a practical device for intervening in complex systems. In addition to illustrating the pivotal importance of system definition, it also demonstrates how systems thinking is a powerful tool for revealing the relationship among factors that at first glance appear unrelated. It also underscores the fundamental significance of mental models in shaping how we perceive the world. It is very simple to make the unconscious mental leap to associate the fish's bland taste with spoilage or inactivity. The process of continually redefining a system helps managers both to learn and to develop a more accurate picture of the true nature of a system. This is a valuable process, yet it suffers from the limitation that it does not help managers to recognize the long-term pattern of a system's behavior. Much can be gained by understanding how a system changes over time.

SYSTEMS OVER TIME

The vast majority of systems that managers are concerned with are dynamic rather than static. The behavior of a system becomes dynamic when it changes over time. Static behavior remains relatively constant over time. Few static systems pose a genuine interest to managers. Most organizations are collections of interrelated dynamic subsystems that cause the organization to experience changes in performance. For example, the sales of skis generally fluctuate throughout the year because in most parts of the world skiing is a seasonal sport. Sales also fluctuate over longer time periods in response to economic changes, demographic shifts, and the general popularity of skiing in relation to other sports. The sales function is just one of many subsystems that behave dynamically. For a moment, try to visualize the influence on an organization of simultaneous fluctuations in cash flow, productivity, quality, demand, inventories, and interest rates. Needless to say, the combined effect of the interaction of these factors on an organization can be complex.

Systems thinking is a useful tool for understanding why organizations change over time and what may be done to manage their dynamic dimensions. Chapters 3 and 15 are devoted to an in-depth examination of the dynamic dimensions of systems. The dynamic behavior of an organization is directly related to its structure. In systems theory, structure is often regarded as a prime influence on behavior.

DISCOVERING THE STRUCTURE OF A SYSTEM

In the United States, the system of primaries used to nominate presidential candidates seems to bring out the worst in the candidates. Grueling travel schedules, incessant campaigning, constant fund raising, and mutual attempts to discredit opponents are the norm. Regardless of which candidates are on the campaign trail, this scenario repeats itself every four years. To some extent, the primary system has evolved in a way that promotes this behavior. The modern roles of big media and big money were unforeseen when the political process was first designed. Should the electoral process be redesigned?

In addition to being holistic and dynamic, systems thinking is also oriented toward explaining the behavior of systems in terms of their underlying structures. **Structure,** in this regard, is the framework or arrangement that links the various parts of a system. All systems have structures; that is, the elements of each system are arranged in a particular configuration that governs their relationships with each other. By changing the alignment of the elements, new forms of relations become possible, and the system is able to generate new patterns of behavior. This concept is one of the most important assumptions of systems thinking—structure is the primary determinant of behavior in systems. The implications of this assumption for managers are enormous. It means that managerial interventions intended to change the patterns of a system must be initiated at the primary level of a system, its structure.

Changing the structure of a system redefines the relationships among the constituent parts that form the system and, consequently, change its behavior. Managers who are unaware of a system's structure are restricted in their capability to create long-lasting improvements. Attempts to manipulate a system in a less profound manner may appear to pay dividends by temporarily shifting or suppressing a problem, but the benefits are short-term. For example, when manufacturers defer routine machine maintenance for lack of funds, their current cash position may improve. But they are simply shifting the problem out into the future. When societies with no plan to dispose of solid wastes rely on landfills to solve the problem, they are shifting the problem into the future. When a firm's problems are blamed on consumers or other external forces as a way to avoid dealing with internal problems, this is suppression.

A system's structure can remain hidden from the awareness of managers. When assumptions remain buried amidst the cultural potpourri of traditions and myths, it is easy to overlook them because they are taken for granted as truth. System structure will be discussed in detail in Chapters 3 and 15.

SUMMARY

Systems-oriented strategies for improving organizational performance are complementary to traditional rational-analytical methods. Managers must understand their own mental models in order to recognize how those models may be based on nonsystemic assumptions. The consequence of not doing so is to become a prisoner of one's own obsolete notions.

In the dynamic setting of today's global organizations, managers have many opportunities to add value by using systems thinking, which helps them continually define what was thought to be true. For managers to use systems thinking, they must find ways to incorporate into their mental models an understanding of and appreciation for systems and systems theory. In particular, such mental models must recognize that system behavior derives from structure and is usually dynamic over time.

KEY TERMS AND CONCEPTS

reframing
rational-analytical thinking
systems thinking
theories of action
linear causality
causal loops
defensive reasoning
mental model
system
systems theory
human systems
purposeful systems
boundary
subsystems
hierarchy
defining a system
static
dynamic
structure

QUESTIONS FOR DISCUSSION

1. Describe several examples of the application of systems thinking and rational-analytical thinking in an organization. Compare and contrast the results obtained with each approach.
2. Describe the health-care system of a particular country. Identify some of the primary causal loops. Use this information to explain why health-care costs tend to rise when left unregulated.
3. Based on the Chapparal Steel case, explain how the managers of a particular organization can change their mental models to gain a competitive edge. You may choose one of the following examples or make up your own: a university, the government of a large

city, an automobile manufacturer, a chain of fast-food restaurants, a start-up, high-tech company in Eastern Europe.

4. Draw a line with the words *static* at one end and *dynamic* at the other. Place the various key subsystems of an organization on the continuum represented by the line where you believe that they belong. Explain why you chose that location.

REFERENCES

Ackoff, R. (1978). *The Art of Problem Solving.* New York: John Wiley & Sons.

Ackoff, R. (1991). *Ackoff's Fables.* New York: John Wiley & Sons.

Ackoff, R., and Emery, F. (1981). *On Purposeful Systems.* Seaside, Calif.: Intersystems Publications.

Argyris, C. (1990). *Overcoming Organizational Defenses.* Boston: Allyn and Bacon.

Argyris, C., and Schon, D. (1978). *Organizational Learning: A Theory of Action Perspective.* Reading, Mass.: Addison-Wesley.

Beer, S. (1966). *Decision and Control.* Chichester: John Wiley.

Beer, S. (1985). *Diagnosing the System for Organizations.* Chichester: John Wiley & Sons.

Capra, F. (1992). "Fritjof Capra: A Physicist Talks Management," *Business Ethics* (January–February 1992), pp. 28–30.

Churchman, C. W. (1968). *The Systems Approach.* New York: Dell.

Churchman, C. W. (1971). *The Design of Inquiring Systems.* New York: Basic Books.

Forrester, J. (1973). *World Dynamics.* Cambridge, Mass.: Wright-Allen Press.

Fuller, B. (1992). *Cosmography.* New York: MacMillan Publishing.

Hannan, M., and Freeman, J. (1977). "The Population Ecology of Organizations," *American Journal of Sociology,* pp. 929–64.

Hendrick, B. (1992). "Ford Exec Uses Game to Prevent Another Edsel," *The Atlanta Journal/The Atlanta Constitution,* January 19, p. G2.

Kim, D. (1990). "Learning Laboratories: Designing a Reflective Learning Environment," working paper #D-4026, Systems Dynamics Group, Sloan School, MIT, Cambridge, Mass.

Liefer, R. (1989). "Understanding Organizational Transformation Using a Dissipative Structures Model," *Human Relations, 42,* no. 10, pp. 899–916.

McKelvey, B. (1982). *Organizational Systematics: Taxonomy, Evolution, Classification.* Berkeley: University of California Press.

Miller, J. G. (1978). *Living Systems.* New York: McGraw-Hill.

Morgan, G. (1986). *Images of Organization.* Newbury Park: Sage Publishing.

Perls, F., Hefferline, R., and Goodman, P. (1951). *Gestalt Therapy.* New York: Crown Publishers.

Quinn, R. (1988). *Beyond Rational Management.* San Francisco: Jossey-Bass.

Richmond, B. (1990). "Systems Thinking: A Critical Set of Critical Thinking Skills for the 90's and Beyond." Lyme, New Hampshire: High Performance Systems.

Senge, P. (1990). *The Fifth Discipline.* New York: Doubleday.

Senge, P., and Sterman, J. (1991). "Systems Thinking and Organizational Learning: Acting Locally and Thinking Globally in the Organization of the Future," in T. Kochan and M. Useem (eds.), *Transforming Organizations.* Oxford: Oxford University Press.

Shannon, C., and Weaver, W. (1949). *The Mathematical Theory of Communication.* Urbana: University of Illinois Press.

Stata, R. (1989). "Organizational Learning—The Key to Management Innovation," *Sloan Management Review,* pp. 63–74.

Tracy, L. (1989). *The Living Organization.* New York: Praeger.

ADDITIONAL READINGS:

Maturana, H., and Varela, F. (1980). *Autopoeisis and Cognition: The Realization of the Living.* London: Reidl.

Sterman, J. (1992). Lecture given at Sloan School of Management, MIT, Cambridge, Mass., February 10, 1992.

2 The Global Systemic Framework

Laudable goals such as "empowering" and "enabling" individuals often prove counterproductive unless managers can act locally and think globally.

Peter Senge and John Sterman (1991)

1. To develop an understanding of the primary factors shaping the global business environment.
2. To recognize the forces that form the structure of the global economic system.
3. To appreciate the value of intangible resources as a means of gaining competitive advantage in the global marketplace.
4. To comprehend how global trade blocs and the global triad influence the flow of global commerce.
5. To become familiar with various world-class management practices.
6. To know the various ways that global organizations are changing their structures to adapt to the demands of global business.
7. To understand why organizations form global alliances.

SECTION I
GLOBAL THINKING

THE IMPORTANCE OF GLOBAL THINKING

The purposes of this chapter are to introduce the concept of a systems perspective for managing and then to demonstrate that many of the global issues that concern organizations today can be understood in terms of such a perspective. The organizations that will prosper in the global markets and industries of the twenty-first century must be able to deal with the complexities inherent in any global system. The systems perspective offers a unique way of viewing the world that includes a broadened scope and is capable of recognizing long-term patterns and trends relevant to business. A growing number of organizations are leading the effort to learn about applying a systems perspective to their own processes for managing within global systems. One organization that has applied this style of thinking quite successfully toward dealing with global strategic issues is the Swiss firm Asea Brown Boveri.

Asea Brown Boveri

Asea Brown Boveri (ABB) currently employs more than 240,000 people around the world, and generates more than $25 billion in revenues, mostly in Europe and the United States. ABB was founded in 1987 as the result of a merger between two well-established firms: the Swedish electrical company Asea and the Swiss industrial equipment producer Brown Boveri. The management of both companies recognized the possible dangers created by factors such as: 1. the relatively small size of both companies when compared with the challenges posed by global markets and evolving trade blocs, of which neither Sweden nor Switzerland was officially part; 2. limited resources and duplication of R&D, production, and sales costs; and 3. intense competition from both Japanese and American firms. The merger was engineered and implemented swiftly to create the new global company. The next five years were spent aligning the firm to strategic trends in the global environment.

Their first strategic moves were designed to make them a more powerful player in the "Triad market" of Europe, the Pacific Rim, and North America. ABB proceeded to acquire more than sixty firms around the world, either in whole or in part. It established a strong foothold in the United States market by buying Westinghouse's transmission and distribution operations and Combustion Engineering. It moved into Asia, where it has over 10,000 employees, and entered into the South American continent. ABB expanded its European base and invested heavily in Eastern Europe, acquiring two Polish manufacturers of steam turbines, transmission gears, and marine equipment.

Second, in order to become more of a knowledge-based company, ABB restructured its operations around the world by creating a complex framework that meshed eight main business units, fifty business areas, and 1,110 local companies. The emphasis of this restructuring was twofold: 1. to integrate common lines across national borders by eliminating needless separation of product lines and production facilities; 2. to focus on the development of a set of core technologies and key competencies that could be easily diffused throughout ABB. Finally, ABB sought to improve its global communications networks on two fronts: 1. by implementing a centralized management information system, known as "Abacus," which tracks the performance of over 4,500 ABB profit centers around the world; 2. by engaging in ongoing discussions and dialogue with all levels of managers from its business units throughout the world.

ABB has sought to create synergies between its business units wherever possible. It has attempted to integrate local autonomy with a global perspective in its business decisions. Percy Barnevick, ABB's CEO, describe the company as "being global and local, big and small, radically decentralized with centralized reporting and control." ABB must balance the benefits of being "local, small, and decentralized" to ensure adaptiveness against the need to stay "global, big, and centralized" to compete in the global marketplace.

Source: Adapted from W. Taylor, "The Logic of Global Business: An Interview with ABB's Percy Barnevick," *Harvard Business Review* (March–April 1991), pp. 91–105.

ABB's use of a systems perspective illustrates the key point that within global systems, organizations must find the appropriate vehicle for meshing local decisions with broader, more global considerations, such as patterns of international trade and world economic conditions.

EARLY APPROACHES TO GLOBAL THINKING

The systems perspective has developed primarily over the past half-century. One of the first large-scale movements to define systems thinking was an attempt led by biologist Ludwig von Bertalaffy. This effort established a set of universal principles that could be used to explain the actions of most types of systems. The research was focused on identifying similarities that exist in different fields and then attempting to relate these similarities in terms of a common cause. It was thought that the common cause could be explained in terms of a universal set of laws that govern the operation of all systems. This set of natural laws, thought to be the most general of all laws, was labeled **General Systems Theory** (GST). Such a broad theoretical perspective would, ideally, be applicable to all types of systems, ranging from a living cell to an organization and from a society to stars, planets, and asteroids in space. The Society for General Systems Theory was founded in 1954 to identify the similarities among the concepts, laws, and models of various disciplines and to determine how these ideas could be imported from one field to another for the greater benefit.

The development of GST and its **open-systems perspective** contributed to many subsequent innovations in the social sciences. The open-systems perspective argues that systems and their environments are interdependent and must be open to each other to encourage exchanges of resources, which they need for survival. By 1966, the concept of viewing organizations as open systems had taken root among both managers and academics alike. Katz and Kahn (1978) were instrumental in proposing to managers that organizations could be seen as entities dependent on their environments for resources (Seeger, 1992). This assumption influenced a whole generation of managers to shift their focus from inside their organizations toward the external environment.

Since this time, systems thinking has continued to evolve, becoming more relevant than ever to managers. The business environment of today is more complex and dynamic than ever, posing a greater challenge to managers of all organizations. Systems thinking is especially meaningful to those managers who recognize that the way they think about global systems may serve as the basis for their future competitive strategies. In the global systems of today, improvements in transportation and communication have caused the world to shrink such that local decisions have reverberating global consequences.

Richmond (1990) has proposed that this is due to the tighter coupling of systems. **Coupling** is the degree of interdependency among elements. If a minor change in one variable causes a large-scale response in another variable, the second element is tightly coupled with the first. Systems such as organizations, whole industries, and national economies are more intertwined, or tightly coupled, than ever before. For instance, the oil crisis of the 1970s forced many nations that were heavy consumers of petroleum to become more aware of the interdependencies among countries and organizations. Similarly, the emergence of Japan and other Pacific Rim countries as industrial powers during the past two decades caught many unsuspecting nations, who had been industry leaders, off guard. Even more importantly, the tight coupling between humans and their environment has become more evident than ever before. Environmental issues such as ozone depletion, deforestation of tropical rain forests, and hazardous and solid waste disposal have all served to focus attention on the many ways that the various parts of the world are interconnected. On the positive side, these interconnections mean that the actions of managers have greater potential to create wide-scale, reverberating effects that improve the functioning of many more levels of global systems than at any prior time in history.

~ GLOBAL SYSTEMS: "A BORDERLESS WORLD"

The many emerging global systems—including economic, political, and communications systems—are distinguished from their predecessors, national and regional systems, by the absence of traditional political boundaries. In many respects, national borders have faded from view and no longer remain useful as a formal perimeter for business activity. This is because economic justifications that formerly supported national limits no longer exist (Ohmae, 1990). National identities are being replaced by newer strategic alliances in both government and business. Examples in government are the European Community and the Commonwealth of Independent States. In business, one of the most successful alliances is Airbus Industrie, a consortium of European airplane manufacturers that is second only to Boeing Corporation in worldwide sales.

Such confederations are forming at an increasing rate between what once seemed to be unlikely partners. Former political adversaries and business competitors have found sufficient common ground to set aside their differences in favor of mutual benefits. The common impetus for many of these business links is the desire to be a survivor in the global marketplace. The assumption underlying such coalition strategies is that size equals power in the global markets and industries of the future. The power desired is to influence industry pricing, to influence customers, and to lead in new product development. Whether size translates into power is not always clear. Let us now examine the arenas of global competition to see if such alliances are sensible strategies.

In every important global industry—automotive, pharmaceutical, banking, electronics, aerospace, and machine-tools—major transformations are taking place. These wide-ranging transformations are reshaping the basic structure and character of these industries. The increasing speed with which new products and services enter the market adds a new dimension of unpredictability to the environment. Patterns of competition and cooperation have rapidly been altered as a result of the action of trade blocs. **Global trade blocs** are groups of countries that act as trading partners, working cooperatively to expand globally and simultaneously protecting their core domestic markets. The coupling between these trading blocs has grown continually tighter, as a result of technological innovations such as computerized communication networks. These linkages permit instant exchange of information and high-speed management decision making.

SECTION 2
CHANGING GLOBAL BUSINESS

FOUR FORCES OF GLOBAL TRANSFORMATION

A systemic view of these global shifts suggests that four mutually reinforcing forces are operating to transform the basic character of global business systems (see Figure 2-1):

1. The development of new economic patterns
2. High-speed communications networks
3. The development of trade blocs
4. Interlocked trends of competition and cooperation of global organizations

Collectively these forces are interacting in a way that transcends other national and regional political, social, technological, economic, and organizational influences. This shift in the global framework is so profound that it is altering the basic fabric of the business environment. These forces have the collective power to catapult many global transformation processes forward with a thrust that is greater than the influence of any one government in the world. We will now examine each of these four developments in detail.

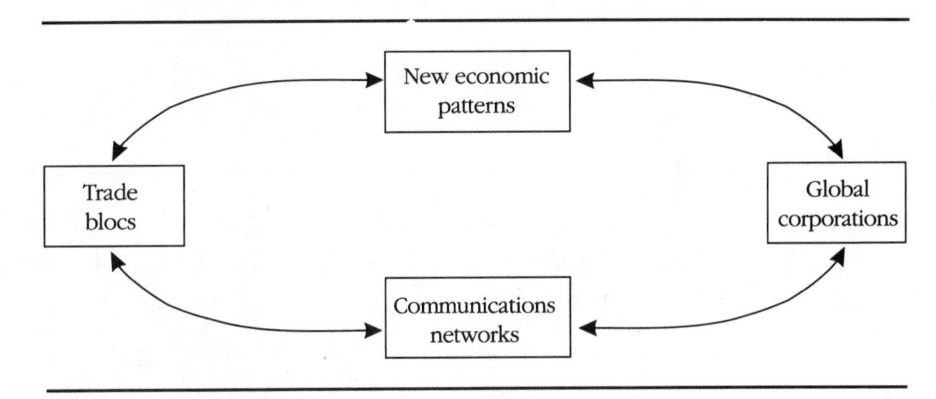

FIGURE 2-1 ***The Four Forces of Global Transformation***

A NEW ECONOMIC PATTERN

The new global economic pattern that has developed over the most recent decade is built on three core themes that have determined its form. First, the idea of a market-driven, deregulated, and decentralized economy has become accepted as the template for economic growth virtually on a worldwide basis. The economic transformation of the countries of Eastern Europe and the republics of the former Soviet Union is but the latest pair of examples of this force in action. Second, rapid innovations in technology have forced the restructuring of many national economic systems. Traditional smokestack industries such as steel manufacturing, shipbuilding, and defense industries have gradually ceased to dominate the landscape of many economies. In some cases these industries have been shifted to less-developed countries, and in others they may be phased out entirely. Although heavy manufacturing will continue to play a pivotal role in many nations, its importance is gradually being offset by the progressive growth of knowledge-based industries. Third, the importance of intangible, informational, and knowledge-based resources is increasing. The impact of these three forces is shown in Figure 2-2.

DOMINATION OF MARKET ECONOMIES

At the beginning of the 1980s, the United States was the only major world economy where state-owned industries and strict regulation of the economic system were absent. In Japan, Asia, Latin America, and Europe, government intervention in national economies played a central role. These interventions took the form of centralized national economic planning. Furthermore, such planning efforts were often implemented through hands-on control of economic performance and the explicit regulation of some industries. These controls were further enhanced by state ownership of many key industries, particularly such industries as automobiles, airlines, aircraft, mining, railroads,

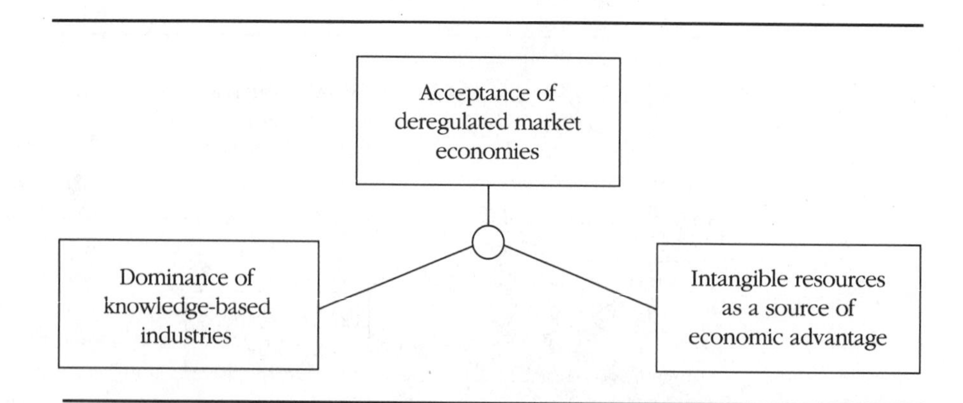

FIGURE 2-2 ***Elements of the New Economic Pattern***

electricity, telecommunications, steel, banking, and insurance. By 1992, the world had changed. China was the sole major country whose economy was controlled by a dominant communist ideology.

Even those economies that had been market oriented in the first place began to change as they abandoned strict regulation. Britain was the first European country to reverse the trend of planned economies. By the early 1980s a broad swing of economic revision had taken place through the deregulation of the economy by initiating the sale of state-owned assets. This sweeping current of changes created a general trend toward the new economic pattern. By the middle of the 1980s, most European countries had deregulated their economies and engaged in a massive privatization of state industries. They were soon followed by many Asian economies.

The movement toward a market economy also struck the communist bloc. In the late 1980s, with the collapse of many of the communist regimes in Eastern Europe, new market economies sprang to life. Poland spearheaded this trend in 1989 by implementing a wide-ranging program aimed at creating a market economy in January 1990. This led the way for other Eastern European countries. Czechoslovakia, Hungary, and Bulgaria also embarked upon similar programs. The movement was capped by the widespread call for economic reforms in many of the republics of the former Soviet Union. The early years of the last decade of this century thus witnessed a series of changes in the structure of economies unprecedented in modern history. Deregulated market economies based on private property rights now prevail throughout the world and serve as the major blueprint for future economic development.

GROWTH OF KNOWLEDGE-BASED INDUSTRIES

For many years futurologists have predicted the coming of the knowledge-based economy. However, only relatively recent developments—such as the widespread use of computers and sophisticated software in the workplace, high-speed telecommunications networks, new technologies (semiconductors and biotechnology), robotics, and automation—have made it finally possible (Toffler, 1990). The escalating economic impact of knowledge-based sectors of the economies is well documented. The portion of the world's gross national product (GNP) created by agriculture and industry has been declining. At the same time the service sector's share of GNP in all developed countries has grown to over 50 percent. This is clearly not to suggest that economies of the future will be supported by fast-food restaurants, law firms, and cleaning services. To the contrary, making things will continue to be important; but the primary resources in making things that serve the needs of customers in continually better ways are information and people.

In knowledge-based industries, organizations will have a distinctly different look than they did in the past. The valued resources will be people and information rather than materials. There will be considerably fewer managers than was the norm in the late 1980s. Some experts predict that the number of managers could be reduced by as much as one half (Drucker, 1989).

Knowledge-based organizations are composed primarily of employees who are specialists; they are self-directed rather than needing to be told what to do by managers. Drucker (1989) has compared such organizations to an orchestra, in which each musician is a competent specialist who is guided by the motions of a conductor as well as commonly shared information (the sheet music). The knowledge-based organization will gain competitive advantage by using information in novel ways and using it with ever-increasing speed.

Even more importantly, knowledge-based organizations will not just use information more effectively to gain competitive advantage, they will create information (Nonanka, 1988). By converting the insights of individuals into opportunities for the entire organization to learn, the value of those insights is multiplied exponentially. Information is created by giving new meaning to an experience—a meaning that offers a fresh perspective. New meaning can be expressed tangibly as new strategies, products, manufacturing processes, or people processes. Groups of organization members can twist these insights into new patterns by discussing them and using them to develop new products and services. Such an example can be found at the Japanese electronics manufacturer, Sharp. (See the box below.)

While knowledge-based industries have thrived, traditional heavy manufacturing industries such as steel, shipbuilding, automaking, mining, and so on—that dominated economies for the past eighty years—have become progressively less important as engines of economic growth. This trend is shown in business literature. The *Fortune* magazine annual "Corporate Reputations Survey" ranks companies on such attributes as quality of management, quality of products or services, and innovativeness. The information-based cor-

▼

Sharp Minds

The Wizard, a 10-ounce, 7-by-3.7-inch organizer based on displays and semiconductors developed for Sharp calculators, began as a product for the Personal Home/Office Electronics Division, fondly called "Phoo-ey." Says Gil DeLiso, now head of the division that markets it in the U.S.: "We saw that lifestyles were becoming busier. People had more information to manage." The gadget, aimed at business travelers, was introduced in 1988.

Bingo. Sales approached $400 million last year. Says Richard Schaffer, editor of *ComputerLetter:* "Sharp essentially created a market that others dismissed as toys. Computer makers think everything is an office machine. The big payoff comes in finding a machine for the people who say no to the question 'Do you need a computer?'" A dedicated sales force is trying to get companies to say yes to Wizards with software customized for their traveling salesmen and such. Among the takers: Prudential Property and Casualty Insurance and PepsiCo.

Source: Fortune (March 23, 1992), p. 108.

porations like Benetton have dominated this list throughout the most recent years.

Another change is that financial services and trading companies have rapidly gained stature in the global economy. This movement has been led by the growth of Japanese banks. Assets of the ten leading industrial giants on the *Fortune* Global 500 list are dwarfed by the assets of the top financial firms. World trade in goods amounts to around $3 to $4 trillion a year, but the financial market turns over $100 trillion a year with credit, trading currencies, and hedging operations. The overall value of international money transactions far exceeds the trade in goods and services (*Fortune,* July 30, 1990, p. 269; *Wall Street Journal,* September 21, 1990, pp. 30–31).

The growing importance of information-based industries and the financial sector reinforces the importance to both economies and firms of one major intangible asset: information. Peter Drucker (1989) proposes that "Knowledge has now become the real capital of a developed economy" (p. 180). It has become apparent to many firms that gains in productivity can be achieved more frequently and rapidly by redesigning systems to more efficiently use

				SALES		PROFITS			ASSETS	
1991	1990			$ millions	% change from 1990	$ millions	Rank	% change from 1990	$ millions	Rank
1	1	General Motors	U.S.	123,780.1	(1.1)	(4,452.8)	490	—	184,325.5	1
2	2	Royal Dutch/ Shell Group	Britain/ Neth.	103,834.8	(3.1)	4,249.3	2	(34.0)	105,307.7	4
3	3	Exxon	U.S.	103,242.0	(2.5)	5,600.0	1	11.8	87,560.0	6
4	4	Ford Motor	U.S.	88,962.8	(9.5)	(2,258.0)	488	(362.5)	174,429.4	2
5	6	Toyota Motor	Japan	78,061.3	21.0	3,143.2	3	5.0	65,178.7	8
6	5	Int'l Business Machines	U.S.	65,394.0	(5.3)	(2,827.0)	489	(147.0)	92,473.0	5
7	7	IRI	Italy	64,095.5	4.3	(254.1)	459	(127.4)	N.A.	
8	10	General Electric	U.S.	60,236.0	3.1	2,636.0	5	(38.7)	168,259.0	3
9	8	British Petroleum	Britain	58,355.0	(2.0)	802.8	55	(70.9)	59,323.9	11
10	11	Daimler-Benz	Germany	57,321.3	5.6	1,129.4	30	8.4	49,811.8	13

FIGURE 2-3 ***Fortune 500 Industrial List (Top Ten)***

Source: Adapted from "The World's Largest Industrial Corporations," *Fortune* (July 27, 1992), p. 179.

information. While the industrial age was known for using machines to extend the physical powers of mankind, organizations in the global economy will seek to use technology to enhance people's thinking or cognitive abilities (Banathy, 1991).

TANGIBLE VERSUS INTANGIBLE RESOURCES

Traditional approaches to business have placed great importance on capital (money, machinery, and plant), raw materials, and labor (workers). These were seen as the major resources that businesses used to generate productivity. Resources of this type are visible and tangible; their costs directly influence the overall cost of operations. While these factors continue to be important for organizations, their relative importance has gradually diminished. Drucker (1980) indicated that three trends emerged from 1975 to 1985 to profoundly mold global economic systems: (1) the decline of raw-material prices, (2) the decline of labor costs, and (3) the shift from material-intensive to less material-intensive production. As the costs of raw material and labor slowly declined, the traditional sources of competitive advantage vanished (Porter, 1990). Among the notable examples of this change is the decline of many industries in the United States.

As tangible resources slowly lose their importance, intangible resources are gaining greater relative value and becoming a competitive advantage for many global firms such as Coca-Cola, Mercedes, Komatsu, BASF, Rolex, and Gillette. **Intangible resources** are the nonmaterial assets of an enterprise such as organizational competencies (creativity, speed) or the products of specific competencies (brand image, goodwill). Intangible resources are valued sources of competitive advantage first because they are difficult to copy and must be developed over time through thoughtful managerial actions and second because they have the flexibility to be applied simultaneously in a wide variety of ways (Itami, 1987).

Intangible resources have the potential to generate multiple gains for organizations that use them (Obloj, 1986). Once a firm's image has been successfully cultivated, for instance, it can be redirected to promote new or developing products. Honda built on its good reputation for quality in cars and motorcycles in its strategy to sell lawnmowers. Its advertising campaign for that new product line suggested that consumers "put a second Honda in your garage" (Itami, 1987, p. 13). Employee creativity represents another important dimension of intangible resources. Businesses with high-performing organizational cultures—such as 3M, Ingersoll-Rand, and Union Pacific—permit the implementation of innovative strategies.

Most intangible resources are not easy to establish. It takes many years to establish a well-known brand name like Rolex, Coca-Cola, Timex, or Sony. Brand image is also a very fragile resource that is easy to lose. Similarly, it is very difficult to develop a work force with sophisticated technical skills and the capacity to be innovative. Consequently, those firms that are dedicated to long-term internal development have risen in the global marketplace. Another advantage of intangible resources is that while tangible resources depreciate with extended use, intangible resources often appreciate.

HIGH-SPEED COMMUNICATIONS NETWORKS

The second of the four forces transforming global economic activity is the widespread use of high-speed communications networks. The incompatibility of national telecommunications networks in a world where high-speed management has become a necessity has forced greater numbers of companies to build their own telecommunications networks linking telephones and computers via satellites (Cushman, 1991). This trend has been led by global service companies, especially airlines and financial institutions. American Airlines established its competitive position almost two decades ago (1976) with the development of the SABRE computer reservations system. Financial institutions (banks, insurers, and securities firms) quickly discovered the advantages of globe-spanning access to computers and telephones in order to deal with millions of daily transactions on a real-time basis. The incredible speed and complexity of such a network is illustrated by Figure 2-4, which portrays the trail of an electronic transaction triggered by the use of a Visa card.

These actions set the trend for other industries. Such technological change has been precipitated in order to facilitate quick response to changing consumer patterns. On the retail front, Benetton, the Italian-based clothing manufacturer, monitors all sales around the globe through computer linkages. This network helps the firm to adjust daily to changes in demand and to respond immediately with new supplies of best-selling items. At Mercedes-Benz dealers, customers can order their own customized automobile (choosing the type of engine, gearbox, color, and so on) through a computer network that connects dealers and the producer.

What are the global advantages of high-speed telecommunications networks? On the local level, they give managers up-to-date views into market demand, production capabilities, inventories, and financial transactions. Access to this information enables managers to adapt quickly to changing conditions in the environment. The time needed to complete the cycle—from identifying a customer need to creating an appropriate product—has been reduced dramatically. Networks enable firms to customize products to meet local tastes and needs while still maintaining standards for efficiency and quality. This information about special needs can be immediately communicated to the production facilities, where flexible manufacturing systems can be configured in such a way to meet the demand. The cycle time of marketing, production planning, production, and delivery is reduced to a minimum. The radical effect this system is having on the manufacturing world is to blur the traditional distinction between customized products aimed at niche markets and the mass production of standardized goods. Even manufacturers who are principally identified as mass producers are now able to customize products and offer short-runs of specialty goods.

THE TRAIL OF AN ELECTRONIC TRANSACTION

1 Customer buys diamond earrings for $895 in Detroit. The clerk passes the Visa card through a credit-verification terminal and punches in purchase data

2 The data travel by satellite, land lines, or microwave to National Data Corp.'s computers in Cherry Hill, N.J.

3 From Cherry Hill the credit query goes to NDC headquarters in Atlanta for processing

The transaction tops $50, so it needs a second opinion. The request is turned over to Visa USA minicomputers at NDC

4 The Visa minis shoot the query to mainframes in McLean, Va., or San Mateo, Calif.

5 The Visa mainframe determines that the card is from a San Francisco bank and sends the transaction to the bank's computer, which checks to see if there is $895 in available credit

The bank's O.K. retraces the path of the authorization request: From the bank to Visa USA to NDC in Atlanta to NDC in Cherry Hill to the merchant in Detroit

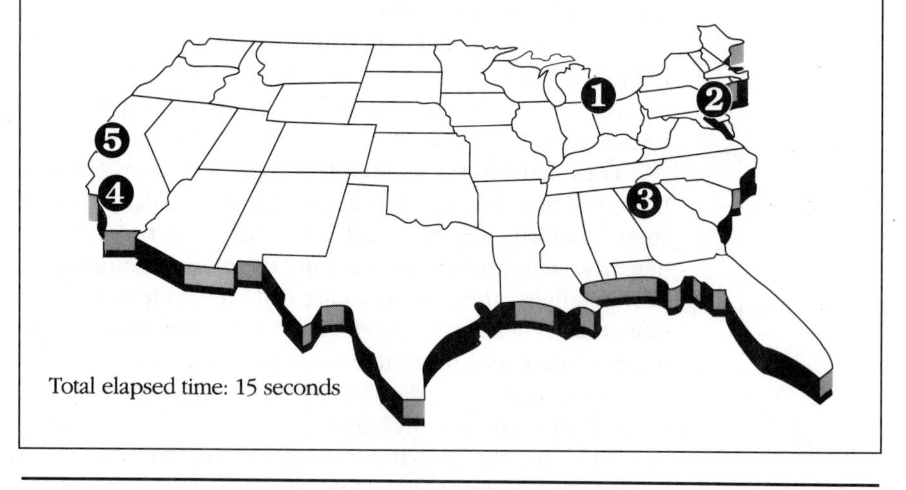

FIGURE 2-4 ***A High-Speed Information Network***

Source: Text from October 14, 1990, issue of *Business Week* by special permission, copyright © 1990 by McGraw-Hill, Inc.

GLOBAL TRADE BLOCS

The new economic pattern and the use of high-speed communications networks have occurred within the context of structural changes in the global economic system, particularly the solidification of global trade blocs and the development of the triad market. Starting around 1950, American companies pioneered low-cost, standardized mass production and exported both this manufacturing strategy and its tangible effects (products and services) to the rest of the world. At the time, other national economies were smaller, less productive, and not as innovative as that of the United States. European countries tried to create a system to counterbalance the power of the United States economy. The European Community, otherwise known as the *Common Market,* formed by six countries in 1957 seemed plausible, but little was actually accomplished, and Europe remained economically fragmented. Its share of global manufacturing and trade continuously declined, falling from 37 percent in 1938, to 27 percent in 1953, 24 percent in 1973, and 22–23 percent by the middle of the 1980s.

While the European countries contemplated the reasons for their decline, Japan was developing a master-plan for its economic rise to power. It lacked all the traditional resources, having few natural resources, a small skilled labor force, and a shortage of usable land. Japan slowly turned these disadvantages into factors supporting its success. It developed lean facilities that concentrated on value-added goods. The automotive industry in Japan is probably the most illustrative of this pattern (Porter, 1990), but similar strategies were employed by Japanese companies in the camera industry (Nikon, Canon, Olympus) and consumer electronics (Casio, JVC, NEC, Sharp, and Sony). The stage was now set for a transformation of the balance of global economic power.

SHIFTS IN THE GLOBAL ALIGNMENT

In the 1980s, two additional trends in global restructuring became more visible. The first trend was the economic integration of Western Europe through the implementation of European Community directives called the **White Paper Plan.** Developed in 1985, this White Paper is a detailed outline for economic unification and a timetable for the implementation of 276 directives and reforms that must take place to achieve the establishment of a single, unified economic system for all twelve members of the EC (see Figure 2-5).These economic reforms were intended to address three main issues:

1. Control of flows of people and commodities across borders.
2. Removal of technical barriers (differing standards, policies, and laws).
3. Removal of fiscal barriers.

Several important matters related to European economic unification remain to be dealt with in subsequent years. They include the opening of cross-border markets for banking, insurance, and transport and the elimination of frontier barriers and controls throughout the entire European Community.

But the changes in Europe were not the only changes in the global economic system. Simultaneous with the EC's initiation of the actions needed to enhance its economic status as a major trade bloc, the countries located on the Pacific Rim were also emerging economically.

THE PACIFIC RIM COUNTRIES

The second major trend in the transformation of the global economic system was the rapid development of the economies of the Pacific Rim, including Japan. The Pacific Rim bloc is the most diversified bloc ethnically, politically, geographically, socially, and in religion. Despite these potential handicaps, the

FIGURE 2-5 ***The Nations of the EC***

area has one common feature—it is the fastest-growing economic region in the world. Its economic growth rate averages 4.3 percent annually, compared to 3 percent for the United States and Canadian bloc, and 3.5 percent for the European Community. Asia's four "little dragons" (Hong Kong, Taiwan, Singapore, and South Korea) became industrialized economies and exhibited impressive continuous annual economic growth at a rate just over 6 percent.

Trailing just behind them in growth were the economies of the "ASEAN-4 Group": Indonesia, Malaysia, the Philippines, and Thailand. Finally, New Zealand and Australia exhibited lower growth rates (over 2 percent), but the economic growth rate still exceeded their annual rates of population growth.

Since the mid 1970s, then, the Pacific Rim bloc has experienced the most dynamic and successful economic development of any region in the world. With the rising economic status of the Pacific Rim region, the global economic structure attained a form that would clearly influence the global business environment for years to come—the Global Triad had arrived.

THE GLOBAL TRIAD

Author and management consultant Kenichi Ohmae (1985) identified the emergence of a triad of key global markets in the early 1980s. This **Global Triad** consists of the European Community, the Pacific Rim countries, and North America (the United States and Canada). Ohmae recognized a fundamental shift in the relative importance of regional markets around the world. Triggered by the impressive industrial strength and rising standard of living in the Pacific Rim, the markets in this region became much more important. Led by the economic resurgence of Germany, the European Community moved forward as a major market. Finally, the United States and Canada still remained as key producers and consumers in the flow of international trade. They responded to the integration of the European Community and the rise of the Pacific Rim economies by creating an economic bloc of their own when they signed a free trade agreement in 1989.

These three geographical centers have gained power by acting as the hubs of activity for the world's dominant businesses and core consumer markets. The importance of the triad is such that any firm hoping to engage in business on a global scale must be able to compete successfully in all three of these markets. Porter (1990) has proposed that an organization's ability to vie effectively in the intense global marketing wars is fortified by its ability to survive in its own domestic market. The organizations that compete effectively on such a global scale are also different in significant ways from typical multinational, international, or regional businesses. It's useful to examine these global organizations more closely and see what makes them different.

~ GLOBAL ORGANIZATIONS: COMPETITION AND COOPERATION

The fourth force at work in transforming the world economy is the growth of competition and cooperation among global organizations. Global organizations are not a new phenomenon. The Roman Catholic Church has survived as a global organization for almost two thousand years. Centuries ago, various British trading companies established commercial links with India and Far East countries. Durant and Durant (1975) vividly described the role of the British traders of the early 1800s:

> If their island was small, the seas that stormed or caressed its shores called them to far-reaching adventures; a thousand liquid roads invited men who could pitch and roll and always stand erect. A hundred lands were waiting with products and markets, to help transform England from agriculture to industry, commerce, and worldwide finance. (p. 340)

The focus of these traders' efforts was on producing goods in a single location and trading commodities or single products. World history has since seen a bold transformation away from the politically oriented, neatly divided, economic subsystems. By the beginning of the 1990s many large industrial firms, such as Siemens and Bayer, established their production facilities and distribution networks outside the countries that were traditionally their home base. Throughout history multinational organizations were the exception; firms with local, domestic orientations were the rule. During the past twenty years this orientation has changed dramatically. Growing numbers of companies have established subsidiaries abroad. Such international corporations have formed through the use of both mergers and acquisitions of foreign businesses. During the 1970s the international corporation was supplanted by the transnational corporation, which was replaced by the global corporation of the eighties. Global corporations of today's world have at least four distinctive features.

1. They have a global perspective on world markets.
2. They are economic superpowers.
3. They employ world-class management practices.
4. They have become interrelated with other organizations through networks of strategic alliances (see Figure 2-6).

A GLOBAL PERSPECTIVE

Traditional multinational corporations were firmly anchored in their home country and expanded by establishing satellites in other countries. They usu-

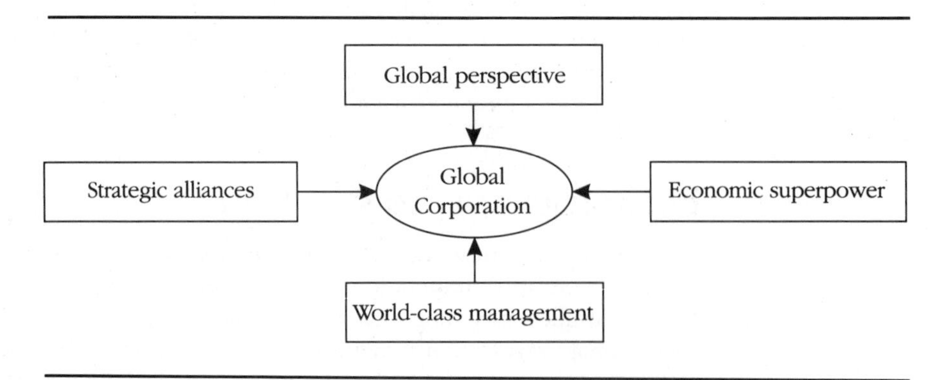

FIGURE 2-6 ***Characteristics of the Global Corporation***

ally had an extensive network of companies that carved their niches in the separate domestic markets of each country or region. The global organization takes a broader, more systemic perspective than a multinational business. To such an organization, the whole world is seen as one large, highly segmented market. The company's strategies encompass the entire world, as expressed in the famous General Electric strategy: "number one or number two globally." Global strategies are oriented toward capturing major shares in all the core markets of the Global Triad by using world-class production and marketing tactics and techniques. Organizations with a global perspective view various markets around the world as being attractive because they can contribute to continued growth or help in reducing costs by providing low-cost production.

These firms raise capital, conduct research, buy supplies, manufacture, and market their products all around the world. Wherever it is feasible—and profitable—these organizations pursue growth, with little, if any, interest in the political boundaries of domestic markets. A typical global corporation might buy its supplies in Israel, raise capital in Europe, produce in the Pacific Rim, and sell in the United States and Canada. The structures of such dominant global corporations as General Electric or Britain's Imperial Chemical Industries are organized on the basis of worldwide business units instead of country-by-country subsystems. Japanese auto manufacturers such as Honda have decided that it is more efficient to produce locally in each major area of the Global Triad. Such regionalized centers for design, manufacturing, and assembly are known as **transplant production facilities** (Womack, Jones, and Roos, 1990). For example, Nissan's assembly plant in the United States (Smyrna, TN) accounted for production of 238,640 units by 1989; Nissan's European plants (in the United Kingdom and Spain) totaled 133,000 units by 1988 (Womack, Jones, and Roos, 1990).

ECONOMIC SUPERPOWERS

Global corporations are economic superpowers with huge sales revenues (see Figure 2-7). Indeed, the sales revenue of some is greater than the GNP of many small countries.

These global organizations are concentrated in the strongest economies of the three parts of the triad. Among the 500 biggest industrial corporations, the United States has 167 companies, Japan 111, Britain 43, Germany 32, and France 29. The picture is similar in the financial services sector. The incredible quantities of assets, revenues, and profits accumulated worldwide by these firms enable them to cross national borders and governments.

Globally competitive organizations gain market power and financial strength over the long term by outperforming their competitors. High-performing organizations are often driven by world-class management practices that provide the margin of competitive advantage, enabling these firms to succeed in the fiercely competitive core markets.

INDUSTRY	COMPANY	COUNTRY	COMPANY SALES ($ MILLIONS)
Aerospace	Boeing	U.S.	$29,314
Apparel	Levi Strauss Associates	U.S.	4,903
Beverages	PepsiCo	U.S.	19,771
Building materials	Saint-Gobain	France	13,311
Chemicals	E.I. Du Pont de Nemours	U.S.	38,031
Computers (incl. office equipment)	IBM	U.S.	65,394
Electronics	General Electric	U.S.	60,236
Food	Philip Morris	U.S.	48,109
Forest products	International Paper	U.S.	12,703
Furniture	Johnson Controls	U.S.	4,566
Industrial and farm equipment	ABB Asea Brown Boveri	Switzerland	28,883
Jewelry, watches	Seiko	Japan	3,070
Metal products	Pechiney	France	3,198
Metals	IRI	Italy	64,096
Mining, crude-oil production	Rubrkohle	Germany	14,902
Motor vehicles and parts	General Motors	U.S.	123,780
Petroleum refining	Royal Dutch/Shell Group	Brit./Neth.	103,835
Pharmaceuticals	Johnson & Johnson	U.S.	12,447
Publishing, printing	Bertelsmann	Germany	9,104
Rubber and plastic products	Bridgestone	Japan	13,226
Scientific and photo equipment	Eastman Kodak	U.S.	19,649
Soaps, cosmetics	Procter & Gamble	U.S.	27,406
Textiles	Toray Industries	Japan	6,666
Tobacco	RJR Nabisco Holdings	U.S.	14,989
Toys, sporting goods	Yamaha	Japan	3,654
Transportation equipment	Hyundai Heavy Industries	S. Korea	6,823

FIGURE 2-7 ***The Largest Global Companies by Industry***
Source: Fortune (July 27, 1992).

WORLD-CLASS MANAGEMENT PRACTICES

Global companies are pacesetters in the implementation of state-of-the-art management practices. Many global firms have been able to create and sustain competitive advantage through the integration of techniques such as Kanban, just-in-time inventory methods, computer integrated manufacturing, robotics, and lean production manufacturing. **Lean production manufacturing** is an organizational strategy that uses teams of multiskilled workers and flexible, automated machines to produce many different products in large volumes (Womack, Jones, and Roos, 1990). Lean production uses fewer resources than traditional mass production to attain the same goals. Producers using this technique strive for continuous improvements in all dimensions of their operation. The method requires having an organizational culture and company-wide learning capability to support it. The use of innovative human resource management practices is also desirable to sustain a lean production system.

Integrated, quality-oriented manufacturing is enhanced by having a well-trained, self-directed, multifunctional work force. Constant retraining of the entire work force, managers and workers, is rapidly becoming a standard feature of such operations. Innovative strategies that combine cutting-edge technology with work force productivity programs have thrived in many global organizations. These organizations thrive on speed and flexibility as the means to respond to dynamic local conditions in markets. Attaining such versatility requires that these organizations continually invest in building the skills, abilities, and learning proficiencies of their employees. To compete successfully in the dynamic core markets of the world demands that global organizations learn to find order in the chaos that such global settings often produce. More importantly, managers of these firms must think in ways that utilize the chaos to create new patterns of information that can then be used to renew and reinvent the organization. Nonanka (1988) has observed that "Chaos widens the spectrum of options, and forces the organization to seek new points of view" (p. 59). One firm that has successfully learned to continually reinvent itself is Italy's Gruppo GFT. The global systems spotlight presents several of the organizational challenges that GFT has faced as it moved to become a global enterprise.

GLOBAL SYSTEMS SPOTLIGHT

Italy's Gruppo GFT

Based in Turin, Italy, GFT is the world's largest manufacturer of designer clothing. It competes at the highest end of the apparel business—"ready to wear" designer collections, one step below made to order haute couture. GFT is the company behind such well-known European designer labels as Giorgio Armani, Emanuel Ungaro, and Valentino. The company makes some of the highest quality and most expensive clothes in the world: men's suits and women's outfits that sell for as much as $1,200.

Until recently, GFT was a relatively small and primarily Italian company. But in the 1980's, it rode the wave of global interest in European fashion to become a billion dollar business, with about 60% of its sales outside Italy and 26% in the United States alone. GFT brought the whole "Made in Italy" (synonymous with high quality and European design) fashion craze to the world. The company almost singlehandedly created the U.S. market for Italian designer apparel. Its U.S. sales grew from $7 million in 1980 to $304 million in 1989. Today GFT's 10,000 employees in the 45 small companies and 18 manufacturing plants under the Gruppo GFT umbrella make, distribute, and market roughly 60 designer and branded collections in 70 countries around the world.

But in the process of becoming a global company, GFT's managers have discovered that being global means something very different from what they had originally thought. They are struggling with a set of paradoxes that more and more companies are facing. For GFT, globalization is not about standardization, it's about a quantum increase in complexity. The more the company has penetrated global markets, the more sustaining its growth depends on responding to myriad local differences in its key markets around the world. In the words of GFT chairman Marco Rivetti, "to be global means to recognize difference and be flexible enough to adapt to it."

But to achieve flexibility requires turning the organization inside out. In a sense, the periphery has become the center—or, at least, the center of top management's attention. Adapting to local differences requires a far more multifaceted organizational structure, one in which innovation occurs at the periphery as well as at the center and where learning flows in many different directions. The centerpiece of GFT's global business strategy is to become an "insider" in each of its major markets.

Becoming an insider, however, has major organizational implications. Put simply, GFT has had to reinvent the way it does business, how it defines its customers, how it develops and manufactures products, how it handles marketing and distribution. What's more, this reinvention is not a onetime event, but an ongoing process. In effect, GFT is trying to create a "designer organization," able to adjust and adapt continuously to differences among markets and changes within markets—much as GFT designer collections change from year to year. The chief role of GFT's global managers is to manage the continuous redesign of the company.

Source: From R. Howard, "The Designer Organization: Italy's GFT Goes Global," *Harvard Business Review* (September–October 1991), p. 29.

STRATEGIC ALLIANCES

The last characteristic of global organizations is their use of strategic alliances. **Strategic alliances** are a pact between two or more organizations intended to enhance each firm's ability to compete. Such treaties were once a rare anomaly in the competitive marketplace. If they occurred at all they were based upon limited relationships such as licensing agreements, joint ventures, or vendor/supplier compacts. Today many global firms are tied together by an intricate network of alliances. For example, Daimler-Benz and Matsushita are linked in multipurpose cooperation. AT&T established a joint venture with Philips to manufacture and market both the model "5ESS" digital switch and transmission equipment outside the United States. AT&T also joined with Olivetti to sell computers and office equipment. Siemens and GTE created a joint venture to get access to each other's markets. Finally, Pratt & Whitney (United States), Rolls Royce (Britain), Japanese Aero Engines, Motoren Turbinen Union (Germany), and Fiat (Italy) have created a joint venture known as International Aero Engines (IAE) to produce a superior jet engine for medium-sized aircraft. Figure 2-8 identifies each part of an IAE engine by country.

These are just a few of the many illustrations of notable strategic alliances

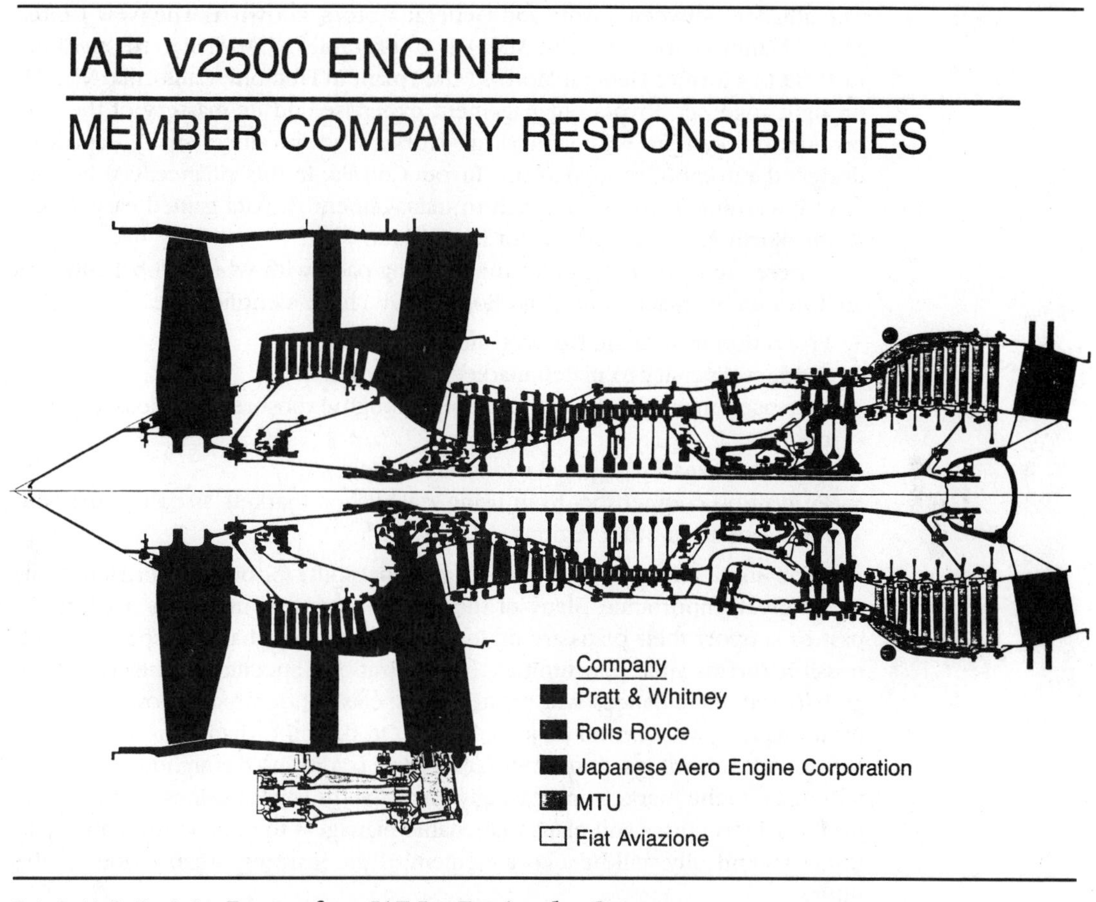

FIGURE 2-8 ***Parts of an IAE Jet Engine by Country***
Source: By kind permission of IAE International Aero Engines AG.

that extend over global markets with their far-reaching networks. The rationale used to justify such relationships is simple: in the global economy, even major corporations cannot compete alone in the dynamic environment. They need each other's support in product development, technological innovation, manufacturing, and marketing. While the task of managing a global organization is quite complex, the formation of strategic alliances greatly increases the intricacy of this process. This greater complexity is often the result of ill-defined, fluid relations between partners.

The effects of alliances are primarily global or regional, rather than local. The joint partners usually envision the world market as their target. As a result, the once clear boundary between competition and cooperation becomes rapidly blurred in such relationships. Strategic alliances also tend not to be stable; they have a relatively short life span. This shortness is generally due to the fact that companies employ them as a tactic for catching up in markets where they lag or for attaining a specific technological breakthrough.

The alliance between Toyota and General Motors, known as The New United Motor Manufacturing Inc. (NUMMI), is an example. NUMMI started operation in 1984 in a former General Motors (GM) plant in Fremont, California. NUMMI brought together Toyota's management expertise and its mastery of the lean production system with American workers to produce small, Japanese-designed automobiles such as the Toyota Corolla. In this alliance, GM benefitted by learning Toyota's approach to management. Toyota gained easy access to the North American market for its product.

There are many reasons for the growing pace with which global alliances are forming. Womack, Jones, and Roos (1990) have identified five:

1. Protection from trade barriers and currency shifts
2. Product diversity to match market fragmentation
3. Increased managerial sophistication through exposure to various environments
4. Protection against regional economic cycles
5. Preventing competitors from using established markets as cash sources to fund growth in other regions

The implications of these transformational shifts for business organizations are of great importance. Many of the assumptions that managers used in the past to support their plans are no longer valid. These changes represent both possible threats and opportunities to organizations. Specifically, the creation of global strategic alliances will create a new class of dominant, powerful organizations that will control major markets. On the other hand, the wide diversity in culture, values, consumer habits, and traditions throughout the world will make niche markets increasingly attractive to small businesses. The organizational structures and plans that enable managers to expand their ability to think systemically will be a core element of the learning organizations of the future.

One increasingly common way for global businesses to cooperate is through the formation of network organizations. **Network organizations** are a group of companies that are loosely coupled by a common interest in a product or market. Each firm contributes to the market only those functions at which it excels. One common type of arrangement is to distribute the design, manufacturing, and marketing functions among various members of the alliance. For example, the Festiva is designed by Mazda, built by Kia in Korea, and marketed by Ford. Such relationships are generally orchestrated by one firm, which takes on the responsibility of designing and preserving the network. That firm is known as a *hub* or *broker firm.* Snow, Miles, and Coleman (1992) recommend that managers who want to place their firms in a favorable position in the next century through the use of networks should follow a few guidelines:

1. Search globally for opportunities and resources.
2. Maximize returns on all assets dedicated to a business, whether owned by the manager's firm or by other firms.
3. Perform only those functions for which the company has or can develop expert skill.

4. Outsource those activities that can be performed quicker, more effectively, or at lower costs by others.

The use of strategic alliances and network organizations are just two of the many ways that global organizations are attempting to become more competitive by modifying their structures. When these approaches are used in tandem with world-class management practices it is easy to envision how global organizations can become formidable competitors in the world marketplace.

SUMMARY

The global systems that have the greatest relevance to managers today involve both global markets and trade blocs. Within both of these systems are numerous dynamic forces shaping the business environment of the future. Chief among these forces are new economic patterns, high-speed communication networks, the triad of trade blocs, and interlocked trends of competition and cooperation of global corporations. In combination, these forces have intertwined to create a new, more complex system. Consequently, the need for a systemic perspective is greater than ever before. The global organizations of today have sought to gain competitive advantage in new ways, such as adopting world-class management practices and managing information more effectively. This new information management means both high-speed processing of information, and creative use of it as a tool of innovation.

KEY TERMS AND CONCEPTS

General Systems Theory
open-systems perspective
coupling
global trade blocs
intangible resources
The White Paper Plan
Global Triad
transplant production facilities
lean production manufacturing
strategic alliances
network organizations

QUESTIONS FOR DISCUSSION

1. How are global corporations different from other types of businesses?
2. What is the importance of the Global Triad for any business that is considering becoming a global corporation?
3. How does the use of world-class management provide firms with a competitive advantage?
4. What are the advantages and disadvantages of strategic alliances?
5. Identify an actual example of a network organization. Explain the benefits that each member receives by being a member of the network.

REFERENCES

Banathy, B. (1991). "Comprehensive Systems Design in Education," *Educational Technology,* March, pp. 33–35.

Cushman, D. (1991). *High Speed Management.* Albany, New York: SUNY Press.

Drucker, P. (1980). *Managing in Turbulent Times.* New York: Harper and Row.

Drucker, P. (1989). *The New Realities.* New York: Harper and Row.

Durant, W., and Durant, A. (1975). *The Age of Napoleon.* New York: Simon and Schuster.

Katz, D., and Kahn, R. (1978). *The Social Psychology of Organizations,* 2nd. ed.

Itami, H. (1987). *Mobilizing Invisible Assets.* Cambridge, Mass.: Harvard University Press.

Nonanka, I. (1988). "Creating Organizational Order Out of Chaos: Self-Renewal in Japanese Firms," *California Management Review,* Spring, pp. 57–72.

Obloj, K. (1986). *Strategic Management.* Warsaw: Warsaw University Press (in Polish).

Ohmae, K. (1985). *Triad Power: The Coming Shape of Global Competition.* New York: The Free Press.

Ohmae, K. (1990). *The Borderless World.* New York: Harper Business.

Porter, M. P. (1990). *The Competitive Advantages of Nations.* New York: Free Press.

Richmond, B. (1990). "Systems Thinking: A Critical Set of Critical Thinking Skills for the 90's and Beyond." Lyme, New Hampshire: High Performance Systems.

Seeger, J. (1992). "Barriers to Entry of New Paradigms in Strategic Systems Thinking," *Proceedings of the International Systems Dynamics Society,* Utrecht, The Netherlands, July.

Senge, P., and Sterman, J. (1991). "Systems Thinking and Organizational Learning: Acting Locally and Thinking Globally in the Organization of the Future," in T. Kochan and M. Useem (eds.), *Transforming Organizations.* Oxford: Oxford University Press.

Snow, C., Miles, R., and Coleman, H. (1992). "Managing 21st Century Network Organizations," *Organizational Dynamics,* Winter, pp. 5–20.

Toffler, A. (1990). *Powershift.* New York: Bantam Books.

Womack, J., Jones, D., and Roos, D. (1990). *The Machine That Changed the World.* New York: Rawson Associates.

ADDITIONAL READINGS

Koch, J. V. (1989). "An Economic Profile of the Pacific Rim," *Business Horizons,* March–April, pp. 18–25.

Nonanka, I. (1991). "The Knowledge Creating Company," *Harvard Business Review,* November–December, pp. 96–104.

Obloj, K., and Joynt, P. (1986). "The Concept of Strategic Reserve," Working Paper of Bedrifstoknomisk Institute, Bekkestua, Norway.

3 Dynamic Dimensions of Systems

Ours is a complex world. But human knowledge is finite and circumscribed. . . . Since theories, like window panes, are clear only when they are clean, and the world does not come as cleanly as all that, we must know where to perform a clean-up operation.

Ervin Laszlo (1972), p. 13

1. To understand the role of perception in recognizing complexity in systems.
2. To see the relationship between system variety and complexity.
3. To be able to differentiate between various system states.
4. To comprehend Ashby's Law of Requisite Variety and understand its significance for managers.
5. To know the effect of feedback loops and coupling in creating system structure.
6. To be able to visualize the action of feedback loops in activating a system.
7. To become familiar with the different types of enigmatic outcomes.
8. To appreciate the role of mental models in relation to enigmatic outcomes.

SECTION I
DYNAMICS

SYSTEMS AND CHANGE

Systems have the capacity for changing both their form and behavior over time. Changes in form may be gradual—like the smoothing effect of water running over the same rocks for centuries or the evolution of life forms. In such cases change can only be observed over relatively long periods of time. For example, if one watches the hour hand on a clock, its movement is imperceptible in the present. It becomes noticed only when seen from time frames of hours rather than seconds or minutes. Sometimes the amount of change is sufficient to permit people to reach a consensus that the system is somehow different. In such cases the change has become recognizable. A change in the system may be noticed, but irrelevant to the survival and prosperity of the system. Some systemic changes that seem trivial, however, end up being significant.

There can also be a metamorphosis in the way a system acts over time. Organizations can leap from relative stagnation to sudden, dramatic growth. Similarly, countries can fluctuate through cycles of global expansion, contraction, disintegration, reformation, and expansion. For example, the political changes in Eastern Europe in the early 1990s were both significant and recognizable transformations of governmental systems. Paradoxically, though, when observed over relatively extended time frames, the behavior of a system may appear stable over time, or changes may follow regular patterns of growth, decline, or fluctuation. In some cases systems may also behave in chaotic and unpredictable ways that do not appear to fit any known pattern of behavior.

Why should managers be concerned with the way a system changes over time? How is this relevant to systems thinking? First, because systems change the management strategies useful at one time may not retain their value during other times. Second, because structure is a primary influence on the behavior of a system, understanding a system's behavior will afford many opportunities to gain insight into its actual structure. Third, the relationship between structure and behavior takes on added importance in the management process because specific types of structures are likely to promote certain patterns of behavior. For example, positive feedback structures promote growth behaviors in systems. Understanding the relationship between specific structures and behaviors is of great importance to managers, then, because it enables them to manipulate the fundamental inputs (structure) of a system to obtain the outputs (behavior, performance) they desire.

In traditional management theory, the ability of a system to behave dynamically over time is often regarded as abnormal. In the systems perspective dynamism is regarded as both expected and natural. Nevertheless, it is clear that systems often behave in unpredictable ways. Many managers have taken logical actions to improve the performance of a system only to witness outcomes that are unexpected, unintended, and contrary to conventional wisdom. As will become clear, this is a natural tendency that occurs when people intervene in systems. As an example, take the success of Japanese auto manufacturers in penetrating American markets. The global systems spotlight examines the unintended consequences of this success.

GLOBAL SYSTEMS SPOTLIGHT

Backfire Problems with Japanese Cars?

The success of Japanese automakers in the United States in the eighties created strong anti-Japanese sentiments among Americans, political pressures in Washington, and economic turbulence among America's auto workers. The solution developed by the Japanese companies seemed very rational: they decided to move production plants to the United States in order to offer new jobs and to Americanize their cars through the use of domestic production and supplies. Honda moved first, building a plant in Marysville, Ohio, in 1982, supplemented by another in Liberty, Ohio, built in 1989. Nissan established a facility in Smyrna, Tennessee, in 1983, which will be expanded in 1992. Finally, Toyota entered via a joint venture with GM in Fremont, California, which began in 1984, and with a plant of its own, opened in Georgetown, Kentucky, in 1988.

It was thought that building these plants would improve the Japanese carmakers' image. First, they established their plants in rural areas that experienced chronic unemployment. Second, they offered long-term contracts to and shared their technological and managerial know-how with United States suppliers. (In many cases this sharing resulted in dramatic increases in the suppliers' productivity and competitiveness.) Third, they built high-quality, affordable cars that, pound for pound, were more American than many American cars. Fourth, in order to Americanize even more, in 1991 the Japanese producers announced plans to increase production and build new plants. Finally, the Japanese companies tried to behave in socially responsible ways. For example, Toyota has donated a $1-million community center to Georgetown, given ten cars to the city and county governments, contributed an amount equal to the school taxes it is exempt from paying, and voluntarily serviced the loan for a new sewer plant.

The unintended consequences of this concerted effort have been multidimensional. The public is still angry and worried by the continued Japanese dominance of consumer markets. American automakers have taken a beating; in 1991 each of the Big Three incurred losses in the American market and saw their market share decline. Toyota, Nissan, and Honda, instead of being just foreign competitors, showed their powerful presence in the United States and built up a production base for further market-share gains. Finally, politicians used the Japanese automakers' actions to generate fear as a springboard for their own gain. For example, in 1991, some legislators introduced bills that would restrict the production level of the Toyota plant to its current level, despite the potential for the creation of more new jobs if production were to expand. Were the unintended consequences predictable?

As the case illustrates, even thoughtfully designed plans succeed in some respects and still produce unintended consequences. This chapter will explore how systems change over time. In particular, we will devote attention to explaining how the behavior of systems can vary in both expected and unexpected ways. We will examine the proposition that unintended consequences are a natural outgrowth of human actions in systems, using both relevant theories and case examples. We will also examine how those results sometimes do not fit prevailing notions of what is likely or possible. These issues will be discussed in relation to several of the major factors that influence human actions in systems: complexity, predictability, and dynamics as well as a key influence underlying them all, perception.

MANAGERS IN COMPLEX DYNAMIC SYSTEMS

Managers often think of systems as though the systems exist independently from their own experience. Such thinking is likely to produce the conclusion that managers' experiences are a property of the system rather than a product of the interaction between themselves and the system. Managers often perceive individual systems as having a particular level of complexity; once identified, that level of complexity is often assumed to be a given characteristic of the system. While systems do have attributes—such as a certain number of elements and a specific number of relationships among these elements—it is difficult to appreciate the complexity of a system based solely on those attributes (Klir, 1985). This is because, as was pointed out in Chapter 1, what is seen depends on what is thought. Thought is the cause, and experience is the effect.

It may be more fruitful to think of a system's complexity from a relative perspective. The complexity of a system depends on the mental model and interests of the person perceiving the system, as well as on the attributes of the system in question. Therefore, complexity is created by the mutual interaction of people with something external to them but within their experience. Since systems can be seen from numerous alternative points of view, their complexity is a matter of interpretation. Figure 3-1 depicts one possible view of activities related to coal mining. Which systems do you see operating in this example? How are the various systems related to each other?

A system's potential for complexity is governed largely by the amount of variety within that system. The **variety** of a system is determined by the number of distinct elements within it (Beer, 1984). (The number of relationships that exist among these parts is also an important contributor to a system's complexity [Klir, 1985].) Variety can be thought of as the sum total of different ways of acting or being that a system is capable of attaining. Organizations that have a lot of variety can exist in many different **system states.** The state of an element is the extent to which it has reached its capacity. The capacities of most elements can be described in terms of two key dimensions: the percent of holding capacity and the pace of increase or decrease in whatever is in holding. For example, a manufacturer's inventory can be described in terms of level—as, say, 60 percent of capacity—or in terms of the rate of change—as a percent of its potential capacity for rate of turnover. The state of a manufacturing firm could be described in terms of its production output per month. This is likely to be a range of values extending from a minimum, which is zero, to a maximum, representing the system's top capacity (Figure 3-2). The actual level of production output is the state of the system—for *that* element. Of course, the system will have other states as well—each corresponding to a different element.

A system with a great deal of variety is often problematic for managers, because such a system is often difficult to regulate using any single approach unless that approach can be responsive to all of the possible states. For example, the materials requirements of a manufacturing process fluctuate as a consequence of the interaction of many different factors. Attempts to plan

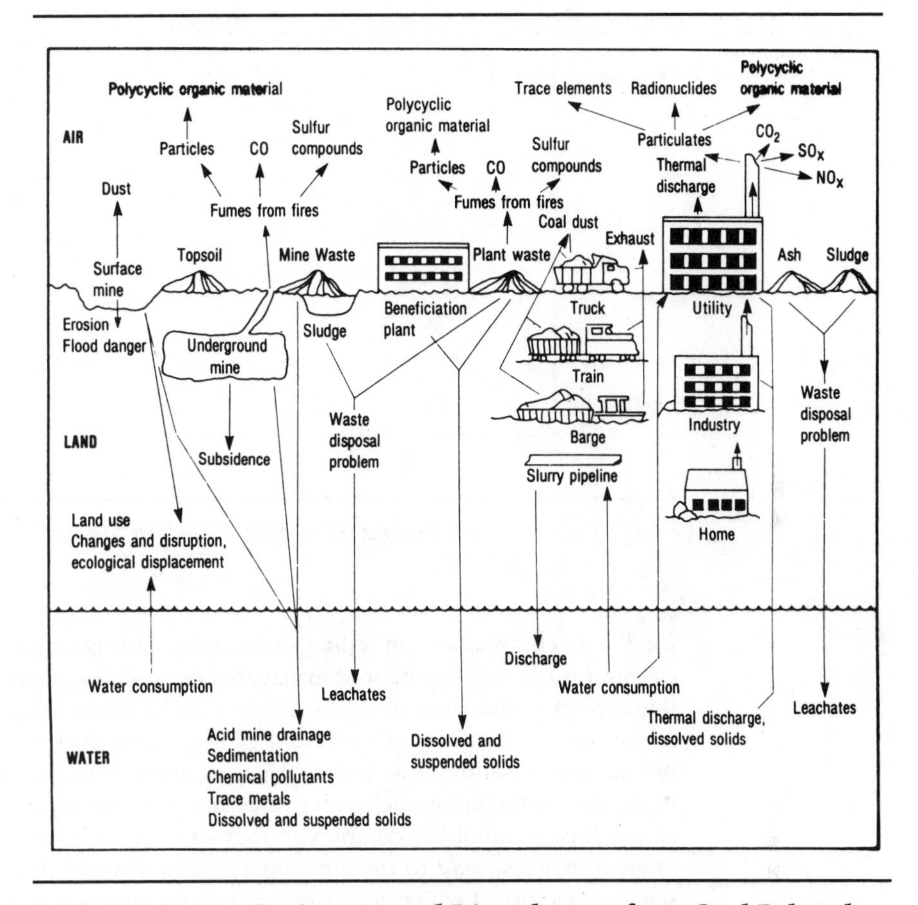

FIGURE 3-1 ***Environmental Disturbances from Coal-Related Activities***

Source: U.S. Congress, Office of Technology Assessment, *Prospects and Problems of Production and Combustion* OTA-E-86 (Springfield, Virginia: National Technical Information Service, April 1979).

materials requirements that do not consider other related system states, such as cash flow or sales, are likely to have limited effectiveness. This is because most simple techniques cannot account for the ability of the system to attain certain system states.

A major school of systems thinking proposes that one of the most important functions of managers is to limit the effects of variety on a system (Ashby, 1960; Beer, 1984). This proposition has been formalized as an important axiom, **Ashby's Law of Requisite Variety.** Ashby's Law suggests that the variety of a system can be reduced only when the means used for reduction has at least as much variety as the system (Waelchli, 1989). The means used to accomplish this reduction can be anything from thinking differently to employing management information systems and controls. Regardless of the approach

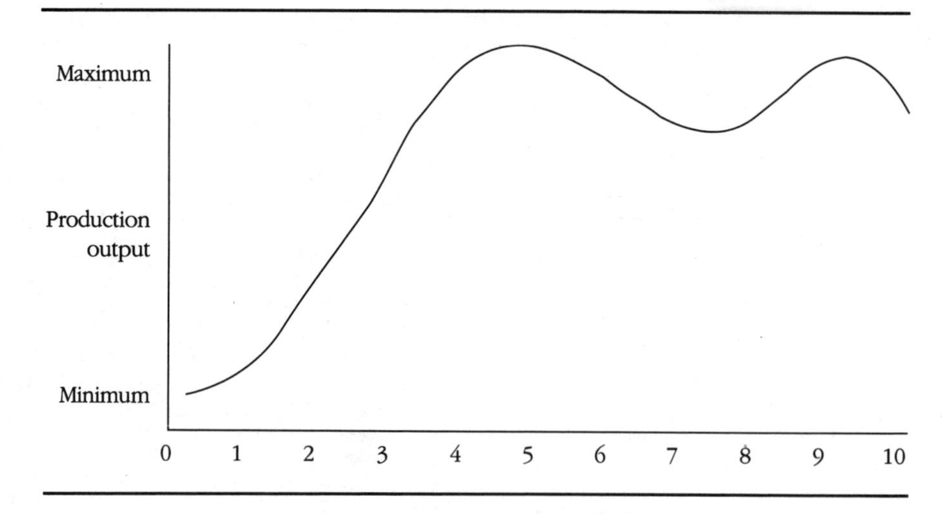

FIGURE 3-2 ***A Range of Possible System States***

used, only complexity can reduce complexity. This principle has major implications for the development of managerial capabilities. Managers who can see things from a number of possible perspectives are more likely to envision ways to improve a situation. Again, what a manager perceives depends on what he or she thinks. Ashby's Law provides strong support to the notion that developing robust mental models can enable managers to see their world in ways that reduce some of the complexity they experience. Conversely, such development may also *add* to the complexity of their world by making them conscious of factors of which they were previously unaware.

The complexity of a system also depends on the level of sophistication and the interests of the person viewing the system. Perception is a selective process, governed by one's needs and interests. The dimensions of a system that the perceiver notices are not always those that are the most relevant for understanding a specific issue. When Isaac Newton saw the proverbial apple fall to the ground, he associated that action with the concept of gravity. Had he not eaten a meal for two days, his interest in the apple may have been quite different.

The implications of the Law of Requisite Variety for systems thinking are great. The law draws a clear line that associates a person's capacity for judgment with success in interacting with a system. The concept of requisite variety is elegant, yet simple. It can be applied to anything from interpersonal relations to strategy formulation in organizations. In systems thinking, the continual process of redefining complex systems in new, more inclusive ways is a prime vehicle for making one's mental models more robust. That is, simply by working to enhance the richness of his or her mental models, a manager builds his or her capacity for variety.

So far complexity has been considered in terms of attributes of the system and the person viewing that system. Though useful as a starting point, this perspective is incomplete because it does not consider how a system changes over time. It is also important to consider whether a system changes over time in a well-defined pattern.

~ PREDICTABILITY

A system's behavior becomes relatively more predictable when it follows the same pattern over time. The longer the period in which it follows the pattern, the greater the predictability. Two primary factors influence predictability: the system's variety and its past pattern of behavior. When a system's behavior is seen as being determined by a given set of initial circumstances, the system is described as being **deterministic** (nonprobabilistic). In such a case, all information about factors that cause system behavior is assumed to be known with certainty and fixed. The behavior of such systems is assumed to be controlled simply by controlling its input, since the initial state determines the outcome. In fact, relatively few systems that are of concern to managers are deterministic. Certain dimensions of mechanical systems may be usefully modeled using deterministic assumptions, but such assumptions fail to work with human systems.

When a system's behavior is the result of the interaction of initial conditions, subsequent conditions, and randomness, it becomes capable of acting in a number of different ways (within limits). Such a system's behavior is **stochastic** (probabilistic). Such behavior is of great concern to managers because it is more difficult to control. This is where systems thinking has great value. When a number of factors act simultaneously to generate outcomes, simple intuition and analytical thinking often are inadequate to predict outcomes. For example, knowing when to reorder raw materials involves stochastic assumptions because many factors influence the decision, including the outflow of finished goods created by consumer demand and the output of the manufacturing system, which will be affected in turn by the impact of absenteeism, materials defects, and natural variations in human energy levels. The simultaneous rise and fall of raw materials, work-in-process, and finished goods inventories and varying rates of flow of inputs and outputs can cause such a system to behave in complex ways.

The predictability of a system's behavior increases when the pattern it follows remains a constant straight line. Such behavior follows a **linear** pattern. For example, the output of an industrial robot may remain constant at x number of units per hour. Over the short run, the profitability of a new company may soar along a straight-line pattern of growth. As far as managers are concerned, these patterns are both rare and inconsequential. The behavior of virtually all systems important to organizations varies over time and does not follow a straight-line pattern. More often systems fluctuate in rates of growth or decline. Such behavior is **nonlinear,** or dynamic. Over relatively long times, most systems exhibit nonlinear behavior.

In rare situations, the behavior of some systems does not appear to follow any discernible pattern. In these situations a system's behavior is characterized as being in **chaos.** Ford says chaos is "systems liberated to randomly explore their every dynamical possibility" (in Gleick, 1987, p. 306).

The vast majority of situations that concern managers involve dynamic (nonlinear) behavior rather than linear or chaotic behavior. Dynamic behavior can be understood as performance that fluctuates (oscillates) over time, often fitting a pattern. There are many reasons why the performance of systems oscillate. Among the most prominent are the influence of time delays between perception and action and the effects of feedback loops (Mass and Senge, 1978). But before learning about feedback and delays, it is necessary to understand more about the patterns of dynamic behavior possible.

UNDERSTANDING DYNAMIC BEHAVIOR

PATTERNS OF DYNAMIC BEHAVIOR

If the behavior of a system is tracked over a period of time, the pattern of performance is usually characterized by either growth, decay, or a combination of the two. Rarely does a system maintain a constant performance level over time. When a system's behavior pattern remains relatively constant, the system is described as being in either a state of **equilibrium** or **near-equilibrium.** Conversely, when a system's behavior fluctuates chaotically, it is termed as being **far-from-equilibrium.** Equilibrium is a stable state where the forces of growth and decay offset each other, creating a degree of relative balance. Some systems are unstable and are unlikely to stay at an equilibrium level very long. In other cases, a system may go through a process of transformation that necessitates moving to a new equilibrium level in order to prosper. When transformation is necessary for survival, it becomes desirable for the system to move rather than remain at a prior equilibrium. Equilibrium will be discussed in greater detail in Chapter 7.

Each basic type of behavior has some specific variations, which are determined by the structure of the system. But the basic behavior pattern of all systems generally fit one of several types of dynamic behavior. There are six fundamental patterns (see Figure 3-3), one of which is equilibrium. Since it is unlikely that a manager will be concerned with any system that remains in a prolonged state of equilibrium, we will discuss the other five types.

- *Growth curves.* The slope, or steepness, of any growth curve may vary, but the curve is always positive. Growth patterns are usually created by the influence of positive feedback within a system. In some systems growth may be exponential; in other words, growth doubles within a given time. For example, a population of rabbits may double every six months or the principal in a savings account may double every eight years at a certain rate of interest. Figure 3-4 shows an example of growth.

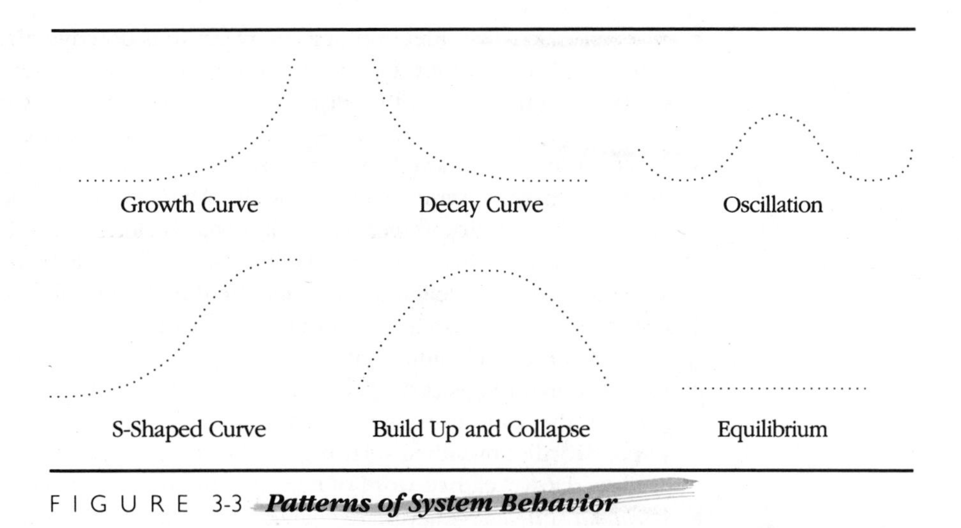

FIGURE 3-3 ***Patterns of System Behavior***

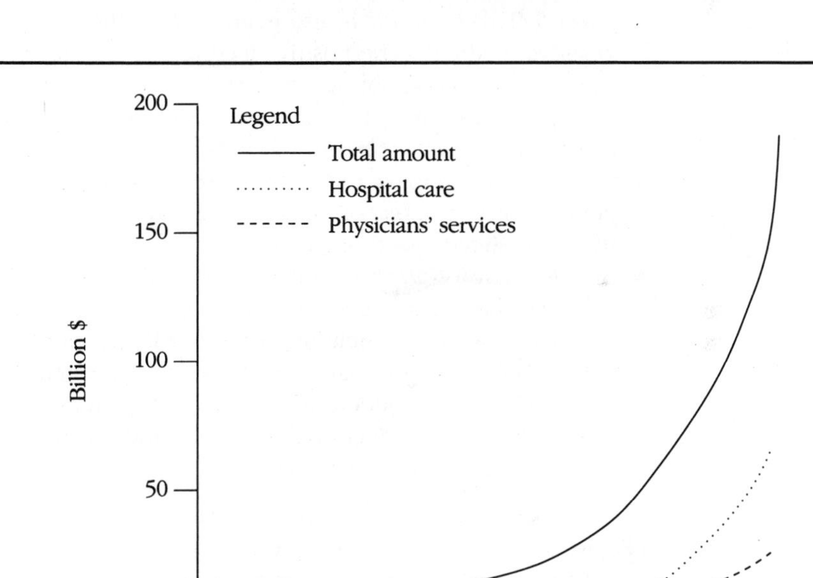

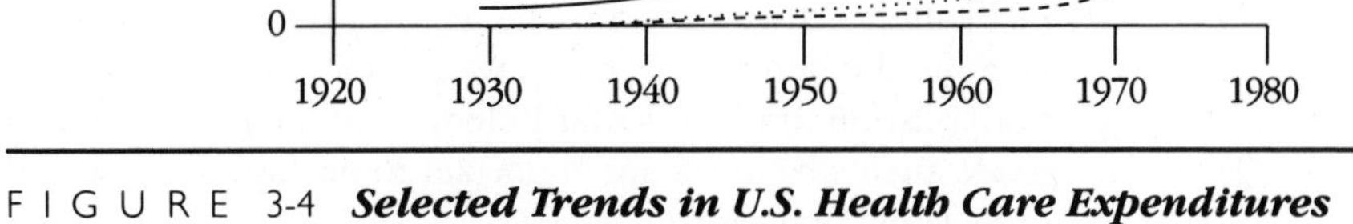

FIGURE 3-4 ***Selected Trends in U.S. Health Care Expenditures***

Source: P. Lawrence and D. Dyer, ***Renewing American Industry*** (New York: Free Press, 1987), p 101.

- *Decay Curves.* The direction of a decay curve is downward, and its slope is negative. Decay curves are also the product of positive feedback within a system. Much like growth, decay can also happen at an exponential rate. For example, as quality declines, market share may decay exponentially.
- *Oscillation.* Oscillation is characterized by its wavy pattern. In some cases the waves may be symmetrical, while in others there may be minimal symmetry and dramatic, atypical spikes in behavior. Oscillation is caused by the interaction of positive and negative feedback as well as by time delays. The tendency toward oscillation may also be due to the inherent stability of a system. Some systems have structures that cause them to repeatedly return to the same equilibrium point. For example, the pendulum on a clock will always return to a vertical position, no matter how hard it is pushed. Oscillation is also caused by the amplification of inputs into the system. For example, word-of-mouth advertising may accentuate promotional efforts for a product. However, bad word-of-mouth communication accelerates decline.
- *S-Shaped Curves.* S-curves are characterized by a pattern that resembles the letter *S.* This behavior is also generated by the interaction of positive and negative feedback. The positive feedback causes steep growth in the early stages of the curve, but negative feedback dampens the growth and flattens the curve. This curve is a common pattern of behavior for new companies or new products. Steep growth is generated as consumers find out about the product and make initial purchases. As company profits rise, new competitors enter the market and the original company's growth rate slows as the pie is sliced into thinner slices.
- *Build Up and Collapse.* This pattern is also a function of the interaction between positive and negative feedback. Build up and collapse results from a number of factors including resource limitations, speculation, and time delays. Many markets—such as those for real estate, commodities, and labor—follow this boom-to-bust cycle. Low prices accelerate demand, but this is offset by the effect of growing demand shrinking supply, thereby driving up prices, which lowers demand.

Oscillation is evident in the cyclicality of U.S. motor vehicle production (Figure 3-5). Four different dynamic patterns are evident in Figure 3-6.

If a manager recognizes these behavior patterns operating in a system that he or she wishes to improve, he or she must look for the structural dimensions creating the behavior. For instance, if one observes exponential growth, attention must be directed toward identifying the positive feedback mechanisms at work. Such structures are important to managers because they are the engines of dynamic behavior.

THE ENGINES OF DYNAMIC BEHAVIOR

Generally, changes in the behavioral pattern of a system reflect a structural response within the system. (Remember that structure plays a major role as a cause of behavior.) While the structure of a system may be purposefully rede-

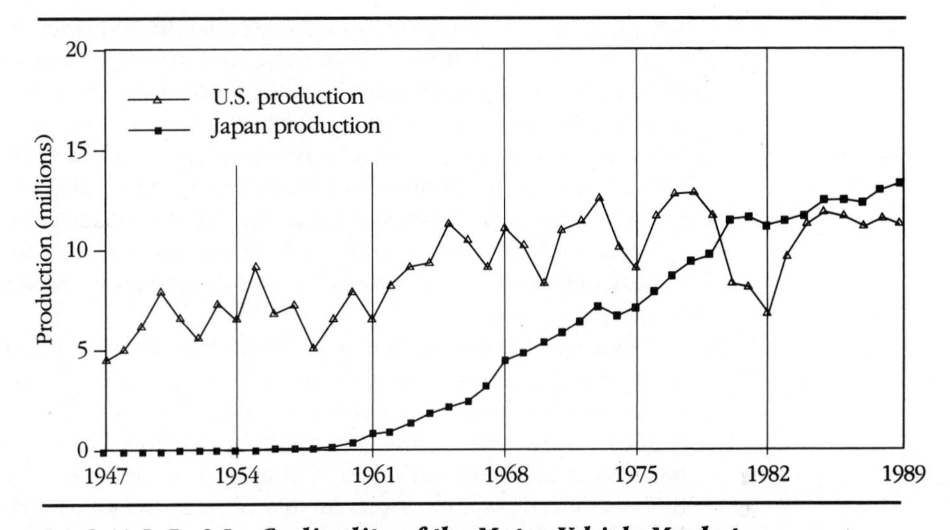

FIGURE 3-5 ***Cyclicality of the Motor Vehicle Market***

Source: J. Womack, D. Jones, and D. Roos, *The Machine That Changed the World* (New York: Rawson Associates, 1990) p. 247.

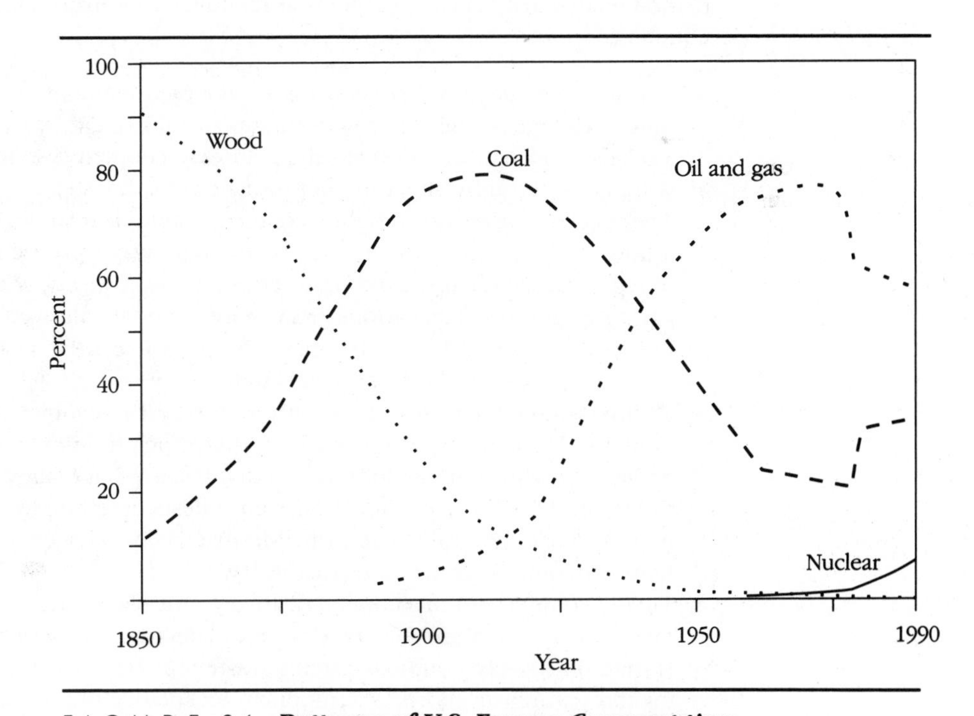

FIGURE 3-6 ***Patterns of U.S. Energy Consumption***

Source: A National Plan for Energy Research, in R. Naill, "A Systems Dynamics Model for National Energy Policy Planning," *Systems Dynamics Review* (Winter 1992), *1,* p. 3.

signed, it can also evolve in response to changing conditions. Structural changes may result from mutual accommodations between various parts of the system. Relations between parts of a system are typically developed over time to reinforce those behaviors instrumental to the survival of a system. For example, organizations such as Apple Computer, Honda, and Marriott Corporation have strong cultures that amplify the effects of their strategies. This positive relationship between culture and strategy allows these organizations to operate with fewer controls and restrictions on people, thereby fostering a climate of innovation, which supports continued growth and success in future strategies.

This type of structure evolves over time and is difficult to re-engineer. Frequently, rather than being the product of a conscious design, the emerging relations within a system are the consequence of an informal process of **self-organization.** Self-organizing processes transform a system's structure by breaking away from prior patterns of behavior to establish a new order on the basis of internal information. In this process, relations between parts may strengthen and the elements may become more tightly coupled, or relations may diminish and the elements become more loosely coupled. The term *coupling* refers to the extent of connection between elements. When parts of a system are loosely coupled, each has greater freedom to express its individual natural tendencies.) Perrow (1984) has identified four major characteristics of coupling:

1. The interdependency between tightly coupled elements is a function of time. A change in one factor necessitates an adjustment by the other within a relatively short time. Conversely, in loosely coupled systems time delays will not significantly affect the performance of the system.
2. The sequence of events in tightly coupled systems is relatively fixed: B must follow A; 11 A.M. must precede 2 P.M.; the only way to score a home run is to tag first base, then second base, and so on. Conversely, when one is preparing spaghetti sauce, various creative freedoms are allowed as to whether the garlic is added before the basil. Because it can be tasty either way, making sauce is a loosely coupled system.
3. Tightly coupled systems allow only one way to accomplish the desired outcome. In many surgical procedures this principle, known as **unifinality,** prevails. On the contrary, loosely coupled systems offer many ways to attain the desired goal. For example, different salespeople employ various techniques to secure a sale. The principle that many ways exist to get to the same outcome is known as **equifinality.**
4. Tightly coupled systems contain relatively little slack, meaning that there is relatively little margin for error without affecting the performance of the system. In loosely coupled systems waste and error can be compensated for through repair efforts or temporary acceptance of lower quality. A bug in a software design may easily cause a program to fail, whereas a performer who is tired may just produce a subpar performance.

COUPLING AND COMPLEXITY IN ORGANIZATIONS

Organizations are designed by managers, engineers, public officials, and consultants, each of whom has a particular mental model of how the parts should fit together. Many of the formal structures created within organizations to achieve specific outcomes do indeed influence performance. However, many of the subsystems within organizations have the potential to behave in ways that extend beyond what was anticipated by the designers' mental models. When subsystems behave in unintended ways and are tightly coupled with other elements, they have the potential to produce unexpected outcomes. These outcomes often defy what systems designers—or managers—believe is possible. The fundamental structures of organizations and technologies greatly influence their potential to produce unexpected outcomes. Figure 3-7 depicts various systems plotted against the interdependency among coupling, linearity, and complexity. (In Section 2 of this chapter we will examine several examples of unintended consequences.)

The minicase that follows illustrates how seemingly unrelated forces interact to shape system behavior.

Dead Cows and Hungry Beavers

Blame it on beavers or balky computers, but one thing is clear: It doesn't take much for the telephone system to go down. And a phone failure means delays in airline flights and problems for air traffic controllers. Two of the thousands of passengers inconvenienced in a September 17 failure were Al Sikes, chairman of the Federal Communications Commission, and panel member Ervin Duggan.

The Federal Aviation Administration, in a report to a House subcommittee, said that from August 1990 to August 1991 there were 114 "major telecommunications outages" across the country that led to flight delays and generated safety concerns. In May, four of the FAA's 20 major air traffic control centers shut down for five hours and 22 minutes. The cause: "Fiber cable cut by farmer burying dead cow. Lost 27 circuits. Massive operational impact." A year ago, the Kansas City, MO, air traffic center lost communications for four hours and 16 minutes. The cause: "Beaver chewed fiber cable." Other causes include lightning strikes, misplaced backhoe buckets, blown fuses, and computer problems. More recently, two technicians in an American Telephone & Telegraph Co. long-distance station in suburban Boston put switching components on a table with other, unmarked components, then put the wrong parts back into the machine. That led to a three-hour loss of service at Logan International Airport.

Telecommunications subcommittee Chairman Edward Markey, D-Mass., has been highly critical of the FCC for its response to the recent phone system failures. "They do not have a plan to deal with modern technologies that can cripple an entire region's telephone service," Markey said. "They've described these outages as unusual or a unique set of circumstances. But the truth of the matter is that this is fairly predictable and it's going to keep happening over and over again."

By the mid-90's the FAA hopes to have in place a new and complex backup system that would turn today's three-hour shutdowns into 20-second interruptions that would go unnoticed by the public. Today, the phone company patrols its cable routes by air, looking for troublesome construction projects. "Ultimately," said the FCC's James Spurlock, "there are no 100 percent guarantees. There is a human dimension," Spurlock said. "Farmers are going to bury dead cows. We can't always prevent that."

Source: Associated Press, "Beavers, Bad Circuits Linked to Phone, Air Traffic Shutdowns," *The Hartford Courant* (November 11, 1991).

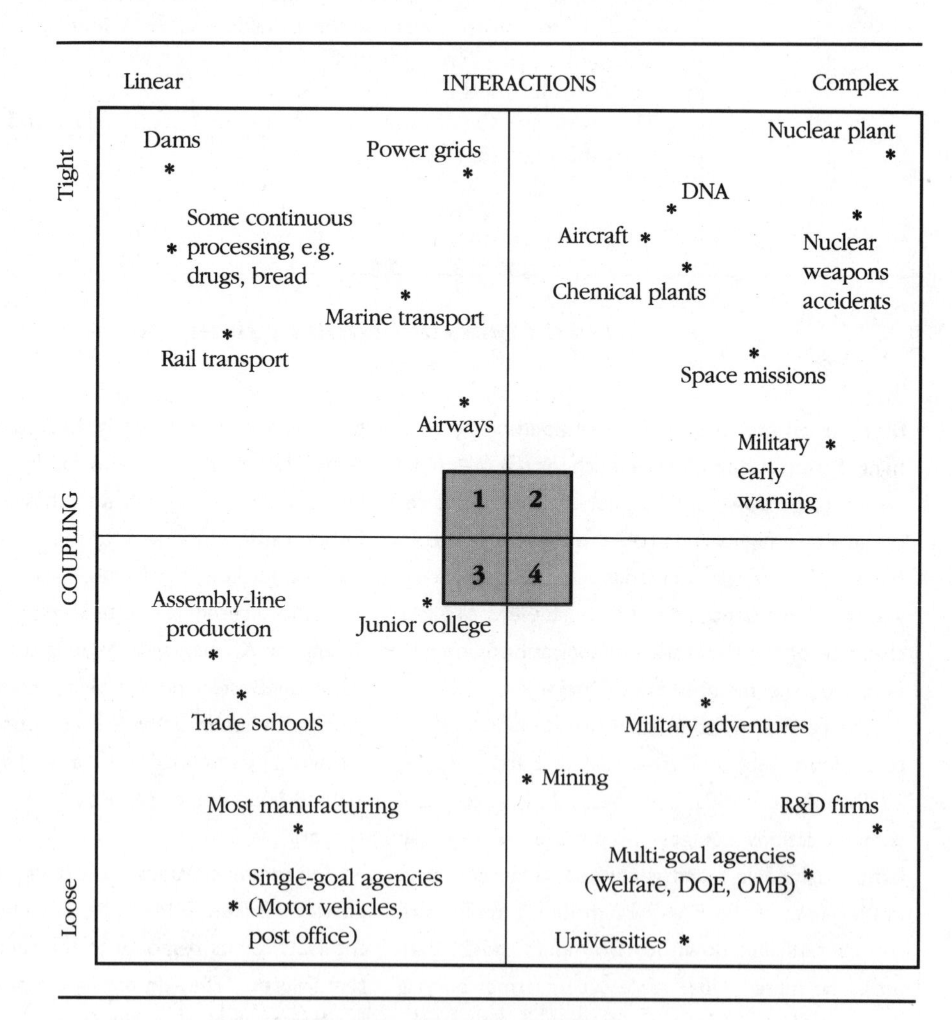

FIGURE 3-7 ***Interaction-Coupling Chart***

Source: C. Perrow, *Normal Accidents: Living with High-Risk Technologies* (New York: Basic Books, 1984), p. 97.

The coupling that exists in a system gains its potential for dynamism through the influence of feedback within a system. Feedback both defines the structure of a system and activates that system to behave dynamically. Developing an understanding of how feedback works to vitalize a system is a critical element of systems thinking for managers.

STRUCTURE AND FEEDBACK

A manager's initial perceptions of a system are often determined by its form or structure. In organizations, structural patterns are often symbolized by formal artifacts such as organization charts. In fact, such artifacts say very little about how the organization actually operates. As discussed earlier, ***structure*** refers to the arrangement among the elements of the system and the relationships that exist among those elements. These relationships can vary on the basis of two main factors: (1) *relative strength* (element B is more sensitive to changes in element A than is element C), or (2) *importance in the system* (the relationship between elements E and F is a primary determinant of the behavior pattern of a system). Structural arrangements of system elements are determined in part by policies, rules, procedures, reporting relationships, and information networks. Such relationships may develop through evolution, or they may be the product of a conscious design. In either case, the structure of the system has a major impact on determining the type of causal relations that will emerge within the system. Those causal relations, in turn, are the engines that drive system behavior. Thus, system structure determines the types and range of performance that a system is capable of achieving. Even though a system may gradually change its goals or tactics over time, its structure provides a degree of continuity that helps to preserve its identity. Structures typically are more resistant to change than are system elements or processes, such as communication.

Interactions among the elements of a system usually establish a pattern over time. Much like rainwater rushing off the side of a mountain, it initially creates furrows that gradually enlarge. The evolution of interdependencies among elements is facilitated by recurring information exchanges. The information flows that trigger changes in the states of elements are known as **feedback.** Feedback is an essential ingredient within systems and a core concept in systems theory. It links system elements and helps to coordinate their actions so that they behave as a whole. From a systems perspective, every interaction can be viewed in terms of an influence process of action and reaction. Feedback is both a reaction to a particular condition that has been recognized and an attempt by an element to influence other elements in the system. For example, improvements in the quality of an organization's products or services can set off an accelerating chain reaction of events that supports further improvements in quality.

The thought of something being both a cause and an effect may seem puzzling. It becomes quite simple to visualize this when one shifts from thinking about cause and effect as linear to seeing it as circular. (See Figure 3-8.)

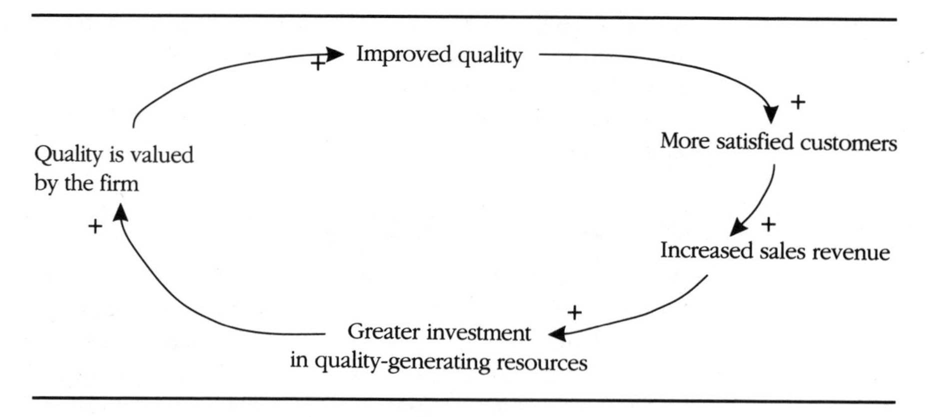

FIGURE 3-8 ***Causal Loop of Quality***

For example, a student may believe that if she studies hard, she will receive high grades. From a systems perspective, however, influence is seen as being circular. High motivation may lead to good grades; which leads to acknowledgment from professor, family, and peers; which causes higher motivation; which causes higher grades. In this situation, the elements of this system are linked by each acting as both cause and effect. This spiral of causes and effects is called a **causal loop.** Causal loops often become self-reinforcing and difficult to alter. Systems driven by powerful causal loops tend to maintain their own momentum; such loops must be recognized before the system can effectively be changed.

Such self-reinforcing cycles are one form of **feedback loop.** Feedback loops that fuel growth and expansion within a system are known as *positive feedback loops.* Those that control and restrict the behavior of a system are known as *negative feedback loops.* Positive and negative loops may interact together in systems to create counterbalancing effects. For example, in market economies the economic laws of supply and demand create both positive and negative loops that tend to offset each other's influence and maintain a general balance of prices over time, when all other things remain equal. The effect tends to stabilize the market over the long term.

The key point for managers is that once the behavior pattern of the system has been identified, it is possible to determine which combination of feedback loops is causing the behavior. Then efforts to change the system can concentrate on modifying the underlying structure that causes the particular feedback.

Over time, through the mechanism of feedback, changes may echo through a system much the same as the water of a pond ripples outward when a stone is tossed into it. When a change affects a system, it has both present and future consequences. Consequently, a whole system may become transformed into a different system than it was the second before the change occurred. The net effect of feedback is to cause continuous change in the state of the system, which in turn alters each of the elements. In a relatively static system, such as a machine, these patterns are less obvious, but over extended periods of time they may become more evident.

DELAYS

Time delays are an important contributor to the dynamic behavior of human systems. **Time delays** occur because there are gaps between the thoughts and actions of people attempting to affect the system. Thoughts and actions are often distant and loosely coupled in relation to time because it is possible to recognize patterns only in the past, not in the present. For example, a commercial airliner is able to remain on course because it continually compares its present position against a preset course. In an organization, goals map out a desired course, but actual consumer demand may differ substantially from the expectations on which the goals were based. Managers try to predict the future by recognizing past patterns. This is like trying to steer the airliner by looking backwards.

Some delays occur quite normally, between the time that necessary information is received and actions are taken (Forrester, 1961). This is because information must be studied and plans of action formulated prior to their implementation. In the meantime, the system is still headed wherever it is going, possibly at an ever accelerating rate. Delays may also occur in the receipt of needed materials, since materials do not arrive at the instant they are ordered (Richardson and Pugh, 1981). Even when kanban ordering systems are used, delays occur between order placement and receipt.

The first section of this chapter has argued that systems may be perceived as being complex dynamic entities that may or may not be predictable. In general, the potential for a system to behave in enigmatic ways is the product of its properties of complexity, predictability, and structure. Even when systems are relatively well understood, they may behave in ways that managers consider to be unexpected and nonsensical. Managerial efforts designed to control or improve a system may, instead, yield results that are unintended and even contrary to what a manager's basic instincts suggest should be happening. Section 2 looks at the role of these enigmatic outcomes such as unintended consequences and counterintuitive behavior.

SECTION 2
ENIGMATIC OUTCOMES

UNINTENDED CONSEQUENCES

When people intervene in complex systems, they may inadvertently cause undesired side effects that may actually hinder their efforts to resolve problems. **Unintended consequences** are effects that cannot be predicted without greater knowledge of the system. Sometimes this information is not available when the original action is taken. At other times, the information is available but not used. This may be due to limitations imposed by time pressures to act or to the perception that the information was not critical. A lack of understanding of the subtle complexities within a system may cause managers to ignore vital signs. For example, there are many indications that people knew that the mirror of the Hubble Space Telescope required further testing and polishing. Reports such as this front-page quote about that telescope—"A scientist who nags colleagues for critical tests of the Hubble mirror is ignored as a company relic" (*The Hartford Courant,* April 1, 1991, p. 1)—are common in many analyses of system failures. The holes in the ozone layer are the unintended consequences of the use of commercial and industrial chemicals, the effects of which were never fully understood. Worker boredom and alienation are the unintended consequences of the misunderstood application of Frederick W. Taylor's scientific management philosophy.

UNINTENDED CONSEQUENCES: NORMAL BEHAVIOR

The tendency for people who intervene in complex systems to have a limited ability to anticipate the consequences of their actions leads to the following proposition: *The radius of the effects of human actions intervening in a system's design or performance is often greater than the radius of human predictions of possible effects.* For a variety of reasons, people are not generally able to envision what effects their actions may precipitate within complex systems (Figure 3-9). In some cases, the complexity of the system simply exceeds people's ability to comprehend all possible outcomes. In other situations, where elements of a system are tightly coupled, chain reactions of effects can be set off by seemingly minor forces. One of the great lessons being learned about high technology is that even relatively simple systems may couple in ways that produce totally unexpected results.

While not all unintended consequences are harmful or unwelcome, they still may have the capacity to disturb the balance within a system. For example, the disaster of the NASA space shuttle *Challenger* was without doubt a personal tragedy for many people. On the other hand, it forced NASA to examine its safety procedures and raised a larger policy issue to the American public and the government: What level of risk is acceptable in space programs?

Furthermore, not all unintended consequences are unpredictable. Many unintended consequences that appear anomalous from a local perspective are predictable when seen from a system-wide perspective. When people interact with large-scale complex systems, the odds of unintended consequences emerging are so common that they are not surprising. As Charles Perrow (1984) states, "If interactive complexity and tight coupling—system characteristics—inevitably will produce an accident, I believe we are justified in calling it a *normal accident,* or a *system accident.* The odd term *normal accident* is meant to signal that, given the system characteristics, multiple and unexpected interactions of failures are inevitable" (p. 5). That unintended conse-

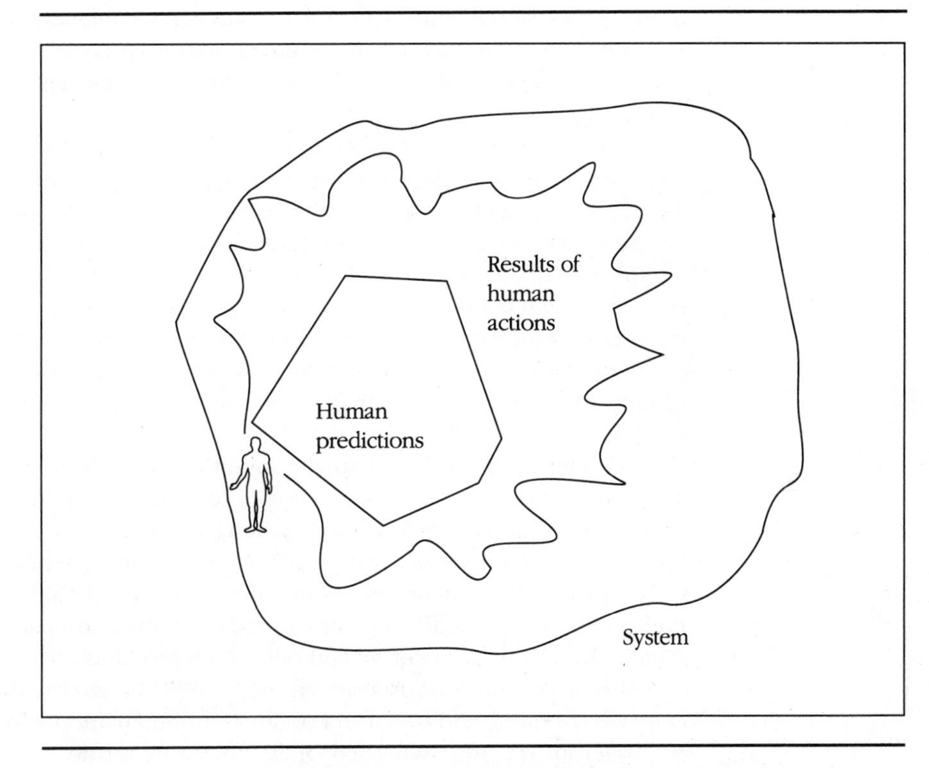

FIGURE 3-9 ***The Origins of Unintended Consequences***

quences are generally unexpected is paradoxical. In complex, dynamic systems, unintended consequences arise with regularity. Although they may not be expected, they are normal. Perrow (1984) has argued persuasively that many high-risk technologies—such as chemical plants, nuclear power plants, dams, ships, and airplanes—are accidents waiting to happen.

UNINTENDED CONSEQUENCES IN AN ECONOMIC SYSTEM

The case of the Organization of Petroleum Exporting Countries (OPEC) shows unintended consequences plaguing an economic system. Peter Drucker (1980) has noted that OPEC has largely failed to achieve its objectives. Drucker has observed, "To the outside petroleum-importing world, OPEC appears a huge success. But from inside OPEC, the cartel must seem a gigantic failure and the years since 1973 a period of bitter frustration and disappointment" (p. 159).

The increase of oil prices by OPEC producers in 1973 found most of the oil importing countries unprepared. The so-called oil crisis began. In many countries, the government imposed restrictions on the purchase of oil and derivate products, mainly gas. The predicted results that oil producers wanted to achieve were attained: their revenues increased considerably and their political influence was duly enhanced. However, despite dramatic increases in the OPEC members' wealth and political influence and the creation of a monopoly dictating conditions on the world market, their real success was short-term and self-limiting.

Several factors that OPEC countries had not taken into account caused the short-term success. First, in the late seventies, oil importers rapidly introduced new technologies that dramatically reduced the demand for oil and derivate products. Second, constantly increasing prices of oil made the search and drilling for oil more economical, even in the most difficult circumstances (such as from the ocean floor). New producers—especially African and Latin American countries—entered the market and consequently increased the overall supply of oil while demand was beginning to stabilize or decrease. Third, the prices of industrial goods—the primary imports of the OPEC countries—increased at a faster pace than the price of oil.

The OPEC countries eventually found themselves in an unexpected and most undesirable trap. The complexity and instability of the situation quickly caused conflicts and strife to arise among OPEC members. Some wanted to limit supply and increase prices; others wanted to increase supply and decrease prices. Proponents of lower oil prices argued that an immediate increase of prices of industrial goods would more than offset additional revenues coming from the export of oil and petroleum products.

After a decade of prosperity in most OPEC countries, the economic climate deteriorated, and social and political turmoil naturally followed. It would be irrelevant to judge whether OPEC countries would have been better off without attempting global domination of the market. What is obvious, however, is that their efforts to dominate oil markets resulted in their isolation, limited political and economic influence, and increased dependency upon Western economies and in some ways contributed to internal political conflicts and wars such as the Iran-Iraq War.

UNINTENDED CONSEQUENCES IN A POLITICAL SYSTEM

The 1991 collapse of the Soviet Union highlights unintended consequences in the political realm.

> An abyss opened for a moment, and black bats flew out. They filled the air with old nightmares, throwbacks to a style of history that the world had been forgetting. The Soviet Union was seized by a sinister anachronism: its dying self. Men with faces the color of a sidewalk talked about a "state of emergency." They rolled in tanks and told stolid lies. The world imagined another totalitarian dusk, cold war again, and probably Soviet civil war as well. If Gorbachev was under arrest, who had possession of the nuclear codes? (Morrow, 1991, p. 20)

This was the beginning of one of the most dramatic periods in modern political history. On August 18, 1991, a group of Communist officials lead by KGB Chairman Vladimir Kryuchkov, Defense Minister Dmitri Yazov, and Interior Minister Boris Pugo, among others, initiated a coup against Soviet leader

Mikhail Gorbachev. The goal of the coup leaders was to forestall the Soviet Union's awkward transition toward a more democratic, decentralized, market-oriented system of government. They hoped for a return to the traditional Communist doctrine that had prevailed for nearly three quarters of a century in the Soviet Union. The coup leaders secured the critical backing of military leaders to place Mr. Gorbachev under house arrest at his vacation home in the Crimea. The military acted swiftly by rolling out troops and tanks.

Little popular support for the coup was found among the populace, and a counter-coup movement emerged led by a Gorbachev adversary, Boris Yeltsin. Yeltsin, a grass-roots leader, quickly gained the loyalty of the Soviet masses and stalled the coup as he built support to overthrow it. Soldiers and tanks in Moscow's Red Square were rendered powerless to act by the pleas and protests of their fellow countrymen.

> Citizens shouted "Fascist" or worse at the troops and scrawled swastikas in the dirt on tanks parked outside the Russian Parliament Building, climbed aboard armored personnel carriers to argue with the commanders and urge them to turn back—all with impunity. When the coup leaders decreed a curfew from 11 P.M. to 5 A.M. the soldiers made no attempt to enforce it. (Morrow, 1991, p. 35)

The coup ended in less than a week, and its leaders were arrested. Yeltsin emerged as the acknowledged savior of the republic and gained the status of folk hero. "As for the surviving plotters, . . . they were facing not only treason trials, but also the knowledge that their mismanaged coup had intensified the move toward democracy and decentralization they had tried to stop. The three days that shook the world were over" (Morrow, 1991, p. 44).

Prior to the coup, Mikhail Gorbachev's agenda for the political reform of the Soviet Union had remained bogged down by factionalism and bureaucracy. The coup unintentionally catapulted the political system into a profound transformation by uniting the splintered populace and circumventing the rusty political machine of the national government. "Then the bats of history abruptly turned back and vanished into the past. By act of will and absence of fear, the Russian people accomplished a kind of miracle, the reversal of thousands of years of autocracy" (Morrow, 1991, p. 20). The coup that was meant to return the Soviet Union to the past inadvertently launched it into the future.

COUNTERINTUITIVE BEHAVIOR

One specific type of unintended consequence is the result of a paradoxical outcome known as counterintuitive behavior. The concept of **counterintuitive behavior** was formulated by Jay Forrester (1975) to explain predictions or conclusions that his computer models made that did not conform to conventional wisdom. In fact, the behavior that results from interventions into a system may be the opposite of what was sought. For example, expanding highways as a means to reduce traffic jams may, in the long run, increase traffic flow by initially making the roads more attractive to commuters. The leading indicator that counterintuitive behavior may be emerging is when a system appears to be generating behavior that is seen as being out of context or out of type.

For example, PeopleExpress Airlines feared being bought out by larger competitors during a period of intense industry consolidation. Managers reacted aggressively by attempting to expand rapidly; they acquired competitors and established a new hub in Denver. Their rationale was that being bigger would make the company less likely as a takeover target. However, their strategy weakened the company financially to the point that it could not sustain losses of market share and profits resulting from competitor counterattacks. As a result PeopleExpress's strategy made the company more vulnerable to internal collapse.

The PeopleExpress case is a classic example of two systemic learning disabilities recognized by Peter Senge (1990), both of which often produce counterintuitive behavior. They are known as "the enemy out there" syndrome and "the illusion of taking charge" syndrome. To some degree, the causes of counterintuitive behavior can be traced to unrealistic expectations that are the byproduct of distorted mental models. "The enemy out there" syndrome causes managers to not see the systemic consequences of their actions; when the consequences occur, they are blamed on external causes. "The illusion of taking charge" syndrome is present when managers attempt to fight the "enemy out there" by being more proactive and aggressive. This stance does not provide insight into ways to beat the system; rather, it reinforces the potential for counterintuitive consequences to occur. Both syndromes are produced by distortions in managerial mental models (Senge, 1990).

The minicases and extended examples presented in this chapter illustrate a common pattern of logic that accompanies virtually all human efforts to intervene in complex systems. When faced with a situation they perceive as problematic, people normally create some sort of plan for action. The intended solution aims at improving the situation by removing the causes of the problem. Frequently, such linear thinking does accomplish this intended goal. Nonetheless, as time passes the effect of the original solutions often diminishes, and unexpected consequences emerge. Once these unexpected consequences become significant problems in their own right, the search for new solutions starts all over again. The new solutions similarly yield results and unexpected consequences, and so on. The cycle tends to reinforce itself unless the system is changed.

~ SUMMARY

Systems have the ability to change over time through processes ranging from slow evolution to rapid transformation. Systems are also dynamic entities capable of behaving in numerous ways, ranging from maintaining equilibrium to following one of five patterns of fluctuation to chaotic, disordered behavior. While a system's environment undoubtedly contributes to its behavior, structure plays a key role in determining how a system will act. Internal structural factors such as feedback loops, complexity, coupling, and linearity all interact to generate system behavior.

Some of these interactions may produce unintended or counterintuitive outcomes. Furthermore, the actions that people take are likely to produce effects that extend beyond what seems possible on the basis of their own mental models. In fact, many unintended consequences are indeed precipitated by the faultiness of our mental models. Clearly, a manager's ability to think systemically is important, and it must be coupled with a capacity to see the world from a dynamic perspective to pierce the fog of complexity in organizations.

KEY TERMS AND CONCEPTS

variety
system states
Ashby's Law of Requisite Variety
deterministic
stochastic
linear
nonlinear
chaos
equilibrium
near-equilibrium
far-from-equilibrium
self-organization
unifinality
equifinality
feedback
causal loop
feedback loop
time delay
unintended consequences
counterintuitive behavior

QUESTIONS FOR DISCUSSION

1. Are organizations basically equilibrium-seeking entities or do they favor continual change?
2. List the five most important causal loops in your life. Compare your list with those of other people. Are there any commonalities?
3. What are the most tightly coupled systems that might be found in the world of business, the environment, and technology?
4. Of the following three points about unintended consequences, choose the one that you think is most important for managers and explain why. (a) There's nothing that can be done about them. (b) Managers need to be better prepared to avoid them. (c) Managers need to learn from them.
5. How can precisely engineered technologies such as airplanes, nuclear power plants, and ships still have normal accidents? Rule out human error as a factor and examine the systemic dimensions of these technologies.

REFERENCES

Ashby, W. R. (1960). *Design for a Brain,* 2nd ed. London: Chapman and Hall.

Beer, S. (1984). "The Viable System Model: Its Provenance, Development, Methodology, and Pathology," *Journal of the Operational Research Society, 35,* pp. 7–26.

Drucker, P. (1980). *Managing in Turbulent Times.* New York: Harper and Row.

Forrester, J. (1961). *Industrial Dynamics.* Cambridge, Mass.: Productivity Press.

Forrester, J. (1975). *The Collected Papers of Jay W. Forrester.* Cambridge, Mass.: Wright-Allen Press.

Gleick, J. (1987). *Chaos: Making a New Science.* New York: Viking-Penguin.

Klir, G. (1985). "Complexity: Some General Observations," *Systems Research, 2,* no. 2, pp. 131–40.

Laszlo, E. (1972). *The Systems View of the World.* New York: George Braziller.

Mass, N., and Senge, P. (1978). "Alternative Tests for the Selection of Model Variables," *IEEE Transactions on Systems, Man, and Cybernetics,* July.

Morrow, L. (1991). "The Russian Revolution," *Time,* September 2, pp. 20–44.

Perrow, C. (1984). *Normal Accidents: Living with High-Risk Technologies.* New York: Basic Books.

Richardson, G., and Pugh, A. III (1981). *Introduction to Systems Dynamics Modeling with Dynamo.* Cambridge, Mass.: Productivity Press.

Senge, P. (1990). *The Fifth Discipline: The Art and Practice of the Learning Organization.* New York: Doubleday.

Waelchli, F. (1989). "The VSM and Ashby's Law as Illuminants of Historical Management Thought," in R. Espejo and R. Harnden (eds.), *The Viable System Model.* Chichester, England: John Wiley, pp. 51–75.

~ ADDITIONAL READINGS

Cain, S. R. (1970). "Anthropology and Change," in *Growth and Change,* #3.

Church, G. (1991). "The Anatomy of a Coup," *Time,* September 2, 1991.

Forrester, J. (1968). "Industrial Dynamics After the First Decade," *Management Science, 14,* no. 7, pp. 398–415.

Hardin, G. (1968). "The Tragedy of Commons," *Science,* December.

Lovelock, J. (1979). *Gala: A New Look at Life on Earth.* Oxford, England: Oxford University Press.

4 The Roots of Systems Thinking

People who described themselves as practical men, proud to be uncontaminated by any kind of theory, always turned out to be the intellectual prisoners of the theoreticians of yesteryear.

John Maynard Keynes

1. To understand how the function of system design is rooted in age-old assumptions and philosophies.
2. To recognize that management, as practiced in Western cultures over the past century, is largely based on one of several possible sets of assumptions.
3. To distinguish between the different types of philosophical positions that can serve as the basis for a world view.
4. To appreciate the power of world views and paradigms in shaping people's perceptions of the world.
5. To see how a person's or organization's choice of a paradigm will influence how business is conducted.
6. To envision the critical role played by information and learning in systems.
7. To comprehend the necessity for continuous improvement in the ways information is gathered and processed in organizations.
8. To evaluate the importance of objectivity in organizations.

SECTION I
SYSTEMS DESIGN AND SYSTEMS THINKING

INTRODUCTION

Systems thinking is based on a diverse set of ideas that have developed in the fields of engineering, organization theory, the natural sciences, and philosophy among others. Clearly, then, systems thinking is an approach to problem solving that has roots in many disciplines. The purpose of this chapter is to introduce and discuss the fundamental elements of the perspectives that have contributed to the development of systems thinking.

This chapter will also introduce several key relevant issues relating to systems research. The role of the inquiring system in systems design will be discussed from a historical perspective. Finally, this chapter will also discuss two fundamental concepts, *world view* and *paradigm.*

The philosophical origins of systems thinking can be traced back to many early traditions in both Western and Eastern culture. Various dimensions of systems thinking are descended from philosophies that date back to the sixth century B.C. Among the oldest of these great writings are Plato's *Republic* and Lao-tze's *Tao te ching.* Both of these seminal works will be discussed later in this chapter.

~ THE EVOLUTION OF SYSTEMS THINKING

A study of how systems philosophy evolved can roughly be divided into two basic areas: the study of how knowledge is acquired (epistemology) and the study of the philosophy of science.

Both of these areas of inquiry have important implications for understanding the evolution of systems thinking. Historically, systems theories have concentrated on issues related to the design of systems and the analysis of problems with these systems. In turn, systems theories have been revised based on the lessons learned from the application of systems analysis and design. Boulding (1956) captured the interrelationship among systems analysis, systems design, and systems practice: "Systems concepts foster a way of thinking which, on one hand, helps the manager to recognize the nature of complex problems and thereby to operate within a perceived environment" (p. 197). These three core dimensions that underlie systems philosophy are depicted in Figure 4-1.

Before attempting to understand the fundamental philosophies that serve as the underpinning for systems theories, it will be useful to briefly review the function of systems design and the role of the systems designer.

~ SYSTEMS DESIGN

Systems design may sound like a high-flying, esoteric activity that is the exclusive domain of people with Ph.D.s in electrical engineering. In reality, systems design is something done by most managers of business organizations. Viewed

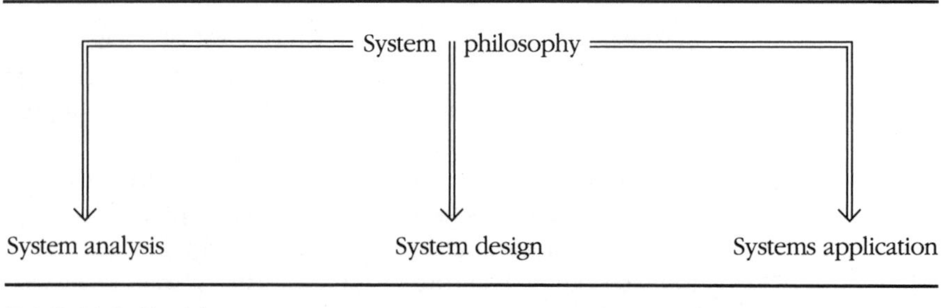

FIGURE 4-1

in the broadest sense, most adults are system designers. They are creators of the life they want to live; as such, they are forced to make the same basic choices that a system designer in an organization must face. **Systems design** refers simply to the process of creating structures to accomplish a specific, planned purpose. Managers may design systems for compensation, information management, strategic planning, or many other functions. These systems all tend to be orderly, precise means for organizing different functions, each with a narrow scope, into an integrated whole. For example, a reward system may include formal peer group recognition, pay increases, and incentive bonuses. All of the specialized functions may be tied together by an overall logic such as reward superior performance or provide incentives while being fair to everyone.

The process of designing a system can be viewed narrowly as a purely mechanical attempt to engineer a framework: in organizations it takes such forms as organization charts, information system flow diagrams, or plant layout blueprints. Viewed more holistically, however, systems design can encompass efforts to engineer human environments so as to foster outcomes, such as better performance. For example, management consultant Michael Hammer (1990) writes, "We should 'reengineer' our businesses: use the power of modern information technology to radically redesign our business processes in order to achieve dramatic improvements in their performance" (p. 104).

The framework of a system should tangibly express the attempt to reconcile the prevailing issues being faced by the system and its designer. One constructive way to frame this process is to see the framework as the means toward achieving an ideal future. Both the functions of design and engineering play a fundamental role in hard systems thinking. In organizations, the focus of design and engineering is to promote stability and enhance certainty by clarifying relationships and making the criteria for decisions more explicit. In organizations, systems design is always purposeful. That is, it is intended to help the organization achieve a predetermined goal such as improving productivity or reducing costs. The specific goal focused on sets up the criteria that will guide the systems design process. Nugent and Vollman (1972) have identified two basic functions that are the essence of the systems design:

1. Clearly define the criteria and objectives of the system.
2. Determine which types of systems are most likely to lead to the accomplishment of those objectives.

This second step concentrates on determining which elements should be included within the system and how they should be related to each other. This essential process of deciding what to include in and exclude from the system depends heavily on the designer's depth of understanding of both the organization and systems philosophy. Understanding the interactive nature of the key elements of a system is a necessary foundation for effective systems design. The case on the Ford Motor Company depicts the redesign of that company's accounts payable system.

Ford Motor Company

In the early 1980's, when the American automotive industry was in depression, Ford's top management put accounts payable—along with many other departments—under the microscope in search of ways to cut costs. Accounts payable in North America alone employed more than 500 people. Management thought that by rationalizing processes and installing new computer systems, it could reduce the head count by some 20%.

Ford was enthusiastic about its plan to tighten accounts payable—until it looked at Mazda. While Ford was aspiring to a 400-person department, Mazda's accounts payable organization consisted of a total of 5 people. The difference in absolute numbers was astounding, and even after adjusting for Mazda's smaller size, Ford figured that its accounts payable organization was five times the size it should be. The Ford team knew better than to attribute the discrepancy to calisthenics, company songs, or low interest rates.

Ford managers ratcheted up their goal: accounts payable would perform with not just a hundred but many hundreds fewer clerks. It then set out to achieve it. First, managers analyzed the existing system. When Ford's purchasing department wrote a purchase order, it sent a copy to accounts payable. Later, when material control received the goods, it sent a copy of the receiving document to accounts payable. It was up to accounts payable, then, to match the purchase order against the receiving document and the invoice. If they matched, the department issued payment.

The department spent most of its time on mismatches, instances where the purchase order, receiving document, and invoice disagreed. In these cases, an accounts payable clerk would investigate the discrepancy, hold up payment, generate documents, and all in all gum up the works.

One way to improve things might have been to help the accounts payable clerk investigate more efficiently, but a better choice was to prevent the mismatches in the first place. To this end, Ford instituted "invoiceless processing." Now when the purchasing department initiates an order, it enters the information into an on-line database. When the goods arrive at the receiving dock, the receiving clerk checks the database to see if they correspond to an outstanding purchase order. If so, he or she accepts them and enters the transaction into the computer system. (If receiving can't find a database entry for the received goods, it simply returns the order.) . . .

Ford didn't settle for the modest increases it first envisioned. It opted for radical change—and achieved dramatic improvement. Where it has instituted this new process, Ford has achieved a 75% reduction in head count, not the 20% it would have gotten with a conventional program. And since there are no discrepancies between the financial records, material control is simpler and financial information is more accurate.

Source: M. Hammer, "Reengineer Work: Don't Automate, Obliterate," *Harvard Business Review* (July–August 1990), pp. 104–111.

THE SYSTEMS DESIGNER

Most organizations have relatively few individuals responsible for designing the basic framework of the entire organization. Numerous other subsystems however, need to be engineered by managers and staff. The Ford minicase offers such an example—the redesign of an accounts payable system. In the process of redesigning this system, a wide variety of people were involved, including both managers and nonmanagers.

Representatives from various functional areas—such as accounting, information systems, purchasing, and shipping and receiving—needed to be involved. Similar activities that occur in many organizations can be improved if managers have a knowledge of basic system design principles. For example, systems that govern compensation, rewards, manufacturing, operations, executive decision support, accounting control, and financial planning all must be designed by someone. Once designed, these systems must be skillfully put into operation, continuously evaluated, and frequently redesigned to align them with the changing needs of an organization.

The process of designing a system is directed toward defining the ideal relations to establish among people, machines, information, and materials. The design sets the boundaries of the playing field and defines specifically how each element will interact with all others. Within that context, other structural factors—such as rules, procedures, and policies—further define the behavior expected from people. This type of structure is relatively static and works in conjunction with systemic structure and processes. In organizations, structure is a critical element in mediating relations with the environment. (Organization structure will be examined fully in Chapter 9.)

The term *organization structure* differs from the concept of system structure discussed in Chapters 1 and 3. System structure is the patterns of mutually reinforcing relations that form within the context of a system's framework and that drive its behavior. Using the human body as an analogy, the systems design process sets the skeletal framework and hooks the various subsystems (digestive, respiratory, and so on) together. The feedback loops that emerge over time within this context will determine how well the system works and which patterns of behavior it will follow.

SCIENTIFIC APPROACHES TO SYSTEM DESIGN

Scientific thinking has often been equated with the ability to reduce complex subjects to their basic elements. This is often regarded as a rigid process built on the use of controlled experimentation to verify the observations or experience of the researcher. In reality, this is a very narrow definition of scientific thinking based on but a single view of what constitutes science, the Newtonian perspective. Newton proposed that the world operated as a rational, mechanical system driven by a set of universal laws that would ensure its consistency. Consequently, the standard he employed to ascertain the soundness of a scientific claim was that the same experimental result must be producible in repeated trials.

Viewed from a broader perspective, natural science is the philosophies and methods we use to understand why the world is as it is. As Einstein and Infeld (1938) suggest:

> Physical concepts are free creations of the human mind, and are not, however it may seem, uniquely determined by the external world. In our endeavor to understand reality we are somewhat like a man trying to understand the mechanism of a closed watch. He sees the face and the moving hands, even hears the ticking, but he has no way of opening the case. If he is ingenious he may form some picture of the mechanism which could be responsible for all the things he observes, but he may never be quite sure his picture is the only one which will explain his observations. He will never be able to compare his picture with the real mechanism and he cannot even imagine the possibility of the meaning of such a comparison. (p. 31)

The systems approach represents one major type of scientific thinking. The balance of this chapter will trace the development of scientific thinking and compare and contrast systems thinking with other types of scientific thinking.

TRADITIONS IN SCIENTIFIC THINKING

For centuries, scientists believed that it was possible to uncover the laws of nature. These scientists argued that a set of universal natural laws—equally applicable to the workings of any field of natural science—existed and could be uncovered. Later, as this perspective gained broader acceptance, attempts to apply these laws to the social sciences and organizations began. These efforts at extending scientific thinking have not lived up to their initial promise, but they have left an enduring legacy. The process of inquiry in the natural sciences has led to the development of conceptual frameworks, and these patterns of thinking have shaped the way people in the West discovered and learned about systems. Over time the various patterns of thinking that emerged within science became defined as proper scientific thinking. Unfortunately, what got lost is the fact that the way scientific thinking is defined at any time is anchored in the culture and dominant views of each era. Without doubt, managerial thinking over the past century has been embedded in the prevailing scientific philosophies, for better or worse.

The technique commonly known as the scientific method is based on a set of philosophies popular in Europe during the 1700s (Capra, 1982). This procedure also serves as the conceptual basis for the rational problem-solving methods used in most organizations today. The rational approach to problem solving is based on three basic sets of beliefs: reductionism, determinism, and positivism. **Reductionism** argues that the best way to understand something complex is to break it into its components and then examine those components. **Determinism** assumes that the universe operates as a mechanical clockwork. It proposes that if the initial conditions of a system can be known and measured, the effect of any changes can be predicted because the system acts in predictable ways according to the laws of nature. **Positivism** assumes that all necessary information about a situation is available, unambiguous, and measurable. Positivism rests on the belief that if something cannot be measured it does not exist. These three assumptions hold up well in mechanical systems, but they are of less value in dealing with complex, dynamic systems such as organizations.

Although organizations may not be directly accountable to the laws of the physical sciences, some aspects of organizations are more relevant to the study of natural science than others. For instance, product development is more related to the physical sciences and engineering than human resource management. As a result, it is not unreasonable to consciously engineer certain aspects of organizations in ways compatible with this approach (Jackson and Keys, 1984). Conversely, highly complex, pluralistic activities—such as strategic planning—are less likely to be successfully engineered (Mason and Mitroff, 1981).

In some cases deterministic models are useful tools for thinking about complex issues because their simplicity enables managers to easily manipulate ideas in their minds. This can lead to breakthroughs in complex issues. But managers must avoid being seduced by a model's simplicity and substitute it for reality. While the reductionist perspective has advantages under certain circumstances, it also has limitations. For this reason it can complement systems thinking but not substitute for it. The following example illustrates this point.

To illustrate the limitations and power of alternative ways of thinking, assume that you are posed with the following problem. There are four sticks arranged in the following pattern:

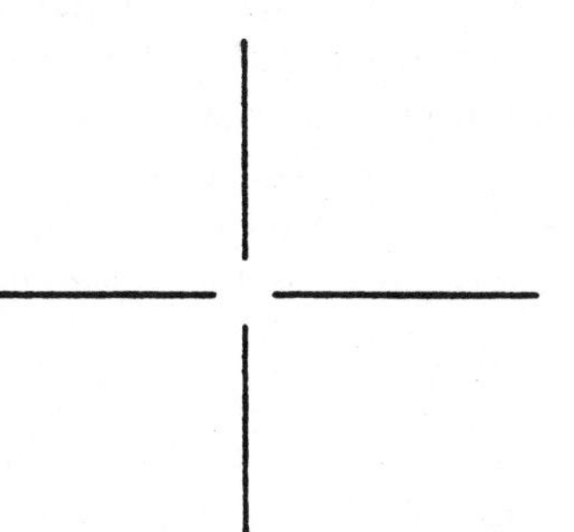

The problem you have to solve is simple: by moving one stick, create a perfect square. How would you attack this problem? Probably, you would follow the typical analytical tactic of trying to disarrange and arrange the sticks many times. You would try to build a square—but what kind of square? Obviously enough (the authors have performed an experiment with this problem in the United States, Poland, Norway, Denmark, Israel, and other countries), you would follow spatial logic. After all, you know what a square looks like:

All the people trying to solve the problem knew after a few minutes that it is *impossible* to build a square moving just one stick. Nevertheless, most of them pursued the logic they considered systematic, despite the obvious paradox hidden in the problem. However, once somebody *accepts* the fact that the problem cannot be solved using spatial logic, what is the alternative? She must adopt a systems perspective and look for an alternate frame of reference and a new set of assumptions. The most effective strategy in this case is to consider using mathematical logic—and then the solution comes instantly.

PHILOSOPHY, NATURAL SCIENCE, AND SYSTEMS THINKING

Systems theory has traditionally been the child of the various philosophies of science. In this regard, it is similar to the relationship between natural science and both engineering and technology. Many scientific laws and principles apply directly to these fields and serve as their theoretical basis. Technology provides the means to manifest this body of theory in a way that is useful to people. Engineering may be viewed as the applied science that unites or integrates the physical sciences and technology. The important point is that both engineering and technology are generally regarded as direct descendants of natural science (Checkland, 1985).

Patterns of thinking from the natural sciences are not as applicable to systems thinking as they are true in the engineering disciplines. Organizations and social systems may be governed by some natural laws, but thinking in the social sciences and in organization theory must be based on a different set of assumptions. Patterns of human behavior are less predictable than natural patterns governed by scientific laws. Geoffrey Vickers (1983) has persuasively defended the idea that human systems are not analogous to physical and biological systems, and therefore the same scientific principles cannot be used to explain them.

Though the laws of physical sciences may not directly dictate how organizations function, scientific thinking has an important attachment to organizations. The dominant philosophies of science within a culture directly influence the world view of people within that society. A **world view** is a conception of reality that a person or group uses to explain how the pieces of their experience fit together. This is another means through which people interpret their world and find meaning amidst potentially confusing circumstances. The consequence of adopting scientific assumptions into the society's world view is that, over time, certain aspects of the collective world view will become incorporated into the technical activities of the society (Prigogine and Stengers, 1984). Indeed, the dominant world view that developed over centuries in Western cultures has had a clear effect on the way organizations are structured and controlled. It is not a coincidence that for most of the past century organizations have been designed as if they were precise machines, not as if they were organisms, societies, or weather patterns.

MANAGERS AND PARADIGMS

The philosophies of natural science are relevant to systems thinking in organizations because they shape the dominant paradigms of the people who create and manage organizations. **Paradigms** contain sets of rules that guide how people will think about specific issues, principles, and concepts that govern cause-and-effect relations in the world. Paradigms are adopted by people to help them simplify how the world operates. People use them as a frame of reference. Managers operate on the basis of paradigms all the time, often without realizing it. Joel Barker (1985) has vividly described the impact of paradigms in the watch industry. When mechanical watches dominated markets, most watch manufacturers in the world never could visualize an electronic watch, except for the Japanese. After hearing about the digital watch, invented by the Swiss in 1967, Japanese companies moved swiftly to capitalize on the opportunity while their competitors kept saying it would never happen. The Swiss watchmakers came to be accused of "inflexibility," "refusing to adjust," and being "tied to traditional technology." They "wouldn't see the opportunities" (*Fortune*, January 14, 1980). Today, brands such as Seiko and Pulsar are household names. Barker (1985) has suggested the following scenario as being likely:

> In 1967, the Swiss watch manufacturers' research arm, The Swiss Watch Federation Research Center at Neuchatel, created the first prototype [digital watch]. They presented it to the Swiss manufacturers, and, while there is no public record of the response, we can guess at some of the comments:
>
> "It doesn't tick!"; "It looks too clumsy!"; "What do you mean, it has no gears? No springs? No jewelled bearings? How can it even be considered a watch?" (pp. 59–60)

Paradigms can be adopted by individuals, professions, and even entire cultures. In organizations and broader social systems, such as nations, paradigms compete to gain the acceptance of the widest audience. For such fundamental ways of thinking to be accepted, they must at minimum only be more successful in meeting people's needs than competing views. As Kuhn (1970) has so clearly recognized, "To be more successful is not, however, to be either completely successful with a single problem or notably successful with any large number" (p. 23). Paradigms become accepted not because they represent truth, or are even close to the truth, but because of their ability to address a human condition.

CHANGING THE WORLD VIEW IN A SYSTEM

World views and paradigms are usually deeply embedded into the collective mind of societies. Carl Jung (1936) proposed that within cultures lies a collective memory, of which people are never conscious and which is acquired through heredity. It would be unlikely that any single group of people could change the dominant world view of a social system. Beckett (1973) argues that throughout history world views that contradicted the dominant views have been largely ignored and that actual changes in world view were more the result of changes in the physical location, size, and organization of a society. Although it would be extremely difficult to change the world view of an entire society, it may be possible to change the world views that exist within organizations.

Systems theorist C. West Churchman (1971) has developed a comprehensive strategy for changing the world view within a system, which will be discussed in detail in Chapter 6. The conceptual foundation for the model is the philosopher Hegel's idea of the dialectic. Burrell and Morgan (1979) see **dialectic** as "a method of analysis [which] . . . stresses that there is basic antagonism and conflict within both the natural and social world which, when resolved, leads to a higher stage of development" (p. 280). In organizations, a dialectic is a communication process similar to a debate, in which two opposing philosophical positions are advocated by proponents and challenged by adversaries. Over time a new synthesis position is identified, only to become the status quo and become challenged by a new ideology (see Figure 4-2). In this view, concepts and ideas continuously compete for acceptance. Each philosophical position contains strengths that may enhance its competitive ability and weaknesses that may lead to its ultimate demise. All philosophical positions contain inherent weaknesses that will eventually lead to their falling out of favor. The concept of the dialectic helps to explain how opposing beliefs interact with each other to influence policy and strategy formulation processes in organizations.

While systems theorists generally agree that the world view within a system can be changed, the extent of change possible is subject to debate. The dominant world view of a system is held in place by more than just philosophy. Other economic, social, and political realities within the system are all likely to be intertwined with the world view (M. C. Jackson, 1982). Furthermore, as Vickers (1983) has concluded, human perceptions and experience often reinforce each other. The result is that the system may become molded to the world view of its members, which in turn can reinforce the perceived validity of that particular interpretation of experience. A similar pattern has been noted in studies on decision making (March and Simon, 1958).

The most important idea to understand regarding the scientific roots of systems thinking is that the lessons learned from the study of the natural sciences shape the patterns of thinking found within any culture or society. New truths can be uncovered in any domain that is relevant to the ways organizations are managed. Since the natural sciences are considered the purest form of inquiry into the physical realm, it is not surprising that scientific philosophies often become the seeds of ideas in the social sciences and the study of organizations.

A major technique for changing the world view of a system is to think creatively. The Global Systems Spotlight about Honda shows how creativity can be fostered.

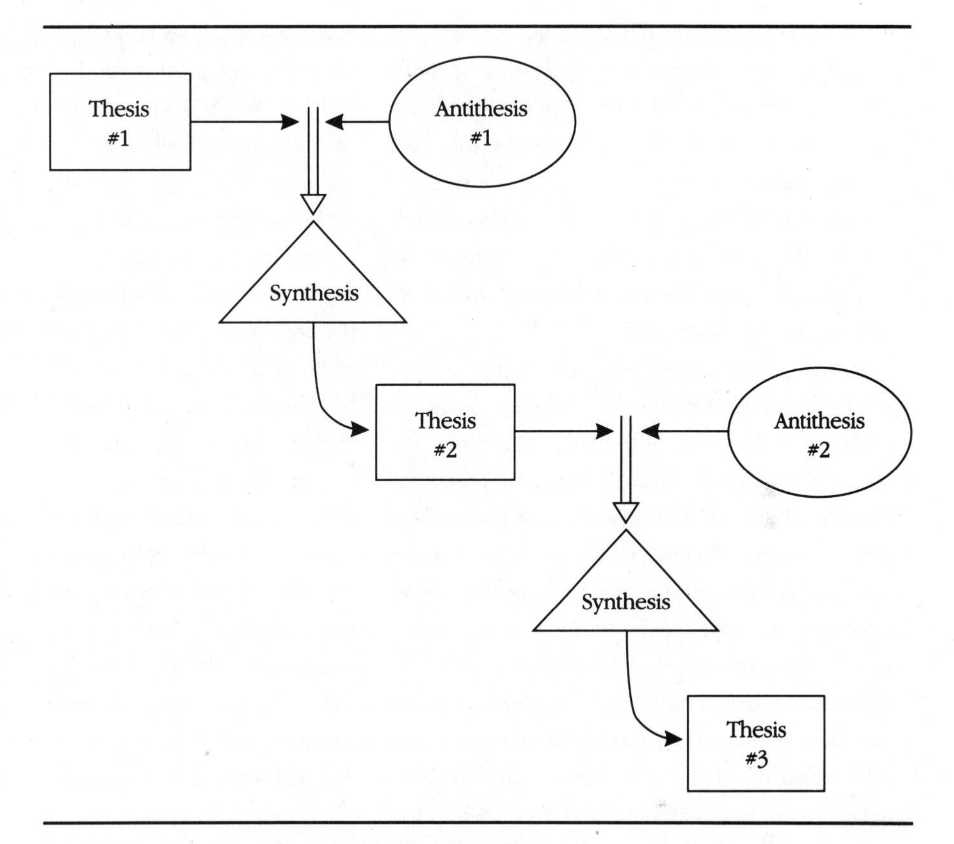

FIGURE 4-2 ***The Dialectic Process***

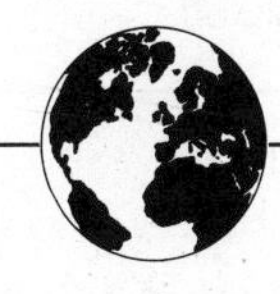

GLOBAL SYSTEMS SPOTLIGHT

Innovation at Honda

In 1978, top management at Honda inaugurated the development of a new-concept car with the slogan, "Let's gamble." The phrase expressed senior executives' conviction that Honda's Civic and Accord models were becoming too familiar. Managers also realized that along with a new postwar generation entering the car market, a new generation of young product designers was coming of age with unconventional ideas about what made a good car.

The business decision that followed from the "Let's gamble" slogan was to form a new-product development team of young engineers and designers [the average age was 27]. Top management charged the team with two—and only two—instructions: first, to come up with a product concept fundamentally different from anything the company had ever done before; and second, to make a car that was inexpensive but not cheap.

The mission might sound vague, but in fact it provided the team an extremely clear sense of direction. For instance, in the early days of the project, some members proposed designing a smaller and cheaper version of the Honda Civic—a safe and technologically feasible option. But the team quickly decided this approach contradicted the entire rationale of its mission. The only alternative was to invent something totally new.

Project leader Hiroo Watanabe coined another slogan to express his sense of the team's ambitious challenge: Theory of Automobile Evolution. The phrase described an ideal. In effect, it posed the question: If the automobile were an organism, how should it evolve? As team members argued and discussed what Watanabe's slogan might possibly mean, they came up with an answer in the form of another slogan: "man-maximum, machine-minimum." This captured the team's belief that the ideal car should somehow transcend the traditional human-machine relationship. But that required challenging what Watanabe called "the reasoning of Detroit" which had sacrificed comfort for appearance.

The "evolutionary" trend the team articulated eventually came to be embodied in the image of a sphere—a car simultaneously "short" (in length) and "tall" (in height). Such a car, they reasoned, would be lighter and cheaper, but also more comfortable and more solid than traditional cars. A sphere provided the most room for the passenger while taking up the least amount of space on the road. What's more, the shape minimized the space taken up by the engine and other mechanical systems. This gave birth to a product concept the team called "Tall Boy," which eventually lead to the Honda City, the company's distinctive urban car.

The Tall Boy concept totally contradicted the conventional wisdom about automobile design at the time, which emphasized long, low sedans. But the City's revolutionary styling and engineering were prophetic. The car inaugurated a whole new approach to design in the Japanese auto industry based on the man-maximum, machine-minimum concept, which has led to the new generation of "tall and short" cars now quite prevalent in Japan.

Source: I. Nonanka, "The Knowledge Creating Company," *Harvard Business Review* (November–December 1991), p. 100.

The innovations described in the case are an example of how new meaning can be created when information is changed into new patterns. This example is important because it illustrates that organizations must take a stance that values such types of activity where the contributions to productivity are indirect. (This is in contrast to direct uses of information as is seen in the use of high-speed communication networks.) For centuries the emphasis in classical Western philosophy has been on thinking scientifically rather than creatively.

SECTION 2
PHILOSOPHY AND SYSTEMS THINKING

THE ORIGINS OF SYSTEMS THINKING IN CLASSICAL PHILOSOPHY

The classical philosophies of the Western tradition have placed considerable emphasis on looking at how people attempt to understand human experience. They have also established the philosophical standard against which all forms of scientific research have been compared for centuries. Most importantly, with regard to systems theory, the rationalist philosophers such as Plato and Descartes have tried to determine the appropriate role and function for the designer of systems. Although the individual philosophies do not constitute a comprehensive systems theory, collectively the various philosophies represent a broad fundamental foundation for systems thinking to build upon. Some of the positions to be discussed in this section reflect views that are antagonistic to systems thinking. However, we believe that in order to understand the value of systems thinking, one must also understand the strengths and weaknesses of its mirror image.

THE RATIONALIST PHILOSOPHERS

Plato. Plato was one of the first well-known Western philosophers to address the philosophical issues that relate to system design. Plato's book *The Republic* examined various issues relating to the design of a system of government. His work revealed that he was aware of the importance of integrating the various parts of the government into a cohesive whole. Plato believed that the key to developing a system was to identify a common purpose that all elements could support.

The main elements of Plato's approach to systems design are:

1. The importance of identifying a common purpose that will unite a system.
2. Understanding the parts of the system with respect to the whole system rather than as isolated parts.
3. Identifying the functions of systems designers.

Plato is also responsible for identifying difficulties inherent in the role of systems designer. Specifically, Plato sought to determine how someone could be knowledgeable about a system without losing his or her ability to see the system objectively. He was also concerned with determining how the systems designer could create a plan that balanced the important systemic needs that the designer recognizes with the issues valued by other people within the system. The challenge that Plato envisioned was for the systems designer to avoid placing priority on personal values at the expense of the beliefs of the constituents of the system. These are age-old questions that managers have attempted to negotiate for centuries. Plato's primary message for systems thinkers appears to be that system designers should be conscious of avoiding conflicts of interest. The challenge for the systems designer is to simultaneously be an objective observer of the system and still remain a part of the system—to be in the system, but not of the system.

The question of whether it is possible to attain any degree of objectivity at all is a matter for philosophical debate. The belief that objectivity is attainable rests on the premise that there is an external world independent of people's perceptions. If one accepts that such a world exists, the objective observer attempts to see that world as if his or her biases and preconceived notions had no influence on the way he or she saw the system. Evidence accumulated by quantum physicists suggests that this is not possible (Zukav, 1979); they remind us that people are part of nature, and nature cannot study itself. Psychologist Carl Jung (1955) expressed the relationship between objectivity and perception this way: "when an inner situation is not made conscious, it happens outside, as fate. That is to say, when the individual remains undivided and does not become conscious of his inner contradictions, the world must perforce act out the conflict and be torn into opposite halves" (p. 173).

Rene Descartes. Rene Descartes continued Plato's rationalist approach but introduced a revolutionary angle. Building on his background as a mathematician, Descartes envisioned what he referred to as the "foundations of a marvelous science" (Capra, 1982, p. 57). Descartes visualized the universe as being one large, logical system that operated with machinelike precision. Mathematics provided the precise tool that allowed him to quantify and measure nature in all its beautiful complexity. He believed that all relationships found within the universe could be quantified and understood through the use of mathematics. Furthermore, Descartes presumed that he could design a precise mechanical model that accurately reflected how nature behaved. This task became feasible because of his positivist view that the laws of nature could become known as matters of fact through the use of the scientific method.

Descartes constructed a mathematical representation of reality that was based on logic and accepted methods of scientific methods of inquiry—analytical geometry. The events that led to this breakthrough in thinking are recounted below in the Systems Profile.

SYSTEMS PROFILE

Rene Descartes

One day at Neuburg, he escaped the cold by shutting himself in a "stove" (probably an especially heated room). There he tells us, he has three visions or dreams, in which he saw flashes of light and heard thunder; it seemed to him that some divine spirit was revealing to him a new philosophy. When he emerged from that "stove" he had (he assures us) formulated analytical geometry and had conceived the idea of applying the mathematical method to philosophy.

It took a proud spirit to dare his scope. Consider what he undertook: mathematics, physics, astronomy, anatomy, physiology, psychology, epistemology, ethics, theology; who would venture today on such a circumnavigation? For this he coveted seclusion, made experiments, equations, diagrams, weighed his chances of escaping or appeasing the Inquisition, and sought to give mathematical method to his philosophy, and philosophical method to his life.

Where should he commence? In the epochal *Discours de la methode,* he announced a first principle that in itself could have brought the world of authority down on his head; all the more so since the essay was written in readily intelligible French, and in animated, captivating, first-person style; here were many revolutions! He would begin by rejecting all doctrines and dogmas, putting aside all authorities, especially of *ille philosophus, the* philosopher, Aristotle; he would start with a clean slate and doubt everything. . . . "The chief cause of our errors is to be found in the prejudices of our childhood . . . principles of which I allowed myself in youth to be persuaded without having inquired into their truth."

But if he doubted everything, how could he proceed? In love with mathematics, above all with geometry, which his own genius was transforming, he aspired to find, after his initial and universal doubt, some fact which would be admitted as generally and readily as the axioms of Euclid. . . . He hit upon it exultingly: *Je pense, donc je suis, Cogito ergo sum,* "I think, therefore I am"—the most famous sentence in philosophy.

Source: W. Durant and A. Durant, *The Age of Reason Begins* (New York: Simon & Schuster, 1961), pp. 638–39.

Descartes's system for understanding the world appeared to many observers of the time to be flawlessly logical. Based on its seemingly perfect consistency, this system became the foundation of a revolutionary intellectual movement. His innovative thinking commenced an era emphasizing rationalism in scientific investigation. Descartes started with a set of assumptions that were very similar to those used by Plato. He saw nature as having two distinctly separate elements. He divided human existence into two categories: the realm of mind (thought) and the world of physical experience. His justification for this position was that the mind was the site of all intelligence, and the body was a mindless slave that followed directives issued by the mind. This analogy was extended outward to explain the rest of the universe. Descartes reasoned that since the physical world failed to manifest any form of intelligence, it must be the object of the mechanical laws of the universe. In essence, Descartes envisioned the physical world as operating as one huge clockwork that moved with both precision and stability.

With this world view as a foundation, it is easy to see how managers would be attracted to adopt this same type of clockwork mentality. The simplicity of the method offered the misguided hope that managers could operate organizations with similar machinelike precision. Henry Ford's vision for mass producing automobiles may not have been as unique as it appeared to his contemporaries. The rational-mechanistic world view seemed to perfectly complement the engineering orientation and the emphasis on technology, rather than people, that dominated Ford's era. Descartes's paradigm was both powerful and persuasive. Contrary to earlier approaches, Descartes's method was supported by reason, and yielded many practical insights into the nature of complexities that eluded the human mind.

In most respects, Descartes's methods were analytical and did not incorporate systems thinking into his paradigm. His world view relied primarily on positivistic assumptions to explain the relationships he observed. Indeed, his approach runs counter to many of the current notions of how dynamical systems operate. Yet it must be remembered that Descartes developed a philosophy that has been the dominant paradigm in Western cultures since the sixteenth century (Churchman, 1971, p. 61).

FOUNDATIONS OF THE RATIONALIST PHILOSOPHY

Rationalist philosophers such as Plato and Descartes adopted a particular world view that generally assumes that the character of the natural world can be discovered through the use of reason. For organizations, the implications are that managers can solve problems by understanding the facts of a problem situation and that the facts are available to be discovered. This philosophical ground has given birth to a general viewpoint called *logical positivism* (Churchman and Ackoff, 1950). Logical positivism is a philosophy that emphasizes the use of logic as a means of seeing the necessary facts relating to a particular situation. In this view, logic and reason replace such methods as intuition or spirituality as the prime vehicle for discerning problematic situations. Positivism implies that the needed facts are available and can be known—indeed, that they need to be uncovered. Logic is the tool for doing the uncovering. Facts are seen as existing as some type of well-formed artifacts that naturally occur in the world.

The rationalist philosophers were convinced that human powers of reason could reveal great truths if applied properly. Positivism is based on a belief in the underlying validity and constancy of the principles that guide the natural sciences—that is, belief in the laws of nature. If one accepts that the world operates on the basis of fundamental laws, then the role of the manager is to uncover the laws through the use of reason. The major assumption that undergirds this approach is one of generalization, that all situations are governed by the same force. Little emphasis is placed on understanding the factors that contribute to each situation's uniqueness. An opposing philosophy—known as **relativism**—differs. Its fundamental assumption is that knowledge is derived from the observation of specific situations. The emphasis of relativism is on learning about the unique aspects of situations rather than trying to generalize across a number of similar situations (Thayer, 1980).

Many of the management techniques in use today are predicated on rational-positivism. For example, many modern organizations still rely on mechanical models as a conceptual framework for designing their structures. Kenneth Boulding (1956) has labeled the least complex of such systems as *frameworks* and *clockworks*. Frameworks are static structures; clockworks are simple, yet dynamic systems that follow a predesigned pattern of movement.

Without doubt, any philosophy heavily biased toward either analysis or synthesis is likely to have limited value for practitioners. This is not to argue against analysis, but rather to promote adoption of an integrated perspective that combines analysis and synthesis. Both are useful conceptual tools for scientific inquiry, in differing environments.

Physicist Fritjof Capra has some interesting thoughts on this issue—his views are highlighted in the Systems Profile.

SYSTEMS PROFILE

A Physicist Talks Management: Fritjof Capra

The road from quantum physics to modern management is shorter than we might think, argues physicist Fritjof Capra. He's known for his eclectic range of thought: comparing physics to Eastern mysticism in the *Tao of Physics,* for example, and surveying new-paradigm thinking in *The Turning Point.* But these days, Capra is turning his thoughts toward business.

Interviewer: When you speak of the situation being critical, are you referring to environmental problems?

Capra: Yes, but more than that. It's the fact that the forests are disappearing, the deserts are expanding, the ozone layer is being depleted—and then there's global warming, population growth, the widening gap between the rich and poor. The nineties are critical. The fate of the earth and humanity is at stake. But there is a way out, and that's what I'm promoting.

You see, all these problems have solutions, some of which are simple. But they require a radical shift in our thinking and acting.

You might say it's a shift from machine to organism. A shift from a fragmented and mechanistic view of the world to an ecological view—and I use the term ecological in a very broad sense: the sense that all phenomena are interconnected, that we are all embedded in larger systems. Along with this goes a shift in values, from domination and control to cooperation and partnership.

When you see the world as a machine, control is the appropriate mode. But if the world is a living system that contains other living systems—ecosystems, social systems, animals—and if it's alive, the best way to understand it is not by controlling it, but by engaging it in dialogue, by cooperating.

Interviewer: Your words certainly ring true, but I think what makes them particularly compelling is that they come from you, a physicist. I have the feeling that underneath your theory you've had a very vivid and personal scientific experience of interconnectedness. Could you tell me about this?

Capra: The paradigm shift in physics happened in the 1920's, and I graduated in 1965, so I'm much too young to have experienced this firsthand. But I was fortunate enough to meet some of the key scientists involved in quantum physics, first and foremost Werner Heisenberg, and I had long discussions with him. What he told me, quite vividly, was how he and other physicists, in the 1920's found themselves in a crisis of understanding—their old ways of viewing the world simply fell apart. And they were led eventually to a whole new view of matter—which is a view of interconnectedness. Then, in the late seventies, I began to see our society in a similar crisis of world view. In our social affairs, it's the same mechanistic Cartesian, Newtonian view that is now at its end. It's no longer useful. We need a new view.

In essence, there are a number of basic shifts underway today—both in physics and in the larger society, particularly in management. One is a shift from structure to processes. In the old paradigm, it was believed that you had fundamental structures—the basic building blocks of matter—and that processes happened through the interactions. In the new view, every structure is a reflection of an underlying process. Structure doesn't pre-exist. It literally comes from process.

In business terms, the new paradigm tells us that when there's a problem in an organization, you don't blame people, you don't blame departments—those would be the structures—but you look instead at the processes. Because you see, physics taught us that structures really have no independent life. They are literally a reflection of energy, or process.... What physics teaches is that there literally are no parts. There are only relationships to the whole. We must shift our focus from objects to relationships.

Source: Excerpted from: "Interview: Fritjof Capra: A physicist talks management," ***Business Ethics,*** Minneapolis, Minnesota (612–448–8864), January-February 1992, pp. 28–29.

PHENOMENOLOGY

Phenomenology is a way of thinking based on the assumption that knowledge and wisdom are properties belonging to people. As the observers of a reality, people give meaning to their experience through thinking. This suggests that meaning is something found within oneself rather than discovered in the environment.

Rationalists believe that through the judicious use of logic, the mysteries of the universe can be deciphered and codified. They believe that once discovered, these universal laws could be generalized uniformly to explain problems within any domain. On the other hand, a relativist might argue that situations should be understood as being the unique products of the interactions among many forces and factors. From this perspective, trying to develop universal laws is nonsensical, since laws cannot adequately explain the unique dimensions of situations. Each situation must be considered on an individual basis.

~ A PHILOSOPHY OF SYSTEMS INQUIRY

GOTTFRIED W. LEIBNIZ

Just as Descartes's philosophy departed from the traditional philosophies of his time, Leibniz's inspiration and direction were clearly different than those of his predecessors. Leibniz believed that the natural world was capable of exhibiting a variety of qualitatively different behaviors. This was quite contrary to Descartes's view, which argued for a mechanical world programmed according to the laws of nature. Leibniz saw the universe as possessing a natural potential for acting in ways that were responsive to the situation.

Although Leibniz tried to prove that mathematics was compatible with the concept that nature could act in a variety of qualitatively different ways, many scientists and theologians of his time were upset by his radical departure from accepted philosophy. Nevertheless, Leibniz made a major contribution to the principles of systems design.

Leibniz is generally credited for developing the concept of an inquiring system. An **inquiring system** is an organized method for collecting information about a system (Churchman, 1971). Inquiring systems help organizations and other systems to learn about their environment and provide a framework for using this knowledge to enrich the organization. Churchman's work suggests that the presence of an inquiring system is instrumental to the systems design process.

The methods used in the design of an inquiring system are based on a set of philosophies that derive from **epistemology**, the study of how people acquire knowledge and use this knowledge to understand the world (Flood and Carson, 1988). Epistemology also considers the design of the methods used to acquire information and the way that information is organized. Put simply, before a system can be effectively designed, it is necessary for the designer to develop a process, or inquiring system, that can explore the potential of the system. Consequently, this process can generate data that will serve the designer by providing a basis for judgments about the design itself (see Figure 4-3). Churchman captured the essence of this approach by stating, "the systems approach itself demands a systems approach to defining it" (1979, p. 30).

The dilemma that confronts a system designer is knowing what type of inquiring system to design without having any a priori assumptions about the system to be designed. **A priori assumptions** are ideas based on previous knowledge of the system or biases about that general class of system. Peter Checkland (1988) has attempted to deal with this question in his innovative soft systems methodology (SSM). (Various forms of soft systems thinking, including Checkland's, will be discussed in Chapter 6.) If the system designer is already familiar with the system, then he or she must also confront the challenges of achieving objectivity that were raised by Plato.

The purpose of an inquiring system is to collect, store, and interpret information to support decision-making efforts within organizations and other systems. In many respects, an inquiring system functions as a **decision support system** (DSS). DSSs are computer-based systems that assist managers to deal with complex problems by facilitating interaction with both data bases and analytical tools (Sprague and Watson, 1986). While an inquiring system plays a supportive role in systems design, it does not actively make decisions or

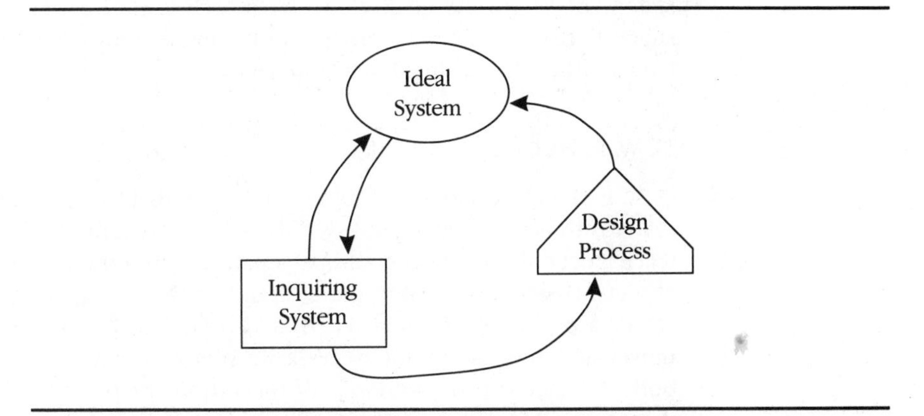

FIGURE 4-3 ***Relationship of Inquiring System to Systems Design***

judgments. It is important to note that the information generated by the inquiring system does not design the system, but how the designer perceives and organizes that information is what matters.

IMMANUEL KANT

Immanuel Kant developed a philosophy that represented a new synthesis; it preserved the positivist framework while acknowledging that nature was more than a simple mechanical clock. Kant concluded that the truth can only be determined through a combination of both observation (phenomenology) and reason (rationalism). Kant's philosophy reversed the traditional explanation for how the natural world could be understood. He argued that relationships in nature were the product of the way people thought about things. This was a dramatic shift from the traditional explanation that knowledge was external to people and was there waiting to be discovered. It also raised the question of whether thoughts and perceptions were accurate representations of the natural world. The positivist belief that analysis and experimentation would unlock the truths of the universe was now seriously being challenged. According to Kant, "science does not engage in a dialogue with nature but imposes its own language upon it" (Prigogine and Stengers, 1984, p. 87).

The issue of how people learn about the world was now framed in a way that made irrelevant the question of whether an objective external reality exits. Kant shifted the focus of philosophy toward defining how a person's prior knowledge of a phenomenon influenced the way he or she actually experienced reality. In many respects, Kant's propositions provided the foundation for a new method for acquiring knowledge. He argued for the necessity of establishing systems that would continually improve the objectivity of information gathering. The process of identifying new information about the external world must include reevaluation of whether the way information is being acquired is consistent with the acquired information. For example, market research not only tells a firm about its environment, but also carries implicit statements about the quality of the research itself. Kant's approach to inquiry suggests that learning is a process of finding a continuous series of unfolding truths rather than a single definitive answer.

G. W. F. HEGEL

G. W. F. Hegel developed a number of ideas that are particularly relevant to systems thinking. Hegel addressed the issue of whether the rational and physical aspects of the world could be separated from the spiritual and emotional aspects. Hegel's philosophy, unlike Kant's, was not a synthesis of the two, but rather a paradigm of coexistence; it recognized that these two fundamental universal forces were not necessarily mutually exclusive. He incorporated both the mechanistic workings of the rationalist perspective with the belief that the world contained a form of innate intelligence. Hegel envisioned nature as being organized into a hierarchy of levels of varying complexity; indeed, he believed that nature was destined to evolve to become more complex. This perspective represented a departure from classical scientific philosophy. Hegel saw nature as dynamic, guided by a form of intelligence that has a spiritual basis. (One modern variation on this view—Eric Jantsch's (1980)—posits that the purpose of nature is to perpetuate its own existence by designing a purposeful process of evolution. This is accomplished through the process of self-organization.)

Another of Hegel's contributions was related to the process of designing inquiring systems. Hegel, like Plato a millenium earlier, questioned the extent to which people can be objective in the process of identifying the needs of a system. Churchman (1971) has provided an extensive interpretation of Hegel's views on objectivity in systems design. This analysis suggests that such objectivity is virtually impossible to achieve. Churchman proposed several strategies to enhance the integrity of observation. One means is to follow a procedure in which a person's observations are continually verified by people outside the system; the assumption is that people outside the system will be biased in ways that cause them to be more objective in relation to the system in question. A second approach is to design a process, called *infinite regress,* in which the goals of the inquiring system are repeatedly critiqued as to the extent of their consistency with the world view of the system. Subsequently, the observations of the system designer and the interpretation of these results are compared with the dominant world view in the system.

Hegel dedicated considerable attention to looking at ways to increase objectivity. He attempted to do this by building on Kant's thesis that "unconstrained reason leads to contradictory conclusions" (Churchman, 1971, p. 170). Hegel's conceptual tool, known as the dialectic, proposes that the biases of a system designer will inevitably create unfavorable consequences for some people within the system. The lack of harmony that results from this inequity will accelerate the level of conflict within the system and trigger the dialectical process, in which opposing views compete for acceptance until a synthesis is formed.

EASTERN PHILOSOPHIES

The concept of dynamic systems is even more richly and elegantly discussed in many Eastern philosophies. This is particularly true of the *I ching* (*The Book of Changes*), and *Tao te ching* by Lao-tze. As the discipline of systems dynamics and nonequilibrium theories evolve, the compatibility with Eastern philosophies grows. In general, the Eastern philosophies such as Zen and Taoism place greater emphasis on the role of flexibility and change in the world than do the Western approaches.

The emphasis in the Eastern approaches is on the dynamic interplay between the fundamental polar opposite forces symbolized as *yin* and *yang.* Normally the *yin* forces are characterized as the passive forces found in nature, and the *yang* are the active influences. A natural flow back and forth between these forces is the underlying source of change in the Eastern tradition. When the relationship between *yin* and *yang* becomes unbalanced, a natural process of readjustment occurs to restore the balance, equilibrium, and harmony of the system. In some respects, the forces of *yin* and *yang* are the counterparts of Hegel's thesis and antithesis in the dialectic. The result of the interplay between the forces of *yin* and *yang* is a natural process of change characterized by a continual ebb and flow between the polarities of activity and passivity.

The Oriental view perceives this ongoing process of change as inevitable and something to be accommodated rather than resisted or overcome. The *Tao te ching* describes the nature of these dynamics:

> Under heaven nothing is more soft and yielding than water. Yet for attacking the solid and strong, nothing is better; it has no equal. The weak can overcome the strong; the supple can overcome the stiff. Under heaven everyone knows this; yet no one puts it into practice. (Lao-tze, 1972, parable 78)

Some Eastern philosophies also deal with the subject of change in terms of the concept of *Tao* (pronounced dow). *Tao* is an essential dimension of life associated with continual flows of change. In this perspective, all forces in the world are governed by a larger dynamic universal enduring force. Because the *Tao* is dynamic and follows a cyclical pattern, all other subordinate forces must follow suit. In the words of one philosopher: "There is no real coming and going. For what is going but coming?" (Shabistari).

The *Tao* acknowledges the polarity of *yin* and *yang* forces, but also adds that all life experiences are natural outgrowths of the interplay between human initiatives and the continual oscillation of the grand force, the *Tao*. "Tao means how: how things happen, how things work. Tao is the single principle underlying all creation" (Heider, 1985).

Systems thinking places great emphasis on recognizing patterns of change. Change is not seen as a reaction to a prior force; rather it is an innate characteristic of all systems. This view is becoming increasingly important in describing organizational transformation processes, and far-from-equilibrium organizational states. These topics will be discussed in detail in Chapter 16.

~ SUMMARY

Systems thinking has roots in philosophy and science. The same issues that surround the debate regarding the merits of various assumptions in science and philosophy also apply to both systems thinking and managing. Traditional approaches to managing are based on assumptions that have been the cornerstone of the rationalist philosophy.

The role of the system designer and the function of systems design has emerged as one of the prime themes of this chapter. The importance of the pursuit of objectivity in order to design a system that is responsive to the needs of its constituents is clear. The process of inquiry to determine the needs of a system raises another major set of issues: What types of information should be collected? Are the facts available to the designer? What is the appropriate blend of qualitative and quantitative data? These can be difficult questions to answer since systems are often a tapestry of structures and processes intertwined with various degrees of complexity. The common theme that emerged in response to these issues was the need for continuous learning and reflection.

KEY TERMS AND CONCEPTS

system design
reductionism
determinism
positivism
world view
paradigm
dialectic
relativism
phenomenology
inquiring system
epistemology
a priori assumptions
decision support system (DSS)

QUESTIONS FOR DISCUSSION

1. Describe a process that is a normal part of business operation such as performance appraisal or strategic planning. What assumptions are these processes based on? In which traditions do these assumptions have their roots? What are the implications of your conclusions for managers?
2. You are assigned to conduct the market research for a major new product release into a market that is largely unknown. Establish a rationale for your approach based on the concept of an inquiring system.
3. You are a manager and have been assigned the task of improving the objectivity of the way you select personnel. Assuming that you have been using traditional selection guidelines, how could the objectivity of the process be improved by incorporating the material from this chapter into the process?
4. A debate takes place on the topics of global warming and ozone depletion. The group of debaters includes a scientist, a philosopher, and a systems scholar. The scientist makes a strong case that, based on hard data, there are no problems. The philosopher has many generalizations, but no data. As the systems scholar, how would you resolve this debate?

REFERENCES

Barker, J. (1985). *Discovering the Future: The Business of Paradigms.* St. Paul, Minn.: ILI Press.

Beckett, J. (1973). "General Systems Theory, Psychiatry, and Psychotherapy," *The International Journal of Group Psychotherapy,* July, pp. 292–304.

Boulding, K. (1956). "General Systems Theory—The Skeleton of Science," *Management Science,* April, pp. 197–208.

Burrell, G., and Morgan, G. (1979). *Sociological Paradigms and Organizational Analysis.* Portsmouth, N.H.: Hienemann.

Capra, F. (1982). *The Turning Point.* New York: Bantam Books.

Checkland, P. (1985). "From Optimizing to Learning," *Journal of the Operational Research Society, 36,* no. 9, pp. 757–67.

Checkland, P. (1989). "Soft Systems Methodology: An Overview," *Journal of Applied Systems Analysis, 15,* pp. 27–30.

Churchman, C. W. (1971). *The Design of Inquiry Systems.* New York: Basic Books.

Churchman, C. W. (1979). *The Systems Approach and Its Enemies.* New York: Basic Books.

Churchman, C. W., and Ackoff, R. L. (1950). *Methods of Inquiry.* St. Louis: Educational Publishers.

Durant, W., and Durant, A. (1961). *The Age of Reason Begins.* New York: Simon & Schuster.

Einstein, A., and Infeld, L. (1938). *The Evolution of Physics.* New York: Simon and Schuster.

Flood, R., and Carson, E. (1988). *Dealing with Complexity.* New York: Plenum Press.

Hammer, M. (1990). "Reengineering Work: Don't Automate, Obliterate," *Harvard Business Review,* July–August, pp. 104–110.

Heider, J. (1985). *The Tao of Leadership.* Toronto: Bantam.

Jackson, M. C. (1982). "The Nature of 'Soft Systems Thinking': The Work of Churchman, Ackoff, and Checkland," *Journal of Applied Systems Analysis, 9.*

Jackson, M. C., and Keys, P. (1984). "Towards a System of Systems Methodologies," *Journal of the Operational Research Society, 35,* no. 6, pp. 473–86.

Jantsch, E. *The Self-Organizing Universe.* New York: Pergamon Press.

Jung, C. G. (1936). *Collected Works: The Archetypes and the Collective Unconscious,* vol. 9. Princeton: Princeton University Press.

Jung, C. G. (1955). *The Interpretation of Nature and the Psyche.* Princeton: Princeton University Press.

Kuhn, T. S. (1970). *The Structure of Scientific Revolutions,* 2nd ed. Chicago: University of Chicago Press.

Lao Tsu (1972). *Tao Te Ching,* New York: Vintage Books.

March, J., and Simon, H. (1958). *Organizations.* New York: John Wiley.

Mason, R. O., and Mitroff, I. I. (1981). *Challenging Strategic Planning Assumptions.* New York: John Wiley.

Nugent, C., and Vollman, T. (1972). "A Framework for the Systems Design Process," *Decision Sciences,* pp. 84–109.

Prigogine, I., and Stengers, I. (1984). *Order Out of Chaos.* Toronto: Bantam.

Sprague, R., and Watson, H. (1986). *Decision Support Systems.* Englewood Cliffs, N.J.: Prentice-Hall.

Thayer, F. (1980). In C. Bellone (ed.), *Organization Theory and the New Public Administration.* Boston: Allyn and Bacon.

Vickers, G. (1983). *Human Systems Are Different.* London: Harper and Row.

Zukav, G. (1979). *The Dancing Wu Li Masters.* New York: Bantam.

~ ADDITIONAL READINGS

Bellone, C. J. (1980). *Organization Theory and the New Public Administration.* Boston: Allyn and Bacon.

Boguslaw, R. (1965). *The New Utopians.* Englewood Cliffs, N.J.: Prentice-Hall.

Briggs, J., and Peat, F. D. (1989). *Turbulent Mirror.* New York: Harper and Row.

Checkland, P. (1981). *Systems Thinking, Systems Practice.* Chichester, England: John Wiley.

Flew, A. (1989). *An Introduction to Western Philosophy,* rev. ed. New York: Thames and Hudson.

Hofstadter, D. (1979). *Godel, Escher, Bach: An Eternal Golden Braid.* New York: Vintage Books.

Morgan, G. (1986). *Images of Organization.* Newbury Park: Sage Publications.

Peters, T. (1987). *Thriving on Chaos.* New York: Harper and Row.

Rosenhead, J. (1989). *Rational Analysis in a Problematic World.* Chichester, England: John Wiley.

Turner, J. H. (1978). *The Structure of Sociological Theory,* rev. ed. Homewood, Ill.: The Dorsey Press.

Vaill, P. B. (1989). *Managing as a Performing Art.* San Francisco: Jossey-Bass.

Vickers, G. (1970). *Freedom in a Rocking Boat.* London: Penguin.

Weiner, N. (1948). *Cybernetics.* Cambridge, Mass.: MIT Press.

Whitehead, A. N. (1969). *Process and Reality.* New York: The Free Press.

~ PART ~

II Systems Thinking: Basic Approaches

In Part I, systems were differentiated from systems thinking. We explained how the basic rules and relations that operate within a system could also be employed as a framework for viewing the world. Systems thinking was shown to be a powerful tool for understanding complex issues in organizations. Systems thinking is not only desirable, but necessary to achieve a balanced view of the world.

Part II proposes that the several discrete, well-defined systems approaches can be grouped as three main approaches: hard systems approaches, soft systems approaches, and cybernetic approaches. Each of these approaches will be discussed in detail. (A fourth main approach, integrative systems approaches, will be discussed in Part IV. This approach is composed of elements of all three primary approaches.)

The hard systems approaches rely on engineering-oriented concepts that emphasize structural design. They also rely heavily on positivistic and determinist assumptions. Mathematics and computers are popular tools for the hard systems practitioner. Mathematics can serve as a language that translates the relationships between system variables into standard units that can easily be manipulated by the use of computers. In organizational settings, hard systems techniques are based on rational, economic models that seek the optimization of desired outcomes in the most efficient way possible.

Soft systems approaches are based on philosophical premises that are antithetical to the hard systems approaches. Relativistic and phenomenological assumptions play a key role in defining this type of thinking. In these approaches, the fuzzy complexities of organizational realities prohibit use of a paradigm that accepts the existence of problems and solutions. Rather than problems, soft systems thinkers see general issues of concern, which necessitate a continual process of improvement rather than definitive solutions. Soft systems approaches generally strive to achieve consensus over the definition of issues and to generate cooperation among organization members to continuously improve situations.

The cybernetic approaches to systems management are also based on many ideas that have been popularized in mathematics and engineering. Cybernetics principles are based on control theories that seek to explain how performance feedback can be used as a barometer to gauge the extent to which performance meets desired outcomes. In its most basic form cybernetics is a limited, somewhat mechanical approach. But the underlying dynamic of this basic model—the ideas that changing inputs inevitably affects outputs and that feedback, in turn, affects the inputs—is a useful way to view systems.

The differences between the hard and soft systems approaches are boldly captured in Ulrich's (1980) mythical debate between Nobel Prize–winner Herbert Simon and C. West Churchman. Simon is cast in the role of the hard systems proponent of decision science and administrative theory. His intellectual adversary C. West Churchman represents the philosophical tenets that are the foundation for many of the soft systems perspectives. The positions each takes in Ulrich's debate are shown in Figure II.1.

The three types of systems thinking identified here are conceptual tools rather than actual entities. They are metaphors for generalized patterns of thought that have appeared throughout systems literature. Their value for managers is twofold. First, they shed light on the many beliefs and assumptions embedded in modern management techniques. Techniques cannot exist without assumptions and basic positions regarding values. Second, they raise the possibility of free choice in management practice. Managers may recognize the need to selectively alter their perspective to adjust to the particular complexity of a situation. This typology of approaches to systems thinking, in itself, provides a crude framework for thinking about systems issues from alternative perspectives.

SUBJECT	SIMON	CHURCHMAN
1. Subjectivity	Avoided	Not avoided
2. Complexity	Hierarchical	Not hierarchical
3. System wholeness	Denied, except in pragmatics	Critically considered
4. Reducing complexity of simple subsystems	A source of knowledge	A source to irrelevance
5. Purposeful nature of systems	Denied	Critically considered
6. Decomposability of systems into parts	Accepted	Rejected
7. Crucial design task to be solved	Problem decomposition	Problem identification
8. The system designer's main tool	Objectivity	Subjectivity

FIGURE II-1 ***Basic Premises of Simon and Churchman Perspectives***

Source: W. Ulrich, "The Metaphysics of Design: A Simon-Churchman 'Debate,'" *Interfaces, 10,* no. 2.

5 The Hard Systems Approaches

Recognition of the distinction between a stable system and an unstable one is vital for management. The responsibility for improvement of a stable system rests totally on the management. A stable system is one whose performance is predictable. It is reached by removal, one by one, of special causes of trouble, best detected by statistical signal.

W. Edwards Deming (1986)

1. To explain the fundamental assumptions on which hard systems approaches are based.
2. To understand the logic of technical rationality.
3. To appreciate the role of uncertainty reduction in hard systems approaches.
4. To describe how the key hard systems assumptions are expressed in the rational problem-solving process.
5. To recognize the importance of complexity and the routine-nonroutine dimension in problem-solving situations.
6. To identify the potential situations in which the various operational research techniques will be valuable.
7. To understand the role played by systems analysis in organizations.

SECTION I
HARD SYSTEMS APPROACHES

INTRODUCTION

In the previous chapter the philosophical roots of systems thinking were identified and discussed. In particular, emphasis was placed on exploring the dominant paradigms of rationalism and positivism. These powerful world views have been expressed in the scientific principles and technological applications of the twentieth century. The impressive contributions of engineering to human life have created interest in the possible benefits of applying engineering techniques to problem solving in organizations. The world views that have dominated scientific and engineering perspectives have had a major impact in shaping the development of hard systems thinking in organizations.

The purpose of this chapter is to introduce hard systems approaches, hard systems thinking, and hard systems applications such as operational research and systems analysis, two techniques that dominated the organizational landscape during the 1950s through the early 1970s. It is also to clearly differentiate the hard systems approaches from the other approaches and forms of systems thinking.

In this chapter, hard systems thinking is viewed metaphorically, rather than as an operational tool. This is because the focus of this book is on how managers think, not on presenting techniques for solving problems. Therefore, this chapter will be directed toward examining the hard systems paradigm and will review only those hard systems techniques that are particularly illustrative of the essence of the hard systems approach.

~ FUNDAMENTALS OF THE HARD SYSTEMS APPROACHES

DEFINING THE APPROACH

The **hard systems approaches** are a paradigm for framing situations that assume the manager's primary task is to identify the most efficient way to solve a clearly defined problem. Solutions in these approaches emphasize structural changes that are based on scientific principles and assumptions of economic and technical rationality. Hard systems approaches are used for problem solving, decision making, and designing systems. **Hard systems thinking** is a perceptual framework that assumes systems are comprised of structures and that relations exist within a mechanical framework which can be purposefully modified to attain desired outcomes.

Although hard systems thinking endeavors to be scientific, it should not be confused with science itself. Science is the pursuit of knowledge in relation to the discovery of how nature operates. Engineering is a practice that seeks to blend scientific insight with technology to reach predetermined goals. Seen from another perspective, engineering pragmatically attempts to employ science to get the job done in the most economic way possible. Science is interested in closing the gap between experience and knowledge. Engineering is interested in using technology to close the gap between the current state of affairs and a desired future state of affairs. The same is true of hard systems thinking. The desired future is not always attained, however, as the global systems spotlight shows.

The global spotlight illustrates that hard systems approaches have limitations and that when their principles are extended to their extremes, they may produce counterintuitive outcomes. There is clearly a need to match the ap-

GLOBAL SYSTEMS SPOTLIGHT

Designing a Soviet System

After the Russian Revolution of 1917, believers in the social utopian system were given an opportunity to try these ideas on a mass scale—an entire nation. Designers of the system thought that all facets of human life could be quantified and connected and controlled. Lenin had read all of the utopian philosophers, as well as the works of social engineers such as Frederick W. Taylor, and believed that such a system could be created. Later, Stalin expanded the scope of this system when he was able to include most of Eastern Europe within this grand design.

The attributes of the system focused on increasing the predictability of everything. The political system included only one party, thereby eliminating the uncertainty of changes introduced by an alternative view. The economy was developed on the basis of state ownership, thereby removing the unpredictability of actions by competing management and owners. A central planning commission was created that would systematically develop a five-year, one-year, and quarterly plan. These plans included all the parameters of production for every factory or means of production, thereby eliminating the vagaries of initiative. A system for balancing resources and finished goods (input-output matrices) was developed as an algorithm that did not consider marketability or human desires but reduced decision situations to a routine basis.

The entire system was reinforced and maintained through the use of a police network and intense propaganda (control of all media, both internal and external sources). Reduction of uncertainty meant that the system had to be almost totally isolated from its external environment. Complex problems were broken into the smallest possible chunks in order to make them routine. They were dealt with sequentially, and there were massive delays in implementing solutions, such that many fixes could no longer be responsive to the original issues. In short, the entire political and economic system of the Soviet Union was reduced to acting like a machine controlled by the First Secretary of the country. In fact, the economic system was the mirror image of the political system, with an analogous hierarchy.

Unfortunately for the Soviets, the designers did not predict that the costs of controlling this massive system would be so high. Nor did they realize that the reduction of uncertainty would be counterproductive and impossible or that the system they created would be terribly inefficient. But, ironically, these results occurred—despite the fact that the original plan for the system was to promote efficiency and rationality.

proach with the situation. Misapplication of the hard systems approaches often stretches the techniques beyond their capabilities. Rational problem-solving approaches have varying levels of effectiveness in reducing the perceived uncertainty in distinctly different types of problematic situations. The effectiveness of hard systems approaches largely appears to be contingent on the "goodness of fit" between the capacity of the approach to the form of the problem (Ackoff, 1979).

Hard systems approaches are composed of a number of specific methods that may help managers improve the efficiency with which a system reaches its goals. This emphasis on designing the most efficient ways to attain goals clearly suggests that hard systems approaches are based on an engineering world view. Hard systems approaches also have an economic orientation and a bias toward action that distinguishes them from the other systems approaches.

CORE VALUES AND BELIEFS OF THE APPROACH

In general, hard systems thinking emphasizes quantification and measurement as tools for understanding systems. This strategy is intended to reduce the level of uncertainty associated with problem solving. The core beliefs of hard systems approaches are that rationalization and systematization of problem-solving processes will lead to the best decisions.

Rationalization is a way of thinking in which experience is interpreted by the use of reason. This approach emphasizes the generality of a problem situation rather than its uniqueness. Rationalization depends on a person's ability to fit the characteristics of a situation to his or her mental model of how it could exist. The tendency toward rationalization is accentuated because mental models act as filters on how people perceive the world. Someone who believes that the world behaves mechanically will interpret most of what he or she sees in that light.

Rationalization is a critical feature of hard systems approaches because one of the key assumptions of these approaches is that all problems can be classified as fitting into a well-known category or type of problem. This process of classification assumes that all important facts are known about the problem and that its unique properties are not as important as what it shares with other problems of the same type. One way that problems can be seen more clearly so that their type can be correctly identified is by quantifying and measuring them.

Quantification and Measurement of Problems. Rationalization of problems occurs by precisely quantifying and measuring phenomena. Hard systems thinkers believe that if it cannot be measured, it does not exist. The benefits of quantification and measurement are twofold. First, a common, well-understood language, such as mathematics, can be used to track and describe the variables being studied. Second, measurement and quantification permit relatively simple comparison and manipulation of the variables. In hard systems approaches, this normally takes the form of either mathematical modeling or computer simulation. In complex systems, there may be many advantages to employing such hard approaches to deal with certain classes of problems. Among these advantages are:

1. The problem solver's level of rationality is increased.
2. Simplifying problems helps managers to organize them more effectively.
3. The relative risks of differing alternative solutions are usually explicitly considered.
4. The continuous decline in the cost of computing (hardware and software) has made it reasonable to conduct extensive "what if" analyses of alternatives. (Lyles and Mitroff, 1980)

Technical Rationality. Miser (1989) has identified a specific type of rationality found within hard systems approaches, technical rationality. **Technical rationality** is driven by two primary values: achieving preestablished goals and minimizing the use of resources while attaining those goals. It is based on the belief that science and scientific principles should be the basis for all attempts at problem solving. Hard systems approaches use both scientific methods and principles as the conceptual underpinning of their various problem-solving methodologies. Technical rationality builds on positivistic tradition, which has coupled science and technology as means of dealing with problems (Schon, 1983).

Technical rationality is expressed by managers who use highly structured methods for investigating and solving problems. A premium is placed on the use of problem-solving methods that are systematic, orderly, and follow a well-established pattern. Managers who follow technical rationality usually deal with problems in a sequential fashion that often appears to follow a lock-step pattern. The underlying rationale for following such a mechanical process is that precision and control add both focus and rigor to the process of inquiry by restricting the manager's attention to the most important considerations. While this logic may be appealing, there is little evidence that it is universally valid.

HARD SYSTEMS ASSUMPTIONS

Hard systems approaches are characterized by the basic assumption that the problem-solving process begins with a definitive problem statement. This clearly defined problem is the cornerstone for all subsequent steps. The endpoint of the process is to change the system in a way that eliminates the problem. Once a problem has been clearly identified, the process that follows focuses on identifying and evaluating alternative solutions.

Most problems can eventually be reduced to a process of judging the relative efficiency of various means for reaching the desired endpoint. In such hard systems approaches as operations research (OR) and systems analysis, mathematical models are used to evaluate which alternatives can solve the problem most efficiently. For such models to be used, all relevant characteristics of a system must be reduced to numerical form. This is accomplished by reducing problems into their basic elements and then measuring each factor.

UNCERTAINTY REDUCTION

Hard systems approaches attempt to deal with the confusion that often accompanies complexity by reducing complex problems to simple elements. This reductionist strategy stems from the view that complex problems are a conglomeration of many simple problems. This building block perspective suggests that complex systems are just a hierarchy of parts aggregated together into progressively larger and larger systems. According to hard systems approaches, if simple systems have simple solutions, the best technique is to break a complex system into its constituent parts and then apply the appropriate technique for that particular class of simple system. The appeal of this logic is enhanced further by seeing simple problems as solvable with a finite array of reliable techniques. Many hard systems advocates thus conclude that the critical task in problem solving is to identify the subsystems of a complex problem.

While this strategy may seem elegant in its simplicity, it may also be misleading. This perspective is limited in many respects.

- Reducing a system into its separate parts deprives the parts of their natural relationship to the other parts. The result is to reduce the likelihood that the part being examined will behave in the same fashion as when it is embedded in the overall whole.
- This strategy assumes that it is only important to know the inner workings of each element and not how the elements work together.
- This process of reduction may reveal nothing about how a system works, only how its individual parts function.

This approach might work well in fixing the engine of an automobile, but the mechanical assumptions also have limitations. Can this strategy work equally well for explaining the workings of the human body, a basketball team, an organization, or a society? The answer is yes and no. The global spotlight on the Soviet economy offered one case where this approach was not fruitful. But some dimensions of systems are mechanical and can be understood through the use of such methods. One example of success in using this strategy is the use of operations research (OR) at American Airlines to manage passenger load.

American Airlines

In its 1987 annual report, American Airlines broadly described the function of yield management as "selling the right seats to the right customers at the right prices." While this statement oversimplified yield management, it does capture the basic motivation behind the strategy. A better description of yield management as it applies to airlines is the control and management of reservations inventory in a way that increases (maximizes, if possible) company profitability, given the flight schedule and fare structure.

The role of yield management at American is analogous to the inventory control function for a manufacturing company. Planning departments determine the airline's flight schedule and fares. The combination of schedule and fares defines the products to be offered to the public. Yield management then determines how much of each product to put on the shelf (make available for sale). American's "store front" is the computerized reservations system, SABRE (Semi-Automated Business Research Environment). All sale and cancellation transactions, whether from American Airlines reservations agents or travel agents, pass through SABRE, updating reservations inventory for all affected flights. New reservations are accepted only if yield management controls permit.

Critical to an airline's operation is the effective use of its reservation inventory. American Airlines began research in the early 1960s in managing revenue from this inventory. Because of the problem's size and difficulty, American Airlines Decision Technologies has developed a series of OR models that effectively reduce the large problem to three much smaller and far more manageable subproblems: overbooking, discount allocation, and traffic management. The results of the subproblem solutions are combined to determine the final inventory levels. American Airlines estimates the quantifiable benefit at $1.4 billion over the last three years and expects an annual revenue contribution of over $500 million to continue into the future.

Source: B. Smith, J. Leimkuhler, and R. Darrow, "Yield Management at American Airlines," *Interfaces* (January–February 1992), 22, no. 1, pp. 8–9.

DEVELOPMENT OF THE HARD SYSTEMS APPROACH

The history of attempts to control organizations similar to hard systems approaches can be traced back to the major projects of ancient times such as the building of the pyramids of Egypt or the Great Wall of China. A more recent impetus came from the Industrial Revolution. Many of the design elements of hard systems thinking evolved as a result of the creation of large, complex manufacturing firms in that era. Examples are the efforts of Eli Whitney in manufacturing muskets in Hamden, Connecticut, and the use of assembly lines by James Nasmyth to produce machine tools in England. Many of the efforts to integrate work designs and technology into rational processes—accomplished by such innovators as Henry Ford and Frederick W. Taylor—are well documented. The research that served as the foundation for hard systems thinking was greatly refined by British and American military researchers who attempted to solve tactical war problems with hard systems approaches during World War II (Morse and Kimball, 1946). The rapid development of operations research and systems analysis in manufacturing firms during the 1950s and 1960s was related to experience gained in producing large quantities of military equipment during World War II.

The hard systems movement gained greater momentum through the innovation of problem-solving approaches that employed mathematical models and computers. This approach became known as operational analysis or operations research (OR). OR may be considered as "a branch of philosophy, as an attitude of mind towards the relation of man and environment; as a body of methods for the solution of problems which arise in that relationship" (Kendell, in Beer, 1959). In more concrete terms, **operations research** is a general label for a set of mathematical techniques useful for solving specialized problems such as inventory management, transportation problems, and waiting or queuing problems. OR is a hard systems approach that attempts to solve problems by controlling the use of resources through the application of scientific methods of measurement, comparison, and prediction.

The recognition that individual specialized approaches could prove effective in solving varied classes of problems provided insight into how problem situations could be framed. Hard systems approaches to public sector problems in the United States were continued in the 1950s and 1960s by the RAND Corporation. RAND, a civilian research organization, originally specialized in solving problems for the military using a technique known as systems analysis. RAND introduced a new problem-solving methodology that emphasized evaluation of the relationship between economic constraints and the achievement of desired outcomes. Its innovative applications of systems analysis signaled the beginning of a new era in hard systems thinking. The simultaneous development of statistics, economics, and computers provided a set of powerful tools to managers. The impressive early results achieved by RAND set the stage for the widespread adoption of hard systems approaches in business and industry after World War II. This eventually would lead to the institutionalization of quantitative techniques of management decision making throughout many organizations during the 1960s and 1970s.

The RAND approach was effective in identifying optimal solutions to problems in complex systems that were too broad in scope to be handled by a single OR technique. In essence, the RAND approach was to achieve optimal allocation of economic resources while still attaining the set goals. The optimal allocation of resources is determined through an evaluation process that seeks to identify those conditions that will yield the most favorable balance between resources used and outcome achieved. Most importantly, the RAND perspective was founded on viewing organizations as a total system. The merger of OR techniques with a systems perspective brought systems analysis closer to the desired goal, a blend of problem-solving techniques and managerial perspective.

It would appear that this blending of technique and managerial thought has never fully been realized, however. A survey of managers investigated the extent to which they used quantitative business techniques. The researcher found that "Management science is an important area in the academic world, but managers—at least in this study—are not using the quantitative models that are its tools" (Coccari, 1989, p. 70). While experience has shown that the appropriate scope of systems analysis is largely limited to mechanical problems, this problem-solving technique has introduced a more systemic perspective into hard systems approaches. Although major advances in problem solving have been made through operational analysis, experience has also shown that OR functions most effectively with highly structured, well-defined tactical problems. Many problems found in government or urban settings—which did not have these characteristics—appeared resistant to efforts to engineer single solutions, however (Hoos, 1979).

THE FUNDAMENTALS OF HARD SYSTEMS THINKING

The ultimate goal of problem solving is to reduce the gap between the current state of affairs and some desired future state. Hard systems approaches provide a framework for managers to accomplish this—the problem-solving process:

1. Identifying the problem (problem definition)
2. Identifying alternative solutions
3. Evaluating the costs and benefits of each alternative
4. Eliminating causes of the problem (implementing the solution)
5. Transforming the system to attain the desired state of affairs (Figure 5-1)

The successful application of this five-step process depends on managers' effective judgment and decision making.

RATIONAL DECISION MAKING

In practice, hard systems approaches are contingent on the use of **rational decision making.** Decision making has been defined by Feldman and Kanter (1965) as "selecting a path which will move the system—individual, computer program, or organization—from some initial state to some terminal state" (p. 614). In general, decision making processes are concerned with the

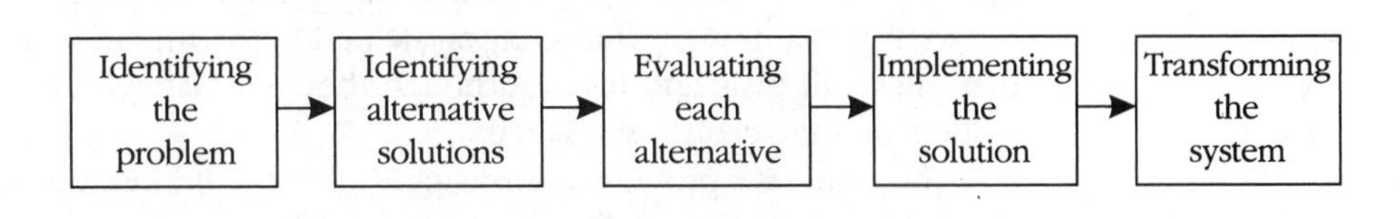

FIGURE 5-1 ***The Problem-Solving Process***

identification of and choice among satisfactory alternatives (March and Simon, 1958). The process of making the final choice of the best alternative is known as **choice making.** Choice making is usually governed by considering the various parameters present in the decision-making situation. **Parameters** are conditions that restrict or limit the range of possible choices available. Parameters typically include limitations on cash flow, human skills, capacity for debt, and availability of information.

Not all alternative courses of action fit the requirements of the desired solution equally well. For alternatives to be truly effective in attaining a desired future state or a goal, they must meet certain minimum requirements to solve a problem. These requirements are the **decision criteria;** they are used to evaluate the fit between the options and the needs of the organization. For example, suppose that a manufacturer's equipment is old and subject to frequent breakdowns, which cause declines in product quality and shipping delays to customers. The firm may decide that it needs to replace a major portion of this equipment; then it must establish a set of decision criteria such as cost limitations for the purchase, the required speed and accuracy of the new machinery (judged in relation to competitors' capabilities), retraining requirements for machine operators, and the amount of time needed to install the new equipment and get it running at full capacity.

Each alternative course of action has the following characteristics associated with it:

- Degree of risk that it will fail to meet the decision criteria
- Costs of supporting its development and implementation
- Benefits it will offer the system

March and Simon (1958) propose that decision-making environments can be categorized into three sets of conditions: certainty, risk, and uncertainty. These conditions are known as *states of nature.* Certainty exists when the exact outcomes of decisions are known. This assumes that managers have complete and accurate information. When decision makers operate under conditions of risk, they are able to identify a finite number of possible consequences to their decision but are unable to determine precisely which one will occur. This condition assumes that managers have accurate knowledge of the odds of a given event occurring. A manager may have tracked product quality in relation to production levels over a year and know the probability that a given production run will yield an output of acceptable quality, but he or she cannot guarantee this outcome. Conditions of uncertainty exist when decision makers are able to conclude that the consequences of a decision will fall into a specific set of results but cannot determine the likelihood of this happening. For example, a new product may show rapid growth, but managers anticipate that sales will peak and level. Although they may say with assurance that this will occur, they cannot say when.

The minicase provides an example of the complexities of the problem-solving process. The option that is most effective in one regard may stretch beyond the parameters of the organization or may be insufficient to meet the decision criteria.

Tilley Video Disc, Inc.

William ("Call me Billy") Tilley, founder, president, and chairman of the board of Tilley Video Disc, Inc., has a very pleasant problem. The market for video discs is expanding rapidly, and he has accepted an attractive offer for his controlling interest in Tilley Video Disc. He will retain his current management responsibilities. The deal is contingent on Billy's development of a plan to expand production from 10,000 to 50,000 units and to lower production costs. Evers, Inc., the company making the offer, feels the growth in sales will be dynamic if costs can be significantly reduced.

Presently, production of the video discs takes place in one plant in Rimer, Oregon. The plant was formerly a slaughterhouse and meat-packing facility. Billy has quickly come to the conclusion that production could be maintained at this facility but that significant on-site expansion is not sound. The equipment in the facility was purchased from a bankrupt company and has been used for several years. It can best be characterized as slow, general-purpose equipment. Following are data on the existing plant for the last twelve months.

Selling price	$6.50/unit
Variable cost	$3.20/unit
Total production	9,824/year
Rejects	644/year
Sales	9,180/year
Estimated capacity	10,000/year
Annual fixed costs	$20,000/year

Ever's vice-president for marketing thinks that the selling price should be reduced to about $5.00 in order to achieve the needed sales growth and to grab market share in the highly competitive consumer market. As a consultant to Billy Tilley, you are charged with developing a plan that will allow the company to increase capacity by 400 percent and reduce variable costs by 20 percent.

Source: M. Vonderemsbe and G. White, *Operations Management: Concepts, Methods, and Strategies* (St. Paul, Minn.: West Publishing, 1988), pp. 206–7.

The case raises a number of conceptual issues that must be dealt with before a technical analysis can be accomplished. What decision criteria should be used to frame the ideal solution? What are the states of nature inherent in this case? What are the parameters? What amount of risk is Billy willing to take? Is this a complex problem that can be broken down into simple parts, or is it a unique problem? This last issue—determining the degree of novelty in a situation—can be expressed in terms of whether a setting is characterized by routine or nonroutine factors.

A FRAMEWORK FOR HARD SYSTEMS THINKING

Although some decisions may initially appear extremely complex, they need not remain this way. Some decision situations may be repetitive and ideal for using a well-established problem-solving method such as OR or systems analysis. Such situations are characterized by consistency in the type of task to be attempted and the existence of a known solution for any version of the problem. Thus, routineness is defined in terms of the extent of task variability and the degree to which the search for solutions is programmed.

Routine situations offer little variability in tasks and include the presence of a programmed path for achieving solutions. Routine decision-making situations offer a greater likelihood of achieving a desired outcome than nonroutine situations because problems fit established categories with proven solutions. Problem solving becomes a simple matter of following a predetermined sequence of steps, a **performance program** (March and Simon, 1958). The specificity of performance programs can range from very precise formulas to general guidelines. Programmed decisions are designed and used in organizations on the assumption that a specific problem will recur in a similar form. This assumption can be made with confidence if the problem area is well understood and there is little uncertainty involved. The use of specific performance programs should, if followed properly, lead to the attainment of the desired outcome.

Performance programs are often comprised of simple **algorithms,** precise formulas that are faultless in their ability to lead to the ideal outcome. For example, a person who desires to make a long-distance telephone call from a public telephone must follow a protocol to complete the call. That protocol contains various algorithms, depicted in Figure 5-2.

In complex, dynamic systems, such as organizations, many decision-making situations are not repetitive; they are nonroutine. **Nonroutine situations** exhibit extensive variability in tasks; further, few existing performance programs are available to provide solutions. In these circumstances, those performance programs that do exist tend to be more general and offer guidance on how to pursue the search for solutions rather than prescribing specific solutions (Figure 5-3). These general performance programs often take the form of **heuristics,** "rules of thumb" invoked to guide the process of acquiring information and identifying solutions (Cavaleri, 1991). Similarly, some situations are not perceived by managers as resembling any situation with which they are familiar. In fact, some situations may be so unique that there may not be a clear starting point for beginning the search to define the problem. The net effect is that these situations are perceived as having a high level of uncertainty. (Of course, what may be complex or uncertain to one manager may be simple to another. What is recognized as a recurring pattern by one manager may go unrecognized by another. Mental models and styles of thinking can have a major influence on how a situation is perceived.)

In such conditions of uncertainty, the decision maker must develop a plan to achieve **nonprogrammed decisions.** Such decisions are customized to meet the unusual needs posed by a nonrecurring situation. For example, the design of a lunar space station is a unique activity that may have few parallels to designs developed for projects on Earth.

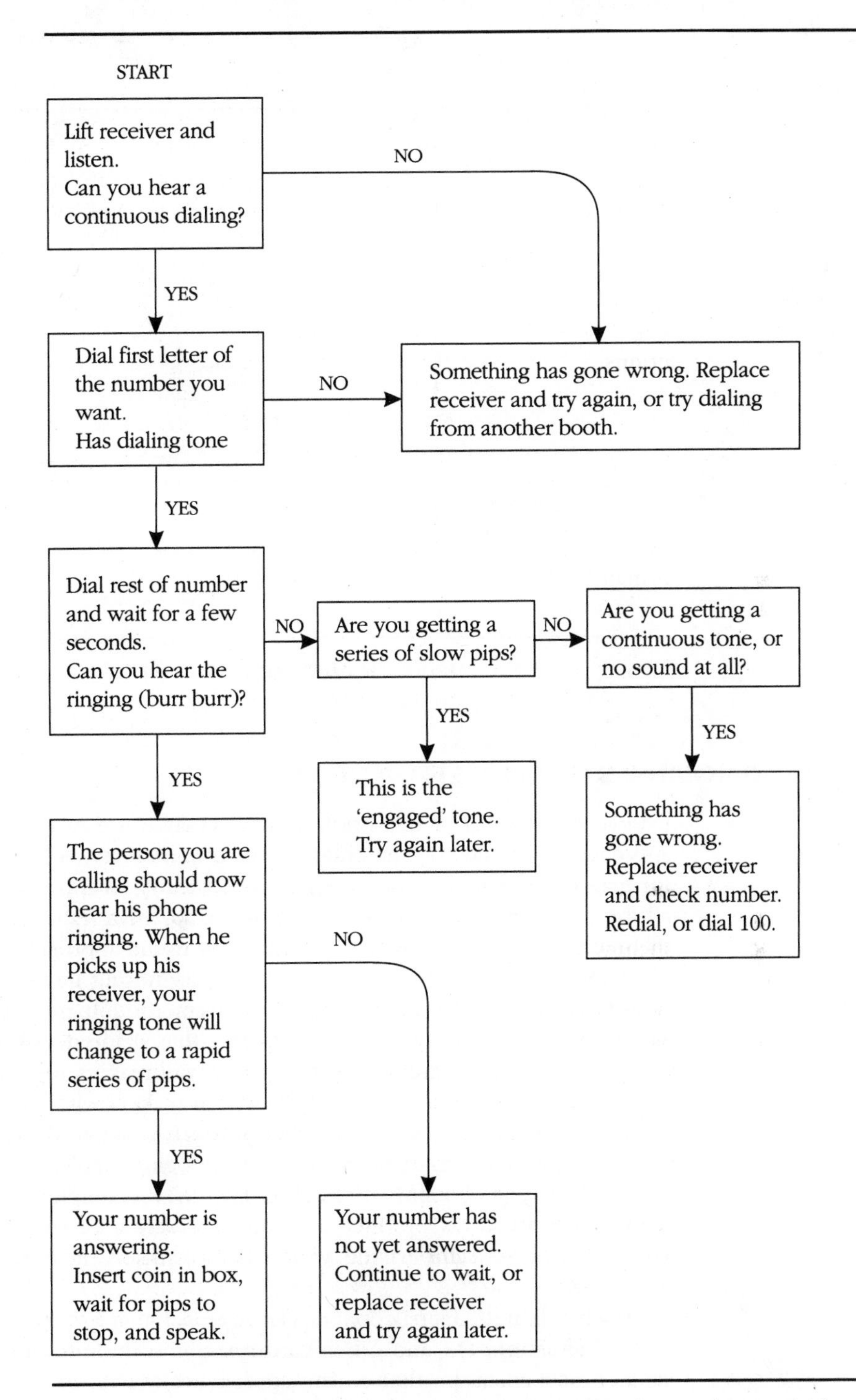

FIGURE 5-2 ***The First Part of an Algorithm for Making a Telephone Call from a Public Telephone***

Source: I. Horabin and G. Lewis, *Algorithms* (Englewood Cliffs, N.J.: Educational Technology Publications, 1978), p. 25.

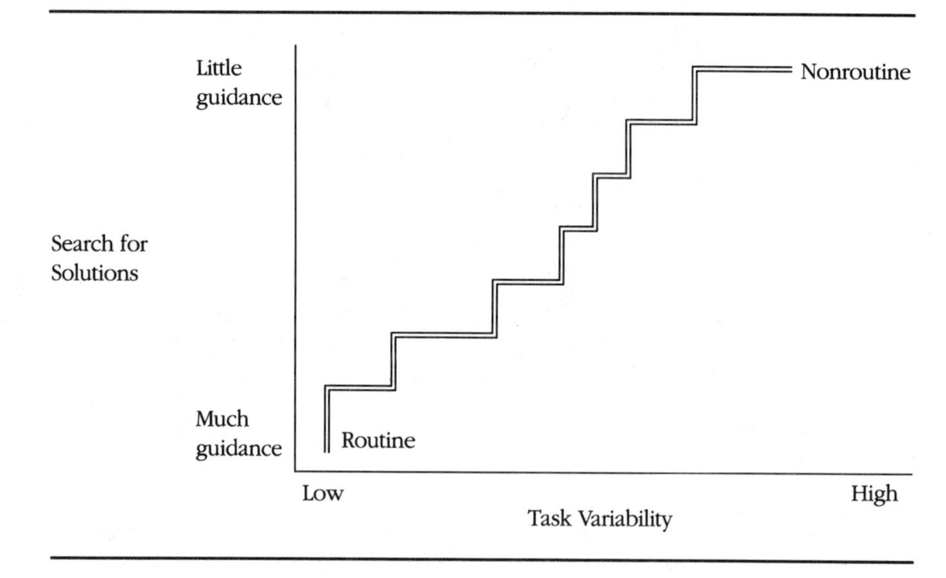

FIGURE 5-3 ***Decision-Making Situations***

DIAGNOSING SYSTEMIC SITUATIONS

One of the themes of this book is that effective use of the various systems approaches depends on matching them to the situations in which they have the greatest potential for providing insight. Clearly, all approaches do not work all the time. By prudently assessing a situation, however, a manager may gain the necessary information to select the most useful approach.

One of the most interesting tools for diagnosing decision-making situations is Jackson and Keys's system of systems methodologies. Their model considers the interaction of two major factors that shape decision-making situations: the extent of agreement among decision makers regarding a common set of goals and system complexity. Decision makers who are unable to agree on a common set of goals are labelled *pluralistic* while those who agree are *unitary.* System complexity can range from being *mechanical* (a simple system with easy problems) to *systemic problems* (complex or difficult problems). The situations produced by the interaction of these forces may be conducive or resistant to the various problem-solving methodologies (see Figure 5-4).

Each cell in the matrix contains an approach that has the best fit with each type of situation. OR and other hard systems techniques are well suited to unitary-mechanical problems. The approaches that fit in remaining cells are the subject of the next two chapters, on soft systems thinking and cybernetics. Now that the hard systems approach has been seen from the broad perspective, we will focus on specific techniques and methods used in this approach.

Decision Maker Agreement	Mechanical	Systemic
Pluralist	Churchman, Inquiring systems Mason, Mitroff, Strategic assumptions: surfacing and testing	Ackoff, interactive planning Checkland, soft systems methodology
Unitary	Operations research	Emery and Trist, Sociotechnical systems
	System Complexity	

FIGURE 5-4 ***Jackson and Keys's System of System Methodologies***

Source: Based on M. C. Jackson, and P. Keys, "Towards a System of Systems Methodologies," *Journal of the Operational Research Society, 35,* no. 6, pp. 473–86.

SECTION 2
HARD SYSTEMS TECHNIQUES AND METHODS

OPERATIONS RESEARCH

Operations research (OR) is an application of hard systems thinking that uses a variety of different mathematical techniques to solve specific types of problems. Operations research approaches problems by using the scientific method of inquiry. Kwak and Delurgio (1980) have developed a seven-stage process that summarizes the application of the scientific method to OR (see Figure 5-5). Ackoff (1967) has interpreted the goal of OR to be the "control of organized (man-machine) systems so as to provide solutions which best serve the purposes of the organization as a whole" (p. 6). In order to gain a systems perspective on the scope of problems, OR includes efforts by interdisciplinary teams, in much the same fashion as systems analysis programs are implemented.

OR employs many of the same fundamentals as other hard systems approaches in that the critical foundation is a well-defined problem. OR places its major emphasis on problem solving through the use of model development.

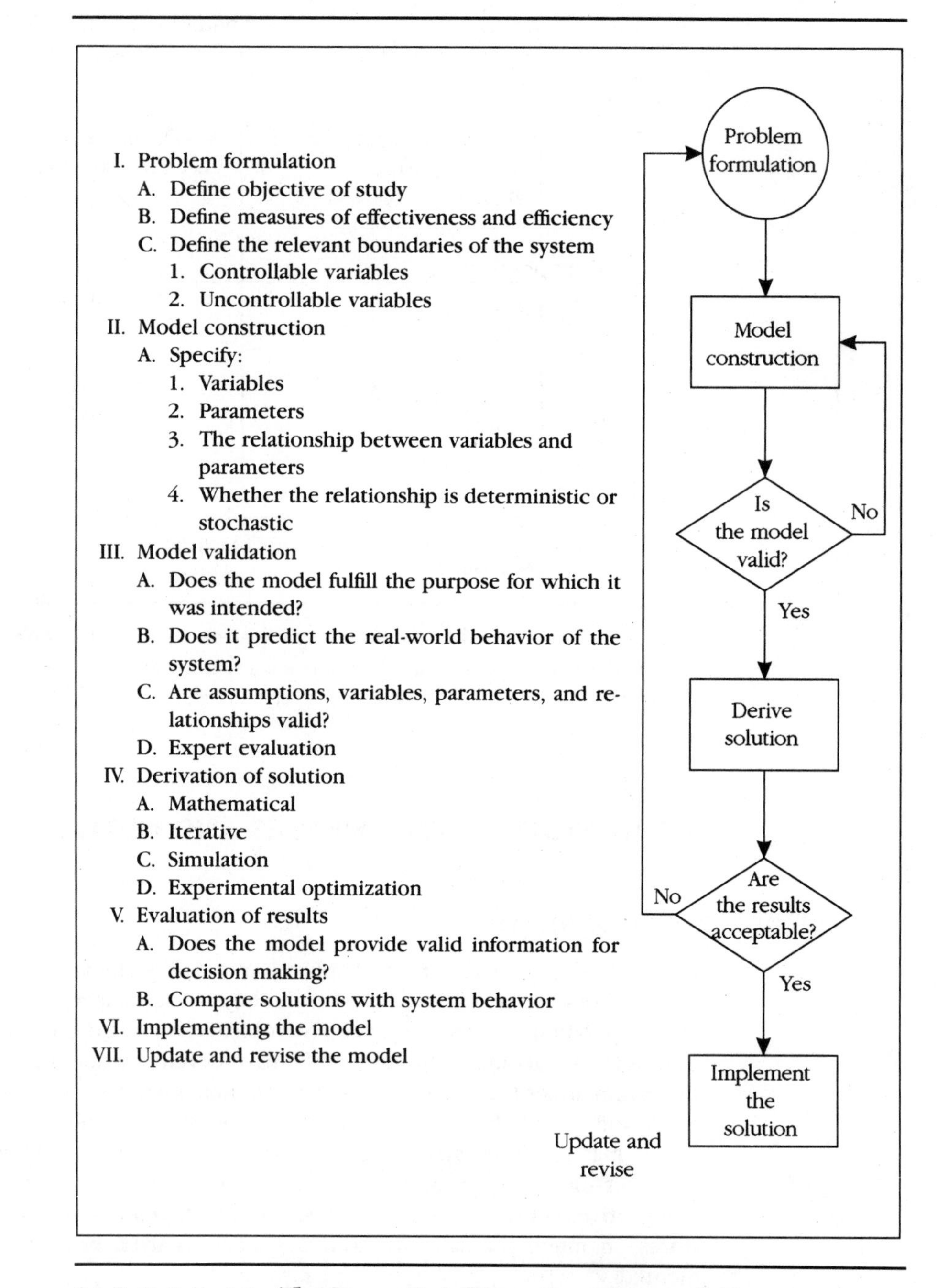

FIGURE 5-5 ***The Seven-Step Operations Research Process***

Source: Adapted from N. K. Kwak and S. A. DeLurgio, *Quantitative Models for Business Decisions* (North Scituate, Mass.: Duxbury Press, 1980), p. 3.

Unlike systems analysis, which is concerned with the generation of alternatives through creative conceptualization of situations, OR relies on mathematical and computer models to generate solutions.

MODELING

In complex systems the number of variables and the interactions between them may result in so many possible variations in patterns of performance that it becomes impossible to clearly define the nature of the system's problems purely through observation. Problem solvers can simplify the complexity of systems by creating models, scaled-down representations of the system. Models reduce the number of variables under consideration to those that are well known and of greatest importance to the behavior of the system. Models provide a wide range of experimental possibilities to researchers because of the relative ease of manipulating variables, cause-effect relationships, and basic system structures. OR models use mathematics to describe the interrelationships among the key variables of system.

Most mathematical models are symbolic representations of reality that attempt to capture the nature of a system in a way that remains consistent with the system over time. As Budnick, Mojena, and Vollman (1977) observed, "A mathematical model explicitly states the mathematical structure which relates the inputs (controllable and uncontrollable variables, constraints, and parameters) to the outputs (values for the criterion as expressed through the objective function)" (p. 11).

In addition to their usefulness in describing the dynamics of complex systems, models can also be used for prediction and explanation. Indeed, the greatest strength of models is their capacity for explanation, whereas their descriptive capabilities are their weakest feature (Chatfield, 1980).

There are a number of different types of models; the two major types are iconic and analog. **Iconic models** are simply reduced-scale representations of what is being modeled such as an architect's three-dimensional model or a globe. **Analog models** use another symbolic representational system to capture the main elements of reality. For example, an organization chart uses various types of lines to represent channels of communication and reporting relationships in an organization. In OR projects, models are continually tested to determine their validity in representing the key elements.

TYPES OF OPERATIONS RESEARCH PROBLEMS

Ackoff and Sasieni (1968) have noted that most OR applications tend to be tactical in nature. Tactical efforts are generally limited to operational, nonstrategic concerns and are focused on improving efficiency. Types of tactical problems can be differentiated on the basis of their form and content. "Form refers to the way in which the properties of a problem are *related* to each other" (Ackoff and Sasieni 1968, p. x). Content refers to the nature or meaning of these properties. Operations researchers are not generally familiar with the content of particular jobs in organizations, but they may be able to recognize that the form of the problem fits into one of the eight classes of problems that OR methods have proven useful in solving.

OR problems can be classified as one of the following types:

1. Allocation
2. Inventory
3. Replacement
4. Queuing
5. Sequencing and coordination
6. Routing
7. Competitive
8. Search

All of these approaches may be categorized as being either deterministic or stochastic. You will recall that deterministic techniques presume that all quantities are known to the decision maker. Stochastic models assume that a range of probable values can describe a particular system. In such cases, probability and chance factors can be introduced into a model through patterns of statistical variation.

Each OR approach is designed to solve problems of a particular form. Problems may have content that varies significantly from the content of other problems but still be of similar form. Problems with similar form can be grouped together in the same class to be solved with the same approach.

APPROACHES TO THE MAIN SIX OR PROBLEMS

Queuing Problems. Since organizations have scarce resources to allocate to many functions, demand for service of one kind or another inevitably accumulates. This accumulated demand takes its form as waiting lines. The lines may be people waiting for service in a fast-food restaurant or to buy a ticket at a movie theater. The lines could be telephone calls waiting to be answered or requests for information from a computerized data base. Such a waiting line is generally referred to as a **queue.**

Organizations face tradeoffs between the cost of building service capacity to avoid queues and the cost of keeping customers too long. Unhappy customers will not return if they are forced to wait too long. But unused service capacity in the form of capital or human resources is inefficient investment. Queuing models attempt to identify the point at which profits are maximized and costs minimized. It can be difficult to measure costs such as waiting in line because people will interpret waiting in line differently.

Allocation Problems. Allocation problems are economic problems in that they involve decisions about how to apportion scarce resources. An allocation model will seek to identify the solution that meets the decision criteria and either minimizes the costs of resolving the problem or maximizes outputs or returns. An example is a brewery with two plants supplying beer to five company-owned warehouses. The allocation model will seek to minimize transportation costs and avoid stock-outs at both the brewery and the warehouses. The economies of the brewing process must also be factored into the model. The L. L. Bean case provides an example of how allocation problems can be solved.

Allocating Telecommunications Resources at L. L. Bean

In the Spring of 1989, L. L. Bean's senior management decided that it needed a better allocation of resources in the telemarketing area. As a consequence, L. L. Bean initiated a project to answer the need for a more structured approach to planning for proper trunking, telephone-agent staffing, and queue management. After an exhaustive shopping expedition showed that no outside sources had taken optimization in the telemarketing context to a level that met the firm's needs, L. L. Bean decided to build an economic-optimization model (EOM) in-house. They needed to apply economic optimization to the simultaneous sizing of the trunks, the agent work force, and the queue capacity.

With annual sales of $580 million in 1988, L. L. Bean conservatively estimated that it lost $10 million of profit because it allocated telemarketing resources suboptimally. Customer-service levels had become clearly unacceptable: in some half hours, 80 percent of the calls dialed received a busy signal because the trunks were saturated; those customers who got through might have waited 10 minutes for an available agent.

As is the case in most operational settings, the types of decisions made in managing a telemarketing call center depend on the time horizon and the nature of the resources involved. . . . In the long term, decisions must be made on the number of telephone lines (trunks) to install, the number of agent positions to establish, the labor markets to use, and such capacity as facilities, buildings, and acquisition of equipment (automatic call distributors [ACD's], work stations, CRT's, and other support equipment. In the shortest term, decisions must be made to schedule staff on a half-hourly basis, 24 hours a day, seven days per week. In the intermediate term, the most critical decisions for the telemarketing operation concern the number of agents to hire and train. [See Figure 5-6.]

While these long-, intermediate-, and short-term decisions might seem quite routine—especially when sales volume remains steady—seasonal build-up, with its big profit potential, changes the stakes dramatically. The three-week peak period just before Christmas makes or breaks the year for the company. For the peak-season build-up in 1989, for example, the number of telephone agents on payroll increased from 500 to 1,275, telephone trunks expanded from 150 to 576, and the overall operational capacity geared up to meet a full 18 percent of annual call-volume in a hectic three-week period.

The model set explicit levels for the following key resources: 1. the number of trunks carrying incoming traffic to telephone agents; 2. the number of agents scheduled; and 3. the queue capacity, the maximum number of wait positions for customers who are successful in seizing a trunk but who are forced to wait for a telephone agent. The fundamental decision variables must be tightly planned and managed, directly and indirectly affecting a variety of important company resources and their associated costs. The modeling effort produced a real home run for the management science project, which was funded at a mere $40,000. The resultant cost savings were estimated as $9 to $10 million per year.

Source: P. Quinn, B. Andrews, and H. Parsons, "Allocating Telecommunications Resources at L. L. Bean, Inc.," *Interfaces* (January–February 1991), *21*, no. 1, pp. 75–91.

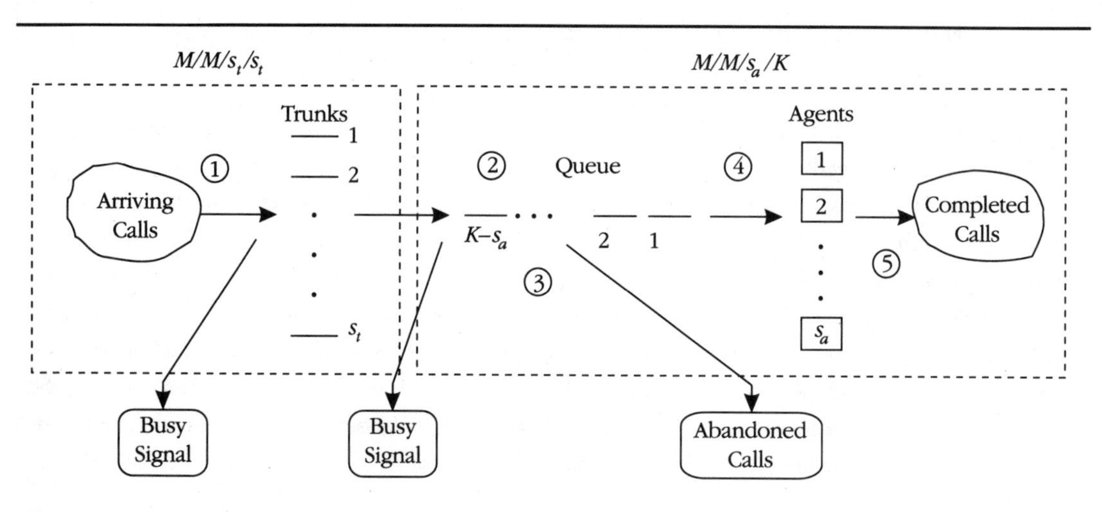

Key:
s_t = number of trunks installed
s_a = number of agents on-duty
K = system capacity for calls waiting and in service
For the L. L. Bean system, the network is characterized by the flow of calls, the points in the system where congestion occurs, and the three key controllables: number of trunks (s_t), number of agents (s_a) and queue capacity ($K - s_a$). Flowing cell processing is used: (1) An arriving call seizes one of s_t trunks if one is available; otherwise, it is routed to a busy signal. (2) The call arrives at an L. L. Bean switch; the switch checks the queue length; if the queue is full, call is routed to busy signal, freeing trunk; otherwise, the call enters the queue, taking up a queue position. (3) Callers may abandon (hang up) while waiting for an agent, freeing trunk and queue position. (4) When one of s_a agents is available, the call is serviced, freeing a queue position. (5) When the call is completed, a trunk and an agent are freed.

FIGURE 5-6 ***The L. L. Bean Call Network***

Inventory Problems. Manufacturing firms invest their finances in the inputs necessary for the creation of goods. But as long as the firm's investment is in the form of raw materials or finished goods in storage waiting for shipment, they are not creating value for the firm. On the other hand, to ensure adequate supply of raw materials to maintain desired manufacturing flows, a sufficient inventory stock must be maintained. Stocks of finished goods must also be maintained to meet customer orders in a reasonable period. There are risks involved with having both inventory stock-outs and excess inventory. The objective of inventory models is to identify the optimum reorder point. Such models take into consideration the following factors: product demand, lead time, inventory carrying cost (insurance and so on), and ordering costs.

Replacement Problems. Most organizations own some form of assets, ranging from manufacturing machinery to computers or vehicles. Because these assets depreciate, both in accounting terms as well as performance ability, they must eventually be replaced. Replacing assets requires an outlay of funds, making it unattractive to do so more often than necessary. To continue using an asset, however, creates maintenance and repair costs, which, as the life of the asset is extended, eventually rise to a point where they are a significant portion of operating costs. Anyone who has owned an old automobile has experienced this situation. There is a clear tradeoff between capital funds outlays for acquisition or lease and operating costs. The goal of replacement models is to identify the point where capital acquisition costs and operating costs are minimized.

Sequencing Problems. Sequencing problems can be divided into two basic classes: waiting-line scheduling and coordination. While waiting-line scheduling involves queues, it includes other issues. In queues, all demands on the system are treated with equal weight. Waiting-line scheduling addresses issues of priority. Thus, a typical waiting-line sequencing problem is a hospital waiting room where patients enter with problems of various severity. The sequencing model must consider the severity of each person's problem and their priority in the queue compared to other patients.

Coordination problems are generally found in complex multitask projects, such as construction of a ship or a factory. In such situations some activities must follow others in a chronological order while others may be conducted simultaneously. The purpose of a coordination model is to identify the particular arrangement of tasks that will allow the project to be completed in the minimum amount of time at the optimal cost. Several techniques are used for this purpose, among them PERT and CPM analysis. PERT and CPM use decision tree analysis to determine which activities are the most important—those on which delays cannot be afforded without delaying the entire project. The sequence of these activities is the *critical path* for the project.

Routing Problems. Routing problems involve determining the route among locations in a network that optimizes some set of criteria. Typical factors considered are the time it takes to get from one location to the other, the cost of travel, and the cost of delays or missed visits to locations. Freight carriers typically face this type of problem. An example is the motor carrier Yellow Freight System.

Yellow Freight System, Inc.

Yellow Freight System, Inc., was founded in 1926 as a regional motor carrier serving the Midwest. By 1980, it had grown to 248 terminals that handled 5.4 million shipments per year. In 1980, after years of debate, Congress passed the Motor Carrier Act, effectively removing a blanket of government controls that restricted market entry and controlled prices. Using this freedom, Yellow Freight has grown to be one of the largest motor carriers in the country, handling over 15.4 million shipments annually in a network of 630 terminals and some 35,000 communities. Yellow Freight serves over 300,000 domestic and international customers; reported revenues in 1990 exceeding $2.3 billion.

As the company grew, terminal managers in the field lacked the information that would allow them to coordinate their activities around the network. A large-scale interactive system called SYSNET was developed to optimize the routing of shipments and design of the network. SYSNET has allowed Yellow Freight to evaluate and implement basic operating strategies that run counter to decades of standard practice. In addition, SYSNET has improved Yellow Freight's overall planning responsiveness in a changing business environment and produced substantial cost savings. But the far more important statistic to the company's customers has been the significant improvement in transit times and service reliability.

Competitive Problems. Competitive situations are characterized by common interest in a goal that is desired by two or more parties (individuals or groups). The competition arises because goal achievement is exclusionary; that is, there will only be one winner—the other competitors become nonwinners or losers. There is an inverse relationship between the processes of winning and losing. As one person's chances for winning improve, the probability of the other's declines proportionately. In this OR approach, competitive situations may be defined as games. Games may be classified in terms of whether they are zero-sum games or non-zero-sum games. Zero-sum games can have only one winner because they are exclusionary. In these games, there is a finite and non-increasing of pool of winnings (prizes, profits, or points) that are available to be won. In a non-zero-sum game more than one competitor may experience gains. Mathematical strategies may be developed to increase the probability that the odds of competitor winning can be identified.

Search Problems. A person may know that he or she wishes to purchase a new automobile but must acquire sufficient information about each type before making a choice. Search is the process of information acquisition that will make the choice of a solution possible. When a match between a class of problem and a class of solution(s) does not exist, then the search process must begin. Search models attempt to minimize both the cost of the search and the risk that the process will make an error. Errors can typically be made when perceptual or observational errors are made. These errors occur when perceptions are inaccurate or incomplete. Errors can also be made in the extent or scope of the sampling process. In shopping for a new automobile, should one sample the dealers in the city, region, or nation? There are a variety of costs associated with the search process:

1. The costs of planning the search, such as site location studies.
2. The costs of doing the search, such as market research studies.
3. The costs of analyzing and reporting the results of the search.

Since all systems have scarce resources, funding for the search process is considered a parameter. The research model then seeks to organize the search in a way that minimizes risk of error and costs in the range allowable within the funding parameter.

As can be seen from the Yellow Freight minicase, OR deals with many highly complex, yet well-defined problems. In general, OR problems tend to be operational rather than strategic or systemic issues. Systems analysis is a hard systems technique that can be applied on a wider scale than OR.

SYSTEMS ANALYSIS

Systems analysis is an integrative problem-solving approach that employs multidisciplinary teams and the tools of mathematics and computer modeling. As was discussed earlier in this chapter, systems analysis is the outgrowth of the efforts by the RAND Corporation to develop an approach that could support decision making in the military and public policy. Systems analysis integrates many of the quantitative and design aspects of systems engineering with several elements of organization theory. The technique recognizes that many systems are complex and have multiple interrelated problems. It also acknowledges that systems are often pluralistic in that they contain many decision makers with diverse and often competing values.

Typically, systems analysis efforts are based on the development of a continuing relationship with a client. The consulting team attempts to present the client with a view of the problem that differs from the perspective generated by the client's management. Through the application of its various quantitative tools, the systems analysis team develops predictions of future consequences of each alternative being considered. Computer modeling has proven to be particularly useful for this purpose.

The five major steps in systems analysis correspond to the five steps of the problem-solving process discussed earlier.

1. *Problem formulation.* This stage—the starting point of systems analysis—is designed to clarify the goals and expectations of the client. At this point the systems team needs to be clear not only about its mission but also about

the constraints and possible obstacles it may face during its engagement. A precise statement of the limitations of the current system must be developed to serve as a platform for the debate that must take place in later stages.

2. *Generating alternatives.* Some aspects of systems analysis remain more an art than a science; such is the case with this second step. Generating alternatives requires the ability to envision how possible alternatives may mesh with the needs of the organization.
3. *Forecasting future environmental states.* It is then necessary to forecast future events to determine whether the alternatives will be compatible with the changes likely to occur in the environment (Quade and Miser, 1985). The underlying rationale for this step is compelling. That this goal may be fully attained is doubtful.
4. *Identifying and evaluating the consequences.* Every alternative has a specific set of costs and benefits, which must be summarized in a way that captures the tradeoffs inherent in each alternative. The future value of the alternative to the system must be projected. Financial techniques such as capital budgeting and other applications of present value theory can be useful in making this evaluation.
5. *Comparing and ranking alternative courses of action.* This is the most difficult step in systems analysis. It is so challenging because each alternative may have its own unique set of costs and benefits, making it unlikely that the team will have a common yardstick to measure alternatives. As a consequence, evaluation of such complex comparisons often requires the expert opinion of someone who has experience in the particular context under consideration.

A major point of departure between systems analysis and other hard systems techniques is the continuous feedback and reality testing built into the design of systems analysis projects. As Quade and Miser (1985) note, some systems analysis models include built-in reiterative processes in which models are continually reevaluated to determine their accuracy (see Figure 5-7). Furthermore, alternative proposals for action are continually compared with any changes in the values, policies, or objectives of the system to ensure their consistency with those overarching characteristics. The inclusion of reiteration of feedback in systems analysis plays two instrumental functions that greatly enhance the validity and value of the approach. First, reiteration of feedback allows the systems team to monitor changes in the system and incorporate them into their assumptions. Thus the solution eventually enacted is more likely to be effective. Second, the feedback mechanism creates a learning environment in which each iteration can yield new information that will help the team to build a more realistic view of the dynamics of the system.

Systems analysis is a rationalized economic approach to problem solving. Its emphasis is on identifying the alternative that offers the optimal future allocation of resources for the system. Although systems analysis may attempt to consider the softer elements of systems, such as values, it does not offer any

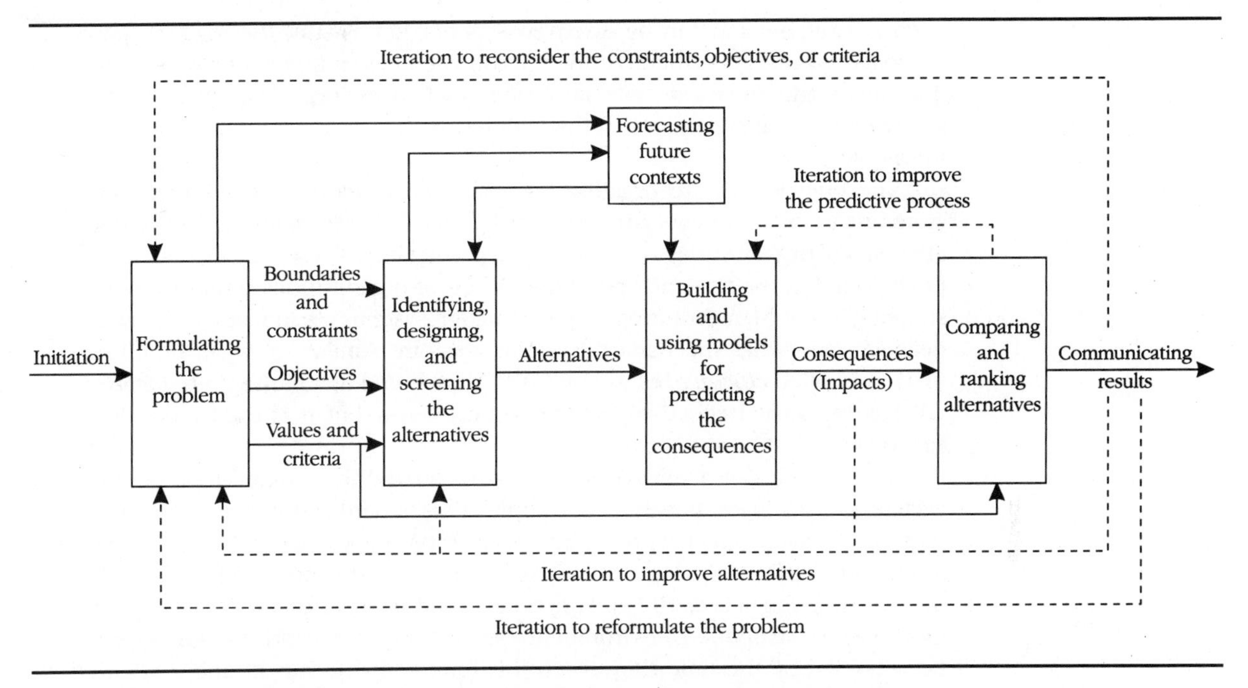

FIGURE 5-7 ***The Systems Analysis Procedure with Iteration Loops***

Source: E. Quade and H. Miser, *Handbook of Systems Analysis: Overview of Uses, Procedures, Applications, and Practice* (New York: North Holland, 1985), p. 123.

insight into how this may be translated into practice. Like any research approach, systems analysis has limitations. In general, its emphasis on identifying the single best alternative to achieve a defined objective may be suited to use in well-defined situations, but it lacks the flexibility and sophistication to address fuzzy problems.

EVALUATION OF HARD SYSTEMS APPROACHES

The beliefs that connect positivism in science, technology, and engineering find their application to organizations in operations research and systems analysis. These two approaches exhibit many of the same strengths and weaknesses identified in the scientific paradigms. The hard systems approaches begin with several important beliefs about both the research process and organizations. They emphasize the formal rather than the behavioral, or emergent, properties of organizations. These beliefs take clear form in the core, but questionable assumption that organizations are able to accurately define problems. (Equally uncertain is the validity of the assumption that organizations can clearly state a set of enduring goals.) Furthermore, in the positivistic tradition, hard systems approaches assume that all goals, parameters, and decision criteria can be measured and quantified. Hard systems thinkers believe that the information they need is available and can be discovered as needed. Finally, the hard systems approach assumes that the practitioner will engineer and implement solutions that meet the decision criteria. If one perceives organizations as well-structured formal systems, then the conclusion that they can be engineered is inevitable.

The question is, are organizations the well-structured, well-understood systems that hard systems advocates see? In the broadest sense, it is true that elements of organizations are orderly and predictable. These elements are generally found at the tactical, operational levels of organizations rather than the strategic levels. Manufacturing and information systems are in many cases well-defined, somewhat mechanical systems that are similar to what Boulding (1956) labelled *clockworks.* Ackoff (1979) has noted that in the United States, OR has been used effectively at the tactical levels, but it is used much less successfully at the strategic levels.

Many aspects of organizations are characterized by extreme levels of uncertainty, even chaos, however. The application of hard systems techniques to these cases seems inappropriate. The hard systems approach has come under intense criticism, reflecting a broad recognition that it is not as effective in ill-defined situations (Lee, 1973). Functions at the tactical level exist in a much more stable, structured environment because many operational goals have been refined by the time they reach this level and can thus be more readily measured. At this level the emphasis is often placed on economic efficiency, so OR approaches are well suited to the situation. Where complexity is more broad-based, systems analysis may be more useful. If the complexity is ill-defined and systemic, however, it may be more appropriate to use the soft systems approaches—the subject of the next chapter.

~ SUMMARY

The hard system approaches represent one of several major components in the systems approach to managing. As with the other approaches (soft systems, cybernetic, and integrative) the hard systems approach is founded on a philosophical grounding that distinguishes it from the other approaches. One of the main features of this philosophical position is the assumption that all necessary information can be obtained. This information is viewed as holding the keys to the development of an economical solution to a well-defined problem. The solution will be one that produces the greatest result with a relatively minimal contribution of effort and resources. Overall, the hard systems approaches are very effective in improving clearly structured situations. Where the assumptions that information availability is present are accurate, this is a powerful way of thinking about organizations. Although the hard systems approaches are often technique-oriented, their contribution to systems thinking should not be overlooked. The hard systems perspective, with its roots in science and engineering, has made great contributions to the way managers represent and attempt to solve problems. Through the use of modeling and simulation, the hard systems approaches provide a useful vehicle for creating simplified representations of situations in organizations that otherwise prove to be too cumbersome for ordinary verbal and conceptual manipulation. Although not all complex situations can be reduced to simple archetypal problems, neither do they need to remain incomprehensible tangles that hopelessly snare managers.

KEY TERMS AND CONCEPTS

hard systems approaches
hard systems thinking
rationalization
technical rationality
operations research
rational decision making
choice making
parameters
decision criteria
routine situations
performance program
protocol
algorithm
nonroutine situations
heurisics
nonprogrammed decisions
iconic models
analog models
queue
systems analysis

QUESTIONS FOR DISCUSSION

1. Identify a situation in an organization that could be improved through the use of hard systems thinking. Which specific applications do you think would prove most useful?
2. Which of the following problems would benefit most from the use of a hard systems approach (explain your answer)? (a) Completing the tunnel between France and England; (b) creating a market economy in Eastern Europe and in the Commonwealth of Independent States; (c) stopping the deforestation of the tropical rain forests of South America; and (d) preserving supplies of fish by preventing overfishing in such areas as the Atlantic Ocean's Grand Banks.
3. You work for a company that manufactures a high-priced radar surveillance device for the military. Your firm has won the bid for a contract to produce three units at a rate of one per year at a selling price of $12 million per unit. The price cannot be renegotiated and has a 5-percent cost overrun provision in the contract. The procurement officer has told you that if the first two units are not of high quality the order for the third will be cancelled. There are virtually no opportunities for benefiting from economies of scale or the learning curve. How will you guarantee the profitability and success of this program by using hard systems thinking?

REFERENCES

Ackoff, R. (1967). "Management Misinformation Systems," *Management Science, 14,* pp. 147–56.

Ackoff, R. (1979). "The Future of Operational Research Is Past," *Journal of the Operational Research Society, 30,* pp. 93–104.

Ackoff, R., and Sasieni, M. (1968). *Fundamentals of Operations Research.* New York: Wiley-Interscience.

Beer, S. (1959). "What Has Cybernetics to Do with Operational Research?," *Operational Research Quarterly,* March, pp. 1–21.

Boulding, K. (1956). "General Systems Theory—The Skeleton of Science," *Management Science, 2,* pp. 197–208.

Budnick, F., Mojena, R., and Vollman, T. (1977). *Principles of Operations Research for Management* (Homewood, Ill.: Richard D. Irwin.

Chatfield, C. (1980). *The Analysis of Time Series,* 2nd ed. London: Chapman and Hall.

Coccari, R. (1989). "How Quantitative Techniques Are Being Used," *Business Horizons,* July–August, pp. 70–74.

Feldman, J., and Kanter, H. E. (1965). "Organizational Decision Making," in J. March (ed.), *Handbook of Organizations.* Chicago: Rand McNally.

Hoos, I. (1979). "Engineers as Analysts of Social Systems: A Critical Enquiry," *Journal of Systems Engineering, 4,* pp. 81–88.

Kwak, N., and DeLurgio, S. (1980). *Quantitative Models for Business Decisions.* North Scituate, Mass.: Duxbury Press.

Lyles, M. A., and Mitroff, I. I. (1980). "Organizational Problem Formulation: An Empirical Study," *Administrative Science Quarterly,* March, pp. 102–119.

March, J., and Simon, H. (1958). *Organizations.* New York: John Wiley & Sons.

Miser, H. (1989). "The Easy Chair: What Did Those Pioneers Have Uppermost in Their Mind, Model Building or Problem Solving?," *Interfaces,* July–August, *19,* no. 4, pp. 69–74.

Morse, P. M., and Kimball, G. E. (1946). *Methods of Operations Research.* Washington, D.C.: Office of Scientific Research and Development, National Defense Research Council.

Quade, E., and Miser, H. (1985). *Handbook of Systems Analysis: Overview of Uses, Procedures, Applications, and Practice.* New York: North Holland.

Schon, D. (1983). *The Reflective Practitioner.* New York: Basic Books.

Schon, D. (1987). *Educating the Reflective Practitioner: Toward a New Design for Teaching and Learning in the Professions.* San Francisco: Jossey-Bass.

Walton, M. (1986). *The Deming Management Method.* New York: Perigee Books.

ADDITIONAL READINGS

Beer, S. (1972). *Brain of the Firm.* Harmondsworth: Allan-Lane.

Checkland, P. (1972). "Towards a Systems-Based Methodology for Real World Problem Solving," *Journal of Systems Engineering, 3,* no. 2, pp. 87–116.

Churchman, C. W., Ackoff, R. L., and Arnoff, E. L. (1957). *Introduction to Operations Research.* New York: John Wiley.

Cyert, R., and March, J. (1963). *A Behavioral Theory of the Firm.* Englewood Cliffs, N.J.: Prentice-Hall.

Hillier, F. S., and Lieberman, G. S. (1974). *Introduction to Operations Research.* San Francisco: Holden-Day.

Horabin, I., and Lewis, B. (1978). *Algorithms.* Englewood Cliffs, N.J.: Educational Technology Publications.

Perrow, C. (1970). *Organizational Analysis: A Sociological View.* Monterey, Calif.: Brooks/Cole Publishing.

Rosenhead, J. (1989). *Rational Analysis for a Problematic World.* Chichester, England: John Wiley.

Ulrich, W. (1980). "The Metaphysics of a Design: A Simon-Churchman 'Debate,'" *Interfaces, 10,* no. 2, pp. 35–40.

6 The Soft Systems Approaches

In the swampy lowland, messy, confusing problems defy technical solution. The irony of the situation is that the problems on the high ground tend to be relatively unimportant to individuals or society at large, however great their technical interest may be, while in the swamp lie the problems of the greatest human concern.

Donald Schon (1987)

1. To distinguish soft systems approaches from the hard systems approach.
2. To understand the basic assumptions that underlie the soft systems approaches.
3. To describe how soft systems thinking can serve as a valuable process for understanding organizations.
4. To define interactive planning, inquiring systems, and the soft systems methodology as models for changing organizations.
5. To understand the differences between organization development and other soft systems techniques.
6. To describe how organizational learning serves as a critical factor for implementing the soft systems perspective in organizations.
7. To appreciate the role of action learning in soft systems approaches.

SECTION I
SOFT SYSTEMS APPROACHES

INTRODUCTION

Soft systems approaches have developed, primarily over the past two decades, as vehicles to deal with the challenges posed by dynamic, ill-defined issues. This chapter will present several managerial perspectives built on a foundation of assumptions quite different from those underlying hard systems approaches. The purpose of this chapter is to explain the essential precepts of soft systems thinking and those differences. This chapter will also explore a number of methods for creating change in organizations that rely on soft systems thinking. Each of these methodologies will be evaluated in terms of its conceptual integrity and its effectiveness as a managerial tool.

Although soft systems approaches have only existed for a few decades, the recognition that a softer approach was needed reaches back to the beginning of this century to the writings of Chester Barnard and Mary Parker Follett. Many of their ideas have various elements that are contained in the current soft systems view. Barnard (1938) is most remembered for his vision of organizations as cooperative systems and his strong belief in the importance of willingness to cooperate, common purpose, and communication in business organizations.

Follett integrated the early works of Frederick W. Taylor and the scientific management movement with the precepts of social psychology, especially gestalt theory. Follett was largely responsible for identifying the potential of working through groups, rather than just as individuals, to accomplish goals. She also discovered what she termed the first contingency principle in management, the "law of the situation" (Follett, 1930). This concept suggested that the structure of the system rather than the managers' relationship with employees should dictate managers' actions. Although neither Barnard nor Follett developed detailed accounts of the methods they envisioned for changing systems, they are still remembered as pioneers, the first to conceive of organizations as social systems rather than just as technical or economic entities.

Hard systems theories assume that a system's goals are well-known, that problems can be identified in an unambiguous, definitive fashion, and that concrete, "cure-all" solutions can be created. Managers who follow hard systems views assume that once the optimal solution has been identified, success is a mere matter of implementing it. In this view the careful use of engineering and scientific methods can accomplish four basic goals: to reduce uncertainty through rationalization, predict future states, prepare for future states, and control the system to ensure efficient operation. The core assumption is that the essence of a system is knowable, and that the system will yield to change brought on by the use of management techniques. The primary benefit of using hard systems approaches in well-defined situations is clear. The complexity of mechanistic systems is usually transparent. The relations between elements follow programmed paths and can be managed independently of other parts of the system.

Although hard systems approaches work quite well in such situations, most of the chronic, difficult problems people must face do not fit this mold. Rather managers' most challenging issues are ill-defined, fuzzy, and messy. These issues range from attempting to gauge customer demand for a new product in development to anticipating the effects of a new manufacturing technology on workers to defining the effects of a new leader's style in a high-performing culture to estimating when a product market will reach its saturation point. In **soft systems thinking,** people play a central role, and they can change things just by thinking differently. The global spotlight that follows compares the use of hard and soft systems thinking in organizations.

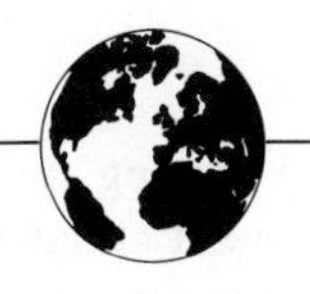

GLOBAL SYSTEMS SPOTLIGHT

Soft Thinking, Hard Results: Japanese Management

Japanese management practices became news to the world in the early 1980s for the simple reason that they appeared to be an important factor for explaining the success of Japanese corporations. The challenge posed by these practices for American and European management became more serious during that decade because their softness seemed so incongruent with the Western management traditions. Instead of the hard management procedures, complex structures, and sophisticated strategic systems found in Western companies, Japanese-style management developed within a framework based primarily upon intangible, soft norms such as subtlety, polarity, empathy, and *kaizen* (continuous improvement). This was the outgrowth of cultural traditions and national experience; more important, it was the result of decades of investment in the country's only major resource—the Japanese people. The climate of mutual trust among employees and managers in Japanese firms means that all work is for the common goals of the firm. This replaces the hard structural and procedural arrangements found in the West, which attempt to clearly define tasks and responsibilities. While such clear divisions of responsibilities may enhance individual productivity, they also significantly increase the complexity of the coordination process.

The emphasis on subtlety in Japanese firms supports the development of learning organizations. Instead of strict formalization and standardization of relations between roles within organizations, Japanese firms are given the freedom to evolve while taking into account complex, dynamic relationships. By rotating their employees among various positions, departments, and locations, Japanese firms enable their managers to learn the rules of the game in their own way, instead of forcing them to learn a set of bureaucratic principles. It may take many years to learn and understand the unwritten rules and subtle practices of the Japanese company, but after they are learned, these socialized principles regulate the behavior and conduct of employees. The power of the soft systems approach in Japan illustrates the importance of managing in ways that are attuned to symbolism, interpretation, learning, and the complexity of interaction that exists between people and technology.

The Japanese use of softer approaches to managing reflects belief in a different set of assumptions about the world than those held by Western managers.

VIEWING THE WORLD ANEW

Organization and management theory traditionally assumes that there is a common reference point for all management decisions in an organization; managers' efforts are directed toward improving the same organization in the same environment and solving the same problems. The implication is that there is common agreement about these fundamental assumptions, which allows them to be taken for granted. Difficulties were always considered to lie in differences of opinion regarding ends and means.

In the 1980s, many organizations began to recognize that the existence of shared fundamental assumptions could be self-limiting. They began to break through those limits by examining the assumptions. Questions such as, "What business are we in?"—which would have been considered foolish in prior decades—became important. What transformed the acceptability of this question? The cause was a change of assumptions, which led to a change in thinking. The realization that an insurance company could define itself as a financial services company provided a new vista for managers. The recognition that corporations could define themselves in very broad, sometimes abstract terms was seen to have certain advantages, including fostering continued growth. For example, the Disney Corporation—already in the theme park, cable TV, film and video, and retail businesses—announced in 1992 that it would enter the magazine publishing business. "The mission is to build a magazine division and have magazines that fit within its realm of family, fun, environment, vacations, technology, and science" (Warren, 1992). New assumptions led to a novel pattern of thinking; managers for the twenty-first century need to manage in novel ways. Many times this change in thinking, or **metanoia,** is externally generated by concepts that become popularized in various media. Imagine the potential power of generating new ways of thinking as both an internally and externally driven managerial process.

SYSTEMS NOTE

A New Way of Thinking in Organizations?

Seventeen years ago sociologist Daniel Bell wrote that for most of human history, reality was nature; then it became technology; and now in the post-industrial age of knowledge work and information science, it's the "web of consciousness." That is what's new about the new paradigm: this focus on human consciousness—not on capital or machinery, but on people. It has challenging implications. "If consciousness is important, then money and profit are no longer that important" argues Michael Ray of Stanford (University). "They're a way to keep score, but if you don't have any vision, you're not going to be successful in the long run."

Source: Fortune, October 8, 1990, p. 164.

FUNDAMENTALS OF SOFT SYSTEMS APPROACHES

Soft systems approaches are based on the belief that because perceptions are subjective experiences, there is no single reality. Accordingly, activities will be interpreted in different ways by various viewers. This makes it extremely difficult to identify a single, well-defined problem in regard to complex, dynamic systems. Therefore, this approach provides little impetus to creating single technical solutions to problems. Soft systems thinking addresses organizational improvement through the vehicles of continuous learning and communication. These are generative tools, which help to build the problem-solving capacity of organizations. **Generative learning** is learning that not only helps through the acquisition of information but also by enhancing the learner's capacity to learn. Learning cannot be effective if a person does not know how to learn.

In soft systems approaches, learning occurs in organizations through reiterative cycles of action, reflection, discussion, and experimentation. The goal is not to solve problems instantly, but to whittle away at problematic areas as the organization members' awareness of the issue related to the problematic area broadens and deepens. Soft systems approaches are educative strategies that enable groups to learn about their own thought processes, individually and collectively, while they explore ways to improve problematic situations. Soft systems approaches center on changing how people think because people are the arbiters of how all situations will be seen. This is a major departure from hard systems approaches with their focus on solving free-standing problems in segmented systems. The **learning organization** is characterized by an emphasis on teamwork, facilitative peer teaching, active experimentation, and thoughtful periods of reflection. Such organizations place a premium on enhancing their own processes for improvement.

Learning organizations are also inquiring systems, discussed in Chapter 4. Continuous inquiry into the nature of organizational deficiencies is a self-refining process driven by organization members' new insights into how the structures that they created influence the situation. In learning organizations inquiring systems create a powerful form of leverage for managers because they simultaneously reduce the perceived complexity of current situations while building the system's capacity to deal with higher levels of complexity in the future.

BASIC PRINCIPLES OF SOFT SYSTEMS APPROACHES

Six principles underlie soft systems approaches and distinguish them from other approaches:

1. Perceptions and experiences are subjective. All events are subject to various interpretation and meanings.
2. There are no problems out there waiting to be solved; problems become enacted through people's conditioning and perceptions.
3. The nature of problematic situations is complex. It is more accurate to identify general issues than to define specific problems.
4. There are no permanent solutions, only improvements. A continuous series of improvements are known as *accommodations*.
5. Systems are projections of the mind, not real objects.
6. System improvements rely on learning and accommodation rather than on engineered and optimized outcomes.

Soft systems approaches are based on the belief that many events that people perceive are not objective factual happenings. The perspective acknowledges that organizations are the mental creations or projections of their members. This process is made possible by the tendency for people to be selective in their attention and memory. Weick (1979) proposed that people in organizations tend to create an **enacted environment.** In effect, they selectively attend to only those dimensions of the environment that they think have meaning for the organization. In this view shared perceptions in organizations are driven by shared experience as well as current thinking. Accordingly, people in organizations develop common world views that explain how cause and effect are part of their world. In essence, an organization, which exists in the shared view of its members, can be seen as if it possessed a single, collective mind (Mitroff, 1984). To understand an organization, then, one must appreciate how its members think about it (Churchman, 1968). From a soft systems perspective, the collective perceptions of the members of a system are the system for the people in that system.

A SOFT SYSTEMS VIEW OF CHANGE

If one accepts the premise that people's thoughts are virtually the same as the system, then externally induced change cannot be lasting change. How can there be commitment to support change that is not understood or does not fit one's mental model? Under these circumstances, the idea of solutions based on change is replaced by a goal of internally generated transformational processes as catalysts to facilitate change. In the soft systems perspective, it becomes impossible speak of the system as separate from its people. Soft systems advocates favor the argument that the primary obstacles to system improvement are self-imposed (Ackoff, 1974). To some extent everyone is handicapped by his or her own lack of awareness, and blind spots to certain issues. Organizations, with their shared mental models, contain **systemic blind spots,** inconsistencies that limit the ability of organizations to act flexibly by imposing artificial restrictions on their thinking.

Internally Generated Systemic Improvements. Weisbord (1989) has made a persuasive claim that complex organizations of the future will have ever-increasing needs for internally generated whole system improvements where everyone participates. He sees the processes used to initiate change as having evolved through the sequence shown in Figure 6-1.

The hard systems model of expert solutions being plugged into organizations clearly is inappropriate for dealing with human systems. If uncertainty is a product of thought, then only changes in thinking can effectively improve a system.

Dealing with Uncertainty. Soft systems approaches, like hard systems approaches, promote the reduction of uncertainty, but not through rationalization. Three other approaches are taken to reduce uncertainty:

Stage I. Experts solving problems piecemeal (Taylorism)
Stage II. Everybody solving problems piecemeal (participative management)
Stage III. Experts improving whole systems (systems thinking)
Stage IV. Everybody improving whole systems

FIGURE 6-1 ***The Evolution of Organization Change Processes***
Source: M. Weisbord, *Productive Workplaces* (San Francisco: Jossey-Bass, 1989), p. 262.

1. Continuous inquiry and learning
2. Clarifying desired future states of affairs and the means to reach them
3. Continuous defining of the organization's world view

The ultimate goal of soft systems approaches is to reduce uncertainty by increasing organization members' insight into how they think and interact with the system they have created.

Soft systems approaches emphasize continuous planning, which is designed to promote learning rather than to predict outcomes (Ackoff, 1974). Preparation for external events becomes secondary to the development of an internal dialogue and discussion, which ultimately will reshape the world view of the members of the organization. Under such circumstances, the need for control is minimized because equifinality is accepted as normal. **Equifinality,** you will recall, proposes that there are many different ways to get to the same outcome from the same starting point. In practice, learning organizations thrive on openness and trust. There must be a vision and organizational setting that members regard as so compelling that they will readily take the risks and share their deep beliefs. In hard systems approaches, people are assumed to be rational, willing to accept solutions because they make technical sense. In messy situations it is virtually impossible to develop solutions that objectively serve as the definitive answer. It is more common to have many definitions of the problem and many opinions of how to improve the situation. Here, consensus building is a more appropriate strategy for organizational improvement than is solution finding.

SECTION 2
MODELS OF CONSENSUS BUILDING

INTRODUCTION

The soft systems approaches have four main goals:

1. Regulation of the system
2. Promoting change in the system
3. Learning
4. Self-renewal

It is not necessary for all four goals to be pursued simultaneously, although it is entirely possible. Regulation occurs in the sense that organizational evolution is directed to a course that runs parallel to the purposes of the system. For that to occur, there must be a process for initiating change. The most common means to initiate planned change in the soft systems approach is **consensus building,** a process that draws the members of an organization into an open dialogue regarding a desired or ideal future for the system. In some cases, this is expressed by comparing current states to future states to determine the gap between the two. In other cases, a debate may be triggered by the consideration of a proposition that runs counter to the current world view. The principles of soft systems methods for change are most clearly identifiable in the conceptual models proposed by Churchman (1971), Ackoff (1974), and Checkland (1979, 1988).

CHANGING THE WORLD VIEW OF A SYSTEM

Churchman (1971) proposes that an organization's prime goal is to make itself capable of redesigning itself so that it may have the capacity to respond to future challenges. This strategy for organizational improvement has two main elements. The first is known as the inquiring system. The **inquiring system** is designed to gather information about a problematic area and assist in the generation of knowledge pertaining to that area. An inquiring system could take the form of an information processing system, a market research function, or an attitude survey of personnel. The ultimate goal of the inquiring system is to produce knowledge that can be used to further efforts to redesign both the organization and the inquiring system itself. The second element of the Churchman model deals with the concept of a world view. A person's **world view** is a comprehensive mental image of the fit among all the relationships within his or her experience.

SHARED WORLD VIEWS

Churchman's model assumes that the world view that a person uses to frame his or her perceptions is, at least initially, not objective and therefore not a good basis for taking action. It needs constant revision to move closer to the truth. Objectivity is increased if a world view is the creation of many points of view rather than the perception of one person who claims to be unbiased (Churchman, 1968). Through continuous clarification of the prevailing world view within a system, the people within that system may develop greater objectivity (Churchman, 1971). Churchman proposes that as the world view is revised, the member's frame of reference for viewing the system will become increasingly more objective. This process of clarification is a continual, ever-refining cycle.

Churchman's perspective is based on the assumption that people's actions depend on the fundamental way that they perceive a system. Ultimately, a world view provides the context for all perceptions and judgments. Since people are selective in how they perceive the world, it is likely that each will recognize those types of information compatible with his or her world view and exclude other information. Churchman's model assumes that perceptions are largely dependent on a person's basic system of beliefs (see Figure 6-2). Therefore any strategy to change the human aspects of a system must modify the world view of the people who make up that system. In essence, Churchman's perspective reiterates the fundamental assumption of soft systems approaches, which is that all situations will be subjectively experienced and have different meaning for each person.

A world view may also exist as a characteristic of a group or system, as a dominant world view that is embedded in the culture of an organization. For example, Kets de Vries and Miller (1984) suggest that the members of future-oriented, high-technology firms are likely to develop shared fantasies around a utopian culture:

> There is a striving for grandiosity, a form of optimism that seeks to improve and inspire, often with little attention paid to present-day reality. We find an attitude that fosters intensive collaboration, and participative, democratic decision making among groups of technical experts, scientists, and managers. (p. 67)

These prevailing views may also be translated into such tangible forms as norms, rituals, or symbols. The shared norms of a social system may be expressed by something as simple as the way people in the system regard customers. In some systems, customers are the raison d'etre of the system, and they are treated specially; in others, they may be seen as a source of uncertainty and are regarded as fickle. At the Seattle-based department store chain Nordstrom's, giving customers anything less than total service flies in the face

---

1. The frame of reference that people use as their guide for perceiving is a set of interconnected beliefs known as their world views. A world view is created by each person based on past subjective individual experiences. The way people see situations cannot change until their underlying world views are altered.
2. People's ability to see things objectively becomes relatively greater when their world views are continually challenged by competing alternative perspectives.
3. The opportunities for a constructive clash of differing perspectives is increased by the creation of a dialectic process within the system.
4. The objectivity of the views held by people within the system can be increased by continually generating information about how the system operates and making this available to people.
5. High-performing systems will be able to use data from inquiring systems to continually redesign themselves to seek continually greater levels of objectivity.

FIGURE 6-2 ***Basic Assumptions of Churchman's Approach***

of all prevailing norms. This world view is symbolized by the hands supporting customers in their organization chart. (Figure 6-3).

Peters and Waterman (1982) have identified seven core beliefs that are common to high-performing organizations:

- A belief in being the "best"
- A belief in the importance of the details of execution, the nuts and bolts of doing a job well
- A belief in the importance of people as individuals
- A belief in superior quality and service
- A belief that most members of the organization should be innovators, and its corollary, the willingness to support failure
- A belief in the importance of informality to enhance communication

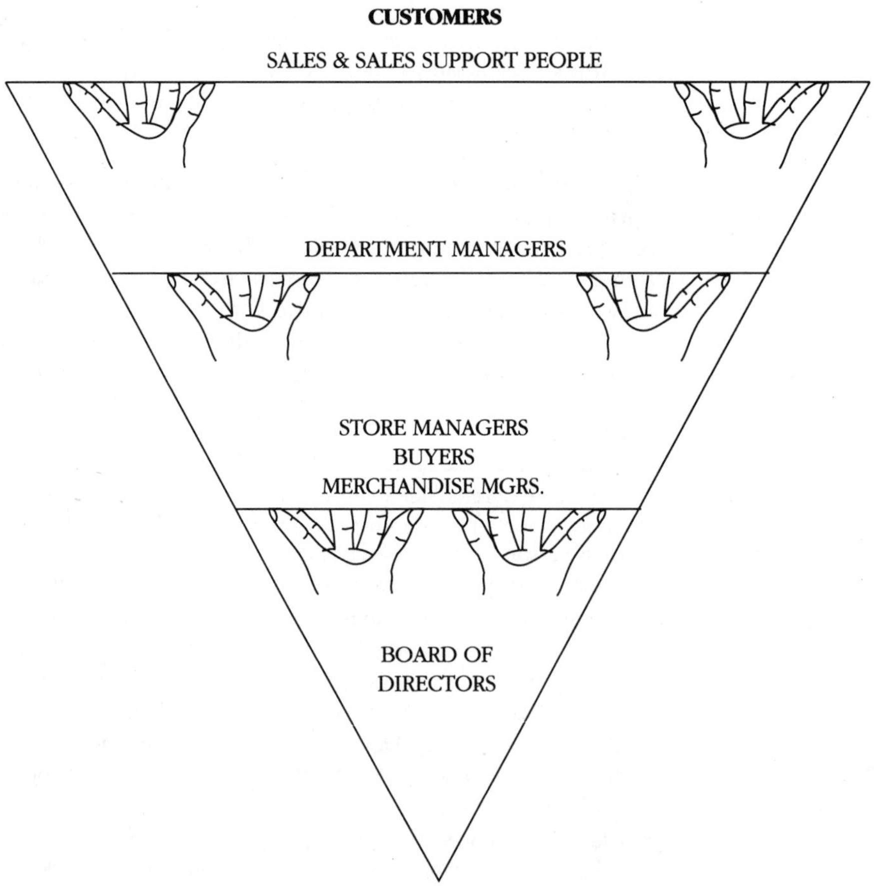

FIGURE 6-3 ***Nordstrom's Organization Chart***

Source: T. Peters, *Thriving on Chaos* (New York: Alfred A. Knopf, 1987), p. 445.

- An explicit belief in and recognition of the importance of economic growth and profits (p. 285)

Churchman's perspective is that change in social systems must originate with fundamental alterations in the shared beliefs of the members of the system. But how is a world view changed, and what should it become?

THE MECHANISM FOR CHANGE

Churchman's method for changing a world view is based on Hegel's concept of the dialectic (see Figure 6-4). The dialectic proposes that ongoing debate between alternative viewpoints (the thesis and antithesis) yields a new perspective that integrates logical elements of the two views. In this model, conflict is temporarily lessened when a third viewpoint, the synthesis, has developed. Since the dialectic is a continuous process, the synthesis eventually becomes the thesis challenged by another antithesis and the conflict reemerges.

Churchman sees the dialectic process as one in which the prevailing world view is exposed and challenged by alternative competing beliefs. As the as-

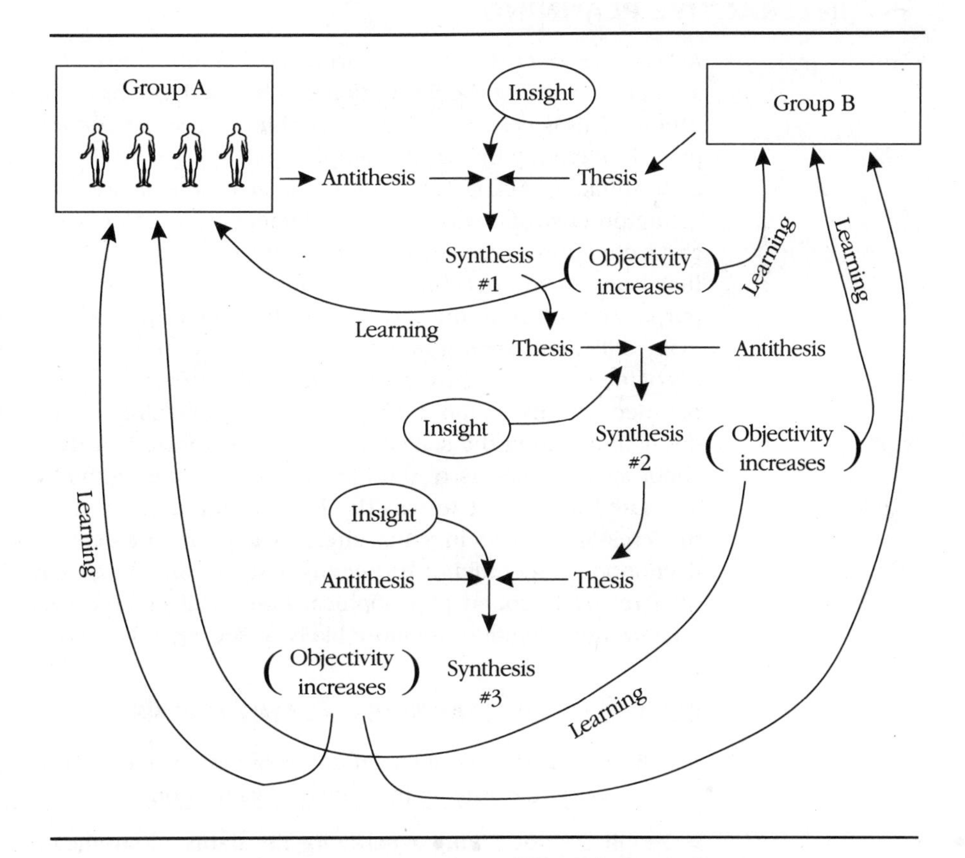

FIGURE 6-4 ***An Interpretation of the Dialectic Process***

sumptions of the world view are put to the test, the objectivity in the system increases. For Churchman, changing the world view is not easy because it is deeply embedded in the culture of the system. Further, the social realities of power and hierarchical structure tend to support the world view in resisting efforts to change.

Churchman proposes that the search for objectivity can be facilitated by employing an inquiring system. The inquiring system is designed to search for objective truths within the system that can eventually be used to redesign the system. The inquiring system continuously compares new information to current data and seeks to reconcile the two. The effect is that learned ideas can be incorporated into the operation of the system. All new information and data are perceived through the world view. In the inquiring system, ideas must constantly be filtered by the world view. While the world view is an element in the inquiring system, the feedback and information in the system is not sufficient to cause modification in the world view alone. In the final analysis, the world view of a system can only be changed by being challenged by a competing world view in the dialectic process.

INTERACTIVE PLANNING

Ackoff's perspective is that the prime goal of interactive planning is to develop a common view of the future that can be used as a basis for action in organizations. This is necessary because problematic situations are likely to be interpreted differently by various people. In keeping with the tradition of soft systems thinking, Ackoff (1974) notes that problems are "products of thought acting on environments, they are elements of problematic situations that are abstracted from these situations by analysis" (p. 21). The prime assumption is that many of the difficulties that cause systems to be unable to attain their purposes result from the way people think about problematic situations. Thus many difficulties are self-imposed; they are due to the subjectivity of the process of interpretation. **Interactive planning** attempts to promote greater levels of objectivity in organizations by creating a planning framework that requires discussion among the stakeholders of an organization. By requiring dialogue about specific aspects of the idealized future, the method ensures that goals are stated in explicit terms. The basic assumption of interactive planning is that creating a forum in which alternative futures are debated will increase the likelihood of identifying erroneous assumptions; thus only those positions with relatively sound philosophical bases will emerge. During such debate organization members are more likely to see any self-imposed constraints.

BELIEFS OF INTERACTIVE ORGANIZATIONS

Ackoff (1974) has identified a number of basic values and beliefs that he feels are necessary to create an interactive organization:

- People are not willing to settle for the status quo of the current state, or for survival or even growth.

- A significant part of the future can be controlled.
- Organization members have a desire for self-development, self-realization, and self-control.
- Ideals can never be attained, but they can be approached with increasing success.
- Learning through experience is sometimes a slow, ambiguous, and imprecise process. As a result active experimentation is the preferred method for learning.
- Planning requires wisdom. Wisdom is the ability to recognize and manage long-run consequences.
- Any aspect of the system can be changed if that improves overall functioning. Nothing is spared from needed examination or change.

The focus of the debate centers on identifying an idealized future for the system and creating a plan to move from the current state to that desired state. Shared perceptions of situations must precede actions. However, since interactive planning is a continuous process, discussion will also follow action as plans are modified or redesigned (see Figure 6-5). The continuous nature of interactive planning provides a fruitful environment for learning to occur. This is because the repetition of action and discussion permits people to progressively note the outcomes of their actions and reflect on this feedback in the group discussions.

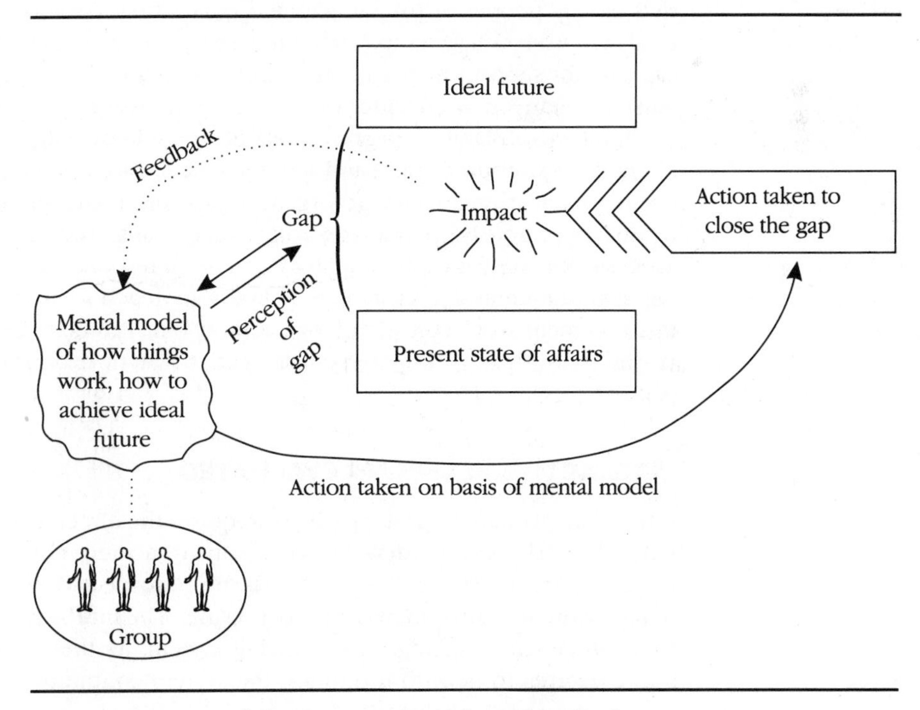

FIGURE 6-5 ***The Interactive Planning Cycle***

THE INTERACTIVE PLANNING PROCESS

The most important benefits of planning are not the outcomes (goals and strategies), but rather the insight gained from participating in the process. To gain these benefits members must engage in the process personally; it is not enough to be in some way associated with the planning process.

The process of interactive planning is based on four principles:

1. Participation
2. Coordination
3. Integration
4. Continuous cycles of planning

Planning efforts that focus on a single aspect of the organization one at a time lose the perspective of how the parts of the system work together. "All aspects of the system must be planned for simultaneously and interdependently" (Ackoff, 1974, p. 29). For example, planning for new products and services must consider the available resources within the organization to support any new product development initiatives. This requires an evaluation of human, technical, and informational resources.

Long-range planning efforts emphasize designing strategies that will mesh with the opportunities, threats, and structural changes of the environment. The goals or ends desired are identified and eventually communicated to the people in the organization. For the ends to be achieved, the means for reaching them must be considered at the same time. Eventually the plans must be carried out by people in the tactical and operational areas of the organization. Goals are useless targets to pursue unless the organization has developed the capacity for supporting the efforts to achieve them. Therefore, planning efforts must simultaneously consider the impact of means on goals and vice versa.

Interactive planning organizations generate knowledge. Relentless cycles of inquiry, experimentation, and action are necessary to create the information needed for learning. Dynamic organizations that thrive in changing environments continuously update and adjust their plans. The continuous planning process ultimately creates a greater capacity to respond innovatively to changing environmental circumstances. Ackoff's approach seems to be matched by this comment from consultant and author Tom Peters: "The essence of successful innovation is, and always has been, constant experimentation" (1987, p. 240).

INTERACTIVE PLANNING EVALUATED

Interactive planning is participative. It requires the direct involvement of stakeholders. It asks stakeholders to make plans to achieve whatever they believe to be important. This is in contrast to the retrospective orientation of traditional planning, which focuses on past results. The interactive system attempts to create a self-designing, self-renewing system. As the concept of the ideal future evolves to fit with the times, the system continuously redesigns itself.

These capabilities are prized in a turbulent environment, which is characteristic of the current global system. The business environment of the 1990s and beyond is what Vaill (1989) describes as "permanent whitewater." If permanent whitewater is the norm of the future, then an equally dynamic management system is needed to keep pace. The power in interactive planning is not in making radical change; rather it is in the collective gains that accrue from continuous improvement, which is a less dramatic but more fundamental type of change.

SOFT SYSTEMS METHODOLOGY

Peter Checkland's soft systems methodology (SSM) is a relatively comprehensive model of change in social systems. It is similar to interactive planning in that it provides a structure to the process of considering the ideal future of a system. This model also incorporates the ideas of the inquiring system and continuous learning proposed by Churchman. SSM is clearly a system for learning to think systematically and for thinking about learning. Learning is focused on complex problematical situations and triggers activity that leads to improved performance (Checkland, 1989). Five major characteristics of SSM are explained in Figure 6-6.

1. *Emphasizes managerial applications.* It defines managing as a process of organized activity. It focuses on the manager's role in recognizing change and taking action in response.
2. *Issues result from differing interpretations.* The same event is likely to be evaluated differently by various people. Facts and situations will have different meaning to different people depending on their world view and past history. This causes situations in which managers must somehow reconcile these differences to coordinate common efforts to achieve goals.
3. *Defining the managerial role from a systems perspective.* The act of consciously defining the manager's role in the context of a dynamic system of events, interpretations, and actions is valuable. It will provide insight into some of the issues that derive from the interconnected nature of human activity systems.
4. *World view must be understood.* The meaning of human activity systems will differ depending on the world view that provides its context. In order to understand the significance of a system for its members, its world view must be identified and its basic assumptions known.
5. *An inquiring system.* SSM helps people in the system to learn. They actively participate in comparing models of human activity systems with their own interpretations of their experiences in actual situations.

FIGURE 6-6 ***Characteristics of SSM***

Source: J. Rosenhead, *Rational Analysis for a Problematic World* (Chichester, England: John Wiley & Sons, 1989), p. 83.

HUMAN ACTIVITY SYSTEMS

SSM assumes that every situation of interest to managers involves bringing people together to achieve a purpose. When people come together to accomplish a task they form a **human activity system.** "A human activity system is a concept of a purposeful 'machine' consisting of activities so linked that they constitute a purposeful whole" (Checkland, 1988, p. 27). The immediate goal of SSM is to explain how such systems operate. This is accomplished by having the members of the particular system design a model that describes how they see the system work. Checkland believes that this modeling, engaged in by stakeholders, will provide them with useful insights that can eventually be applied to the actual situation.

The first step in modeling a human activity system is to identify which world view most accurately captures the essence of the activity in the system. For example, attending a class in a university might be interpreted variously by students as a great way to meet friends, an opportunity to learn, a chance to work, or a vehicle to achieve the credentials for increasing one's wealth. Human activity systems can be visualized as groups of events, processes, and transactions that are connected to each other within the context of a particular world view. Practically speaking, the world views used to make decisions in organizations generally remain unexamined. Clear descriptions of each person's world view are essential for the creation of **root definitions,** which specify the elements that are the building blocks of the situation in focus. For example, the relationship of customers to the system in question is a root definition.

Once the model has been designed, it serves as the basis for a comparison with the dynamics of the actual situation. The process of comparison uses a dialectical framework similar to that proposed by Churchman. This comparative process is meant to provide a framework for discussion of alternative courses of action for closing the gap between the model and the actual situation. This creative brainstorming will not only generate practical potential changes but will also serve as an experience-based learning process.

The longer term goal of SSM is to create an opportunity for learning. A learning environment is built on both modeling and debate. Modeling creates situations in which people refine their appreciation of the environments in which they participate. The mere action of designing a model presumes that the most fundamental elements and their relations in the system are understood. Debate increases people's awareness of alternatives and has the potential to increase the likelihood that people will feel committed to working for the implementation of whichever course of action is selected. As is true with the Churchman and Ackoff models, Checkland's methodology seeks greater objectivity through continuous debate and eventually some form of consensus.

THE STAGES OF SSM

Checkland's SSM has evolved progressively since 1972. It is currently constructed as a seven-step sequential model. However, Checkland has noted that it is not critical to follow the seven steps in precise sequence. The model is reiterative; that is, it is designed to be used in continuous cycles of research, learning, and action.

Stages 1 and 2: Finding out About the Situation. SSM begins by identifying the stakeholders of the problem situation, the social system, and the political aspects of the situation.

1. Analyze the participants and their roles. The focus is on identifying those people who are likely to take ownership of the changes being fully implemented.
2. Analyze the customary social roles, norms, and values characteristic of the problem situation.
3. Analyze the power structure that governs the situation. Identify the forms that power takes in the system and how it is enacted.

Stage 3: Forming Root Definitions. Root definitions typically examine the relationship of the following relevant subsystems: customers, actors (people who would perform the activities), transformation process (how the activity will function in this capacity), world view, owner (how those in power will respond to the activity), and environment (the limitations that external systems present).

Stage 4: Building Conceptual Models. The goal of this stage is to illustrate the relationships among the subsystems. Once the subsystems have been identified, the next step is to select the appropriate action words that describe how these elements are related and in what sequence. For example, a conceptual model is presented in Figure 6-7. Checkland's model is an example of a broader class of models designed to guide change in organizations. Known as **action-research** models, these models are conceptual tools for participant-directed experiential learning. (Action-research will be discussed in greater depth later in this chapter.)

Stage 5: Comparing Models and Reality. After having created the conceptual models, participants may have a new perspective on the problem situation. In order to use this new awareness, a discussion occurs; it focuses on the differences between the model and the real situation. Its purpose is to challenge assumptions and search for ways to employ the new ideas as actions for improvement.

M-14

Stage 6: Defining Changes. The previous stage generated some possible alternatives for changing the system but did not determine the extent to which these changes were feasible. The purpose of stage six is to make that deter-

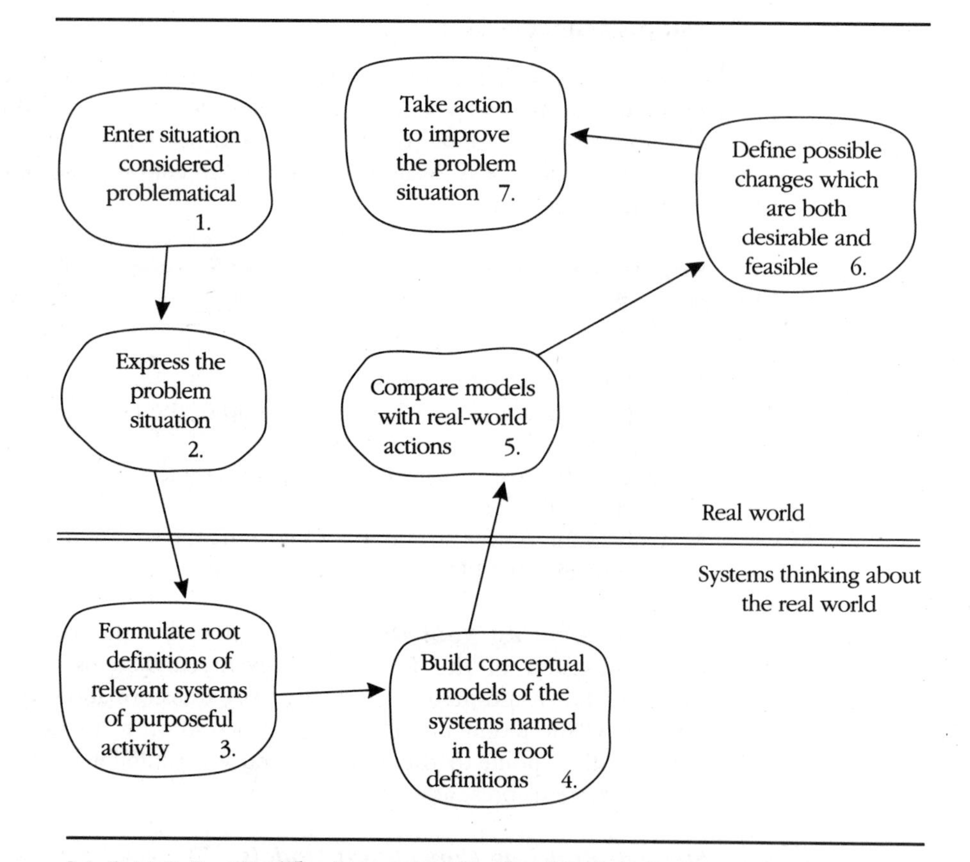

FIGURE 6-7 ***The Learning Cycle of Soft Systems Methodology***
Source: J. Rosenhead, *Rational Analysis for a Problematic World* (Chichester, England: John Wiley & Sons, 1989), p. 84.

mination in terms of two major criteria. The first standard is termed ***systemic desirability***. This factor examines the technical merits of the proposed accommodations. The second dimension, which looks at the social acceptability of the proposal, is termed ***cultural feasibility***. Not all technical solutions will be embraced by people; regardless of its logical merits, a solution will be shunned if it is perceived as threatening.

Step 7: Taking Action. The implementation of the recommendations that were both acceptable and feasible requires action that is guided by the new awareness generated by the learning process. As soon as these new programs take effect, the problem situation is for all practical purposes a new situation. It has a new structure and new dynamics that result from the changed structure. The action-research cycle then begins anew and continues through the seven steps in a continuous process of refinement.

AN EVALUATION OF SSM

As developed by Checkland, SSM attempts to structure the dialectic process in a way that leads to greater objectivity and learning. It is a relatively precise model that includes steps for defining the issues as well as considering the social factors that could impact on efforts to implement new plans (see Figure 6-8).

Unfortunately, it does not address the social and political factors that may inhibit modeling and debate. In general, SSM meets the three goals of soft systems approaches: regulation, learning, and the power to create change. However, much of the capacity for change in SSM depends on the skill of the person directing the change effort (Jackson, 1982). In a broad perspective, SSM represents one of the balanced and integrated strategies for systematically guiding change in organizations. Its perspective differs from the vantage point found in models based on social psychology, such as organization development (OD). Systems can also be changed through improvement in the way insight is integrated into decisions that affect the entire system and by improving the social system. The next section of this chapter examines these issues.

SECTION 3
CREATING A LEARNING ORGANIZATION

ORGANIZATION DEVELOPMENT

There are two action-research methods for creating a learning organization. The first perspective that will be discussed is organization development (OD). Beckhard (1969) has defined **organization development** as an effort planned, organization-wide, and managed from the top to increase organization effectiveness and health through planned interventions in the organization's processes, using behavioral science knowledge.

The basic premise of OD is to improve the organization's problem-solving abilities on a system-wide basis. This is generally accomplished by providing an action-research framework that people can use as a guide for experimentation and discovery. OD is behaviorally oriented; it places high value on changing the social and interpersonal aspects of organizations to improve their overall functioning. French (1969) identified seven basic objectives of OD (see Figure 6-9).

The form of action-research employed in OD is called the survey-feedback method. This technique was developed by Rensis Likert at the University of Michigan. Survey feedback is the most popular system-wide process intervention in OD. It consists of collecting data from an organization or department through the use of a questionnaire. The data are analyzed, fed back to organization members, and used by them to diagnose problems of the organization

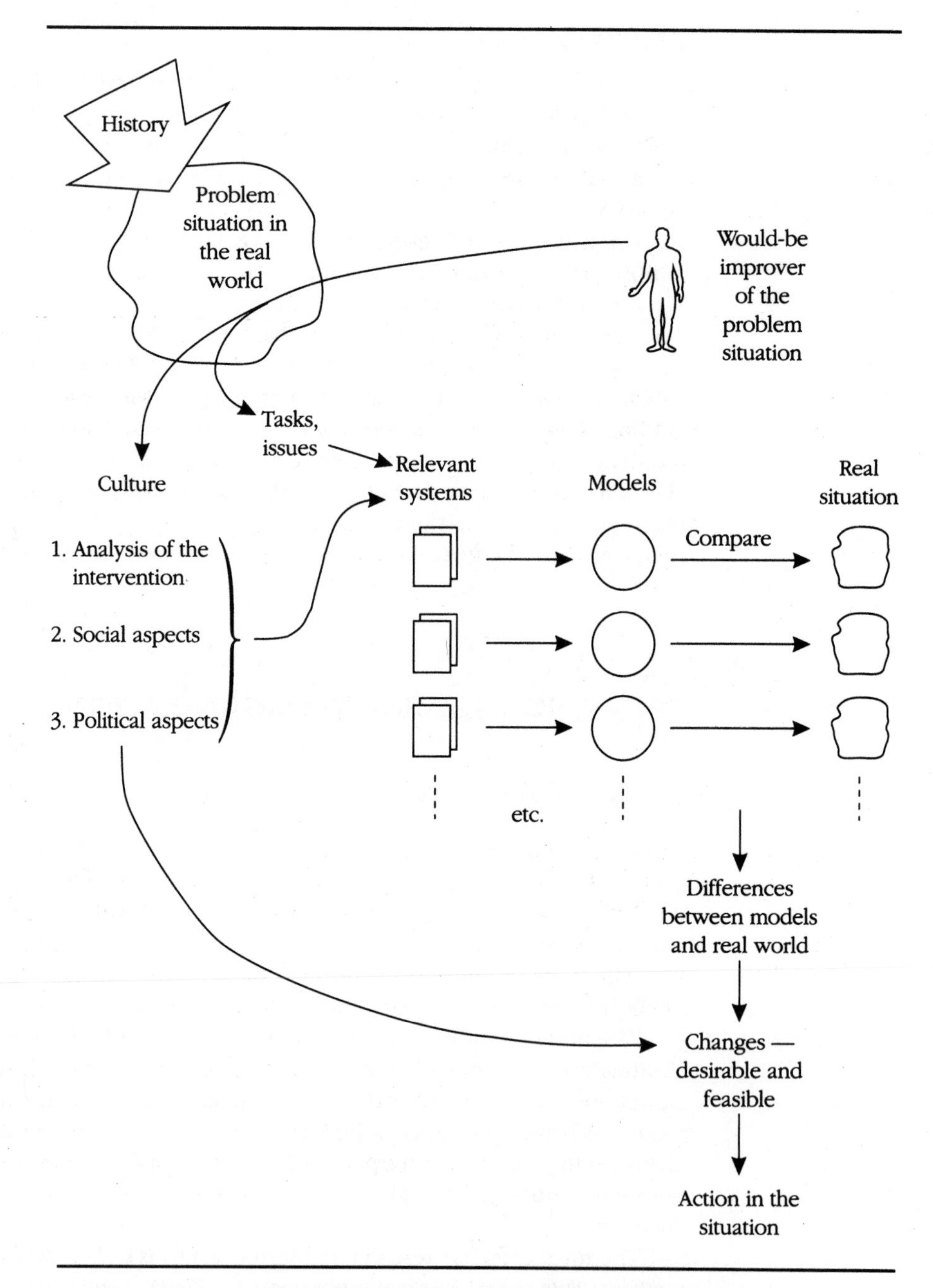

FIGURE 6-8 ***The Process of Soft Systems Methodology***

Source: P. Checkland, "Soft Systems Methodology: An Overview," *Journal of Applied Systems Analysis,* 1988, *15,* p. 29.

1. To increase the level of trust and support among organization members.
2. To increase the incidence of confrontation of organization problems, both within groups and among groups, in contrast to "sweeping problems under the rug."
3. To create an environment in which authority of assigned role is augmented by authority based on knowledge and skill.
4. To increase the openness of communication laterally, vertically, and diagonally.
5. To increase the level of personal enthusiasm and satisfaction in the organization.
6. To find synergistic solutions to problems with greater frequency.
7. To increase the level of self and group responsibility in planning and implementation.

FIGURE 6-9 ***Objectives of Organization Development***

Source: Adapted from W. French, "Organization Development: Objectives, Assumptions, and Strategies," *California Management Review,* 1969, *12.*

and to develop interventions to improve it. The results of the surveys are analyzed and then fed back to the people who originally responded to the survey. The results for various subsystems are compared and then used as the basis for discussion of areas for improvement. Since OD has a social-psychological emphasis, the areas targeted for change are often in the domains of leadership, communication, trust, involvement, and so on.

Fundamentally, OD uses the same action-research cycle as do the models of Churchman, Ackoff, and Checkland discussed earlier. These models for change, however, do not focus exclusively on behavioral issues. Also, the comparisons are initiated by the consultant in OD rather than being an internal function, as in the other models. In OD the survey feedback becomes a sort of status report on the current state of the system. This is eventually compared with what the people of the subsystem would like the state of the system to be; the results of that comparison become the basis for planning future changes.

Although the OD perspective generally centers on improving the problem-solving abilities of the system, it is nevertheless a soft systems approach. It does not focus on designing freestanding solutions in the hard systems sense; rather it is oriented toward creating the capacity for responding to change. Bennis (1969) has stated that organization development is "a response to change . . . a complex educational strategy intended to change the beliefs, attitudes, values and structure of organizations so that they can better adapt to new technologies, markets, and challenges, and the dizzying rate of change itself" (p. 2). Bennis's definition of OD may sound suspiciously like Churchman's model. Both use feedback cycles to change basic beliefs (what Churchman calls *world view*). The fundamental difference between the two is that Churchman's goal is to change the way people view situations in general so as to be closer to the truth whereas OD attempts to reduce dysfunctional beliefs that will limit interpersonal relations, such as mistrust, and to develop greater capacity for change. Furthermore, the methods applied to reach these ends in OD are straightforward and practical, while Churchman's methods are abstract and difficult to effectively put into action.

In general, OD differs from other soft systems perspectives in being highly prescriptive and more explicitly value-laden. OD interventions include a conscious effort to transform systems toward adopting a humanistic orientation. OD efforts are generally open-ended, in that OD accepts many alternative avenues to arrive at the ideal future. But the process of defining an ideal future is governed largely by a set of basic values (see Figure 6-10).

The specific way that OD is practiced depends not only on the values of the OD consultant but also on the culture in which it is practiced. The global systems spotlight highlights differences in the practice of OD that are related to culture.

GLOBAL SYSTEMS SPOTLIGHT

OD Viewed from a Cultural Perspective

Singapore. The application of OD in the Asian context is illustrated by a study of forty-nine companies in Singapore. The study reveals that only the largest and most well established firms have OD programs. In Singapore, OD has come to be associated with training and vice versa. Organizational change efforts tend to use the top-down approach, with middle managers waiting for direction from executives. Some OD programs have created fear and skepticism because middle managers are not skilled in their implementation, and some companies lack the open climate needed to support any OD effort.

Source: J. Putti, "Organization Development Scene in Asia: The Case of Singapore," *Group and Organization Studies* (September 1989), *14*, pp. 262–70.

In any program of intervention designed to promote change, there is a need for sensitivity to the larger society's values and beliefs. OD and the other soft methodologies are ethnocentric in the sense that they all have been developed in Western Europe or the United States; their efficacy in other cultures is not well-established.

Lewin's original model of OD (1951) included efforts both to remove the social barriers to change and to identify an ideal future. The ideal future could become progressively more apparent to organization members as they work through continuous cycles of assessing the current situation and jointly planning changes to the systems. This perspective reflected Lewin's basic frame of reference, which saw change as the product of continuous interaction between driving and restraining forces. These are the core elements of Lewin's

1. Replace the production model of organizations with a social model. See people as humans with a dynamic potential for growth, not as static components of a production process.
2. Provide people with opportunities to develop their potential. Help them to grow as people through choice of work assignments and matching to career paths.
3. Create jobs and a general work environment that people experience as stimulating.
4. Empower people so that they can influence their work environment.
5. View people as whole persons with an entire spectrum of abilities, limitations, life beyond the workplace, and complex, sometimes competing needs.
6. Consider human factors and feelings as legitimate.
7. Develop better methods for resolving conflict.
8. Start processes of organization change at the top of the organization to demonstrate commitment to OD values.

FIGURE 6-10 ***Basic Values in OD***

Source: Adapted from N. Margulies and A. Raia, *Organization Development: Values, Process, and Technology* (New York: McGraw-Hill, 1972); and W. Bennis, *Organization Development: Its Nature, Origins, and Prospects* (Reading, Mass.: Addison-Wesley, 1969).

force field model. This way of viewing change is analogous to the dialectic process discussed earlier.

All soft systems perspectives are marked by the polarities between the present and the future states of the system, as well as competing values and beliefs. Soft systems prescriptions for change generally attempt to resolve these conflicts by educational strategies designed to increase the flow of information to system members. The assumption is that given the right conditions, over time, the members of the system will collectively gain insight, and the organization will move to a more sophisticated level of equilibrium.

The role of learning in soft systems approaches should not be minimized. If in fact the organization's ability to change is dependent on the collective ability of its members to learn, then the importance of designing a system that effectively promotes learning is self-evident. Specifically, the speed at which an organization can learn appears to be a significant source of competitive advantage in the global economy of the twenty-first century. De Geus (1988) argues that the best way to speed the learning process is to "change the rules that managers live by" (p. 73). This can be accomplished by helping managers learn to learn and to think systemically.

ORGANIZATIONAL LEARNING

The second basic perspective to creating a learning organization falls within the domain roughly known as organization learning. The concept has its roots in the work of Argyris and Schon, who said that **organizational learning** occurs "when individuals, acting from their images and maps, detect a match or mismatch of outcome to expectation which confirms or discomfirms the organizational theory-in-use" (Argyris and Schon, 1978, p. 19). Clearly, this is something that is a fundamental conceptual base of the soft systems approach to change. Although learning occurs at the individual level, it becomes organizational learning when individuals act as learning agents for the organization.

They propose that learning is required at each of two levels in order for the organization to have self-generating capabilities. *Single-loop learning* occurs in second level cybernetic systems (see the discussion of a room thermostat in Chapter 7). The system works exclusively by processing single loops of feedback, one at a time. Although thermostats do not learn in the classic sense, their programming and capacity allow them to identify which temperatures are acceptable and take action in response to unacceptable temperatures. *Double-loop learning* is more sophisticated; double-loop systems may actually modify their internal operations to adjust to environmental changes. Double-loop learning may require that an organization accommodate to policies, objectives, or even norms (Argyris and Schon, 1974). The outcome of this process is that accepted ways for approaching issues for that specific organization become modified.

Despite the central role of organizational learning in soft systems approaches, there is no widely accepted model for promoting it (Fiol and Lyles, 1985). Although Argyris and Schon's description of single-loop and double-loop learning are broadly accepted, they do not represent specific methods for promoting this type of learning. Senge (1990) and Kim (1990) have experimented with creating **learning laboratories** to promote organizational learning. Learning laboratories are "workshops that blend system dynamics principles and repeated simulation game trials with ongoing conceptualization and feedback sessions" (Kim, 1990, p. 1).

~ ACTION RESEARCH

The approaches discussed in the second section of this chapter focused on creating planned change in human systems. A theme of that discussion is the concept of learning through experience. Experience was established not only as the basis for learning but also as the source of new information that must be integrated in order to understand how things actually work. This understanding may take the form of shared mental models or jointly created descriptive models used as guides for further activity. The linkage of action, learning, and concepts is of critical importance to the understanding of efforts to promote system-wide change.

The recognition that action, learning, and concepts were linked has its roots in the thinking of the psychologist Kurt Lewin. Much like Mary Parker Follett, Lewin was strongly influenced by the gestalt movement in psychology. Lewin's grand discovery was that the involvement of people in action plans that would affect them increased the likelihood that they would behave so as to support the implementation of the plan. Consequently, each change would require renewed participation from employees, which in turn would yield added insight into the change process. This realization suggested that expert opinion from outside the group may not be as influential in creating change as internally generated ideas (Weisbord, 1989).

Action learning principles have great significance for change of structure, processes, and behavior in systems. **Action learning** supports regulation and change efforts in two basic ways (Mohrmann and Cummings, 1989). First, learning provides useful ideas that will assist people in putting new plans into effect. Second, as these plans require fine-tuning, the learning helps provide a context for people to make decisions about modifications to the current situation. There are four basic steps in the cycle of action learning:

1. Planned activity, putting ideas into action
2. Collecting information, researching the current situation
3. Comparing the actual situation to the intended state
4. Planning for future action to change the situation.

The action learning cycle provides continual opportunities to learn and experiment with new learning. The sequential nature of this cycle is illustrated in Figure 6-11.

SUMMARY

Soft systems approaches to managing regard systems as the mental creations of people. Consequently, any attempts to change a system must be channeled through people's thought processes. There are five fundamental methods for initiating change in organizations via the systems approach: designing an inquiring system to change world view, interactive planning, soft system methodology, organization development, and organizational learning. These methods of achieving systemic change are all based on a set of assumptions about the nature of human experience that differ markedly from those at the foundation of hard systems approaches.

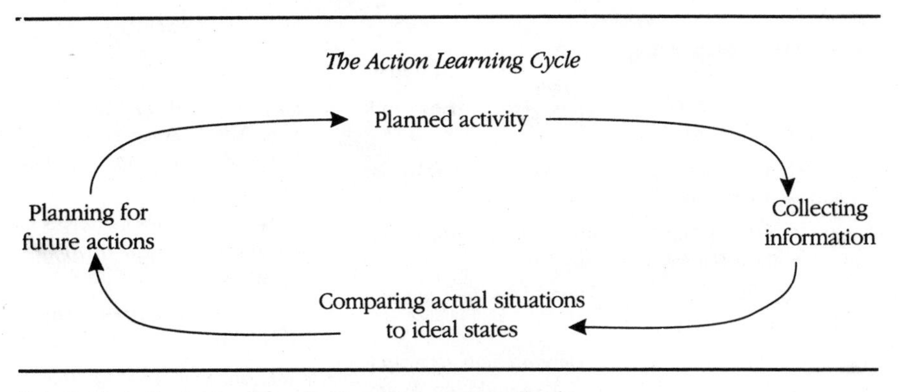

FIGURE 6-11 *The Action Learning Cycle*

KEY TERMS AND CONCEPTS

soft systems approaches
soft systems thinking
metanoia
generative learning
learning organization
enacted environment
systemic blind spots
equifinality
consensus building
inquiring system
world view
interactive planning
human activity system
root definition
action-research
organization development
organizational learning
learning laboratory
action learning

QUESTIONS FOR DISCUSSION

1. Soft systems approaches are a continuous process of refining and redefining issues over the long term. Can such approaches be useful in global organizations that thrive on getting things done quickly?
2. Imagine a scenario where Churchman's inquiring system, Ackoff's interactive planning, Checkland's SSM, and OD are all used in an organization concurrently. What challenges could be anticipated in using them all together?
3. How can managers deal with the political realities of organizations in implementing a soft approach to change? For example, those people who already have power are likely to have more influence in the discussion of a future vision for the company than those people with relatively less power.
4. When would it be more appropriate to use OD than one of the other soft systems approaches?
5. Is it necessary to have any preexisting conditions in an organization before beginning the use of a soft systems approach? For example, how much trust or communication must be in place prior to starting the program?

REFERENCES

Ackoff, R. L. (1974). *Redesigning the Future*. New York: John Wiley & Sons.

Argyris, C., and Schon, D. (1974). *Theory in Practice*. San Francisco: Jossey-Bass.

Barnard, C. (1938). *The Functions of the Executive*. Cambridge, Mass.: Harvard University Press.

Beckhard, R. (1969). *Organization Development*. Reading, Mass.: Addison-Wesley.

Bennis, W. (1969). *Organization Development: Its Nature, Origins, and Prospects*. Reading, Mass.: Addison-Wesley.

Checkland, P. (1979). "Techniques in 'Soft' Systems Practice, Part 1: Some Diagrams—Some Tentative Guidelines," *Journal of Applied Systems Analysis, 6,* pp. 33–40.

Checkland, P. (1988). "Soft Systems Methodology: An Overview," *Journal of Applied Systems Analysis, 15,* pp. 27–30.

Churchman, C. W. (1968). *The Systems Approach*. New York: Dell.

Churchman, C. W. (1971). *The Design of Inquiring Systems*. New York: Basic Books.

De Geus, A. (1988). "Planning as Learning," *Harvard Business Review,* March–April, pp. 70–74.

Fiol, C. M., and Lyles, M. (1985). "Organizational Learning," *Academy of Management Review, 10,* no. 4, pp. 803–13.

Follett, M. P. (1930). "Some Discrepancies in Leadership Theory and Practice," in H. Metcalf (ed.), *Business Leadership.* London: Pitman.

French, W. (1969). "Organization Development: Objectives, Assumptions, and Strategies," *California Management Review, 12,* pp. 23–24.

Jackson, M. (1982). "The Nature of 'Soft' Systems Thinking: The Work of Churchman, Ackoff, and Checkland," *Journal of Applied Systems Analysis, 9,* pp. 17–29.

Kets de Vries, M., and Miller, D. (1984). *The Neurotic Organization.* San Francisco: Jossey-Bass.

Kim, D. (1990). "Learning Laboratories: Designing a Reflective Learning Environment," Working Paper #D-4026, Systems Dynamics Group, Sloan School of Management, MIT. Cambridge, Mass.

Lewin, K. (1951). *Field Theory in Social Science: Selected Theoretical Papers,* D. Cartwright (ed.). New York: Harper & Row.

Mitroff, I. (1984). *Stakeholders of the Organizational Mind.* San Francisco: Jossey-Bass.

Mohrmann, S., and Cummings, T. (1989). *Self-Designing Organizations.* Reading, Mass.: Addison-Wesley.

Peters, T. (1987). *Thriving on Chaos.* New York: Alfred A. Knopf.

Peters, T., and Waterman, R. (1982). *In Search of Excellence.* New York: Harper and Row.

Schon, D. (1987). *Educating the Reflective Practitioner: Toward a New Design for Teaching and Learning in the Professions.* San Francisco: Jossey-Bass.

Senge, P. (1990). *The Fifth Discipline: The Art and Practice of the Learning Organization.* New York: Doubleday.

Vaill, P. (1989). *Managing as a Performing Art.* San Francisco: Jossey-Bass.

Warren, J. (1992). "Disney Making Imprint in Print," *Hartford Courant,* April 12, 1992, p. C1.

Weick, K. (1979). *The Social Psychology of Organizations,* 2nd ed. Reading, Mass.: Addison-Wesley.

Weisbord, M. (1989). *Productive Workplaces.* San Francisco: Jossey-Bass.

~ ADDITIONAL READINGS

Checkland, P. (1981). *Systems Thinking, Systems Practice.* Chichester, England: John Wiley & Sons.

Checkland, P. (1985). "From Optimizing to Learning: A Development of Systems Thinking for the 1990's," *Journal of the Operational Research Society, 36,* no. 9, pp. 757–67.

Fearon, D., and Cavaleri, S. (1991). "Concurrent Learning: Theory and Practice," unpublished, working paper.

Flood, R., and Carson, E. (1988). *Dealing with Complexity.* New York: Plenum Press.

Jackson, M. (1985). "Social Systems Theory and Practice: The Need for a Critical Approach," *International Journal of General Systems, 10,* pp. 135–51.

Margulies, N. and Raia, A. (1972). *Organization Development: Values, Processes, and Technology.* New York: McGraw-Hill.

Nadler, D. (1977). *Feedback and Organization Development: Using Data-Based Methods.* Reading, Mass.: Addison-Wesley.

Quinn, R. (1988). *Beyond Rational Management.* San Francisco: Jossey-Bass.

Rosenhead, J. (1989). *Rational Analysis for a Problematic World.* Chichester, England: John Wiley & Sons.

Sterman, J. (1989). "Modeling Managerial Behavior: Misperceptions of Feedback in a Dynamic Decision Making Experiment," *Management Science,* 35, pp. 321–39.

Vickers, G. (1965). *The Art of Judgment.* New York: Basic Books.

7 Cybernetic Approaches

Control theory shows us how to analyze a closed circle of causation involving a behaving system that can sense and act on its environment.

W. T. Powers (1990)

1. To distinguish between the two primary functions of feedback: control and change.
2. To appreciate the power of feedback loops in shaping behavior in human systems.
3. To see human systems as networks of interconnected feedback loops.
4. To understand the process of managing in terms of feedback loops.
5. To understand the various contributors to system complexity.
6. To recognize the role of variety in creating system complexity.
7. To comprehend Ashby's approach to reducing complexity.
8. To realize the relationship between cybernetic controls and the maintenance of equilibrium.
9. To understand the mechanisms of a simple cybernetic control system.
10. To explain how cybernetic thinking is the basis for total quality management.

SECTION I
CYBERNETIC THINKING

INTRODUCTION

This chapter explores how cybernetic approaches are used to regulate and change dynamic systems. All systems have a need to regulate their own behavior to achieve balance. They also need to change to support growth, expansion, and experimentation. Both regulation and change are established and maintained through the action of feedback loops, which form the underlying structure of a system. Feedback loops are able to form only when there is an effective transmission of information—when the various elements of a system can communicate with each other. Using feedback, systems can coordinate their activities and act with a greater degree of wholeness. Coordination is designed to synchronize the various elements of a system in accordance with an overall purpose. Although each element within a system may have its own sense of timing and purpose, there is usually an underlying theme or an overarching purpose that unifies their actions. Effective coordination ensures that the functioning of one part of the system is enhanced by the functioning of related parts.

Traditionally, cybernetic approaches have been associated with the control of systems. There is a need to ensure that a system is continually moving closer to attaining its purposes. Cybernetic control processes are designed to govern the behavior of a system to continuously increase the probability that its performance will remain purposeful. Cybernetics has come to imply the use of negative feedback loops for purposes of control. However, this view ignores the many influences of positive feedback. We will examine cybernetics from the perspectives of both positive and negative feedback, since in practice the two are inseparable.

~ CYBERNETICS DEFINED

The original concept of a cybernetic approach was developed by the American mathematician Norbert Wiener in 1948. Wiener defined *cybernetics* as "the science of control and communication in the animal and the machine." The term *cybernetics* was based on the ancient Greek word *kybernetike,* which refers to the "art of steersmanship." Wiener's intent was to demonstrate that cybernetics, like piloting a ship, was concerned with keeping a system on the right course, moving in the ideal direction.

In current usage, the concept of cybernetics is much broader in scope. It also includes such processes as information transmission, processing, and storage (Klir, 1965). Cybernetics has also developed into a discipline that relies heavily on mathematics and computers to create programmed control systems. These dimensions of cybernetics are clearly related to the ideas of control and regulation contained in the original concept. Cybernetics can also be understood as a specific form of systems thinking. Robb (1985) has argued that cybernetic systems thinking has served as a common unifying theme that connects the classical, scientific, and organization theory schools of thought in management. Cybernetic thinking can be found in classical management in the emphasis placed on informal communication patterns throughout organizations. Scientific management has applied cybernetic concepts in its concern with feedback loops and closed systems. Organization theories are related in their clear emphasis on integrating organization culture with strategy.

The purpose of this chapter is to examine the underlying assumptions of cybernetic thinking and its basic methods. The definition of cybernetics used in this book focuses on its relation to thinking and to managing human systems. **Cybernetics** is an approach to managing information to regulate and promote change in human systems by promoting the attainment of a wise balance wherever appropriate. Cybernetics assumes that feedback loops are the primary mechanism through which communication, control, and change occur. Positive feedback loops are particularly important; they can accelerate the pace of change. If not carefully managed, they can drive a system beyond its resource limits into a precipitous, often unexpected, period of decline. The

GLOBAL SYSTEMS SPOTLIGHT

A Cybernetic Perspective on Political Conflict

Several ongoing conflicts concerning ethnic, political, economic, and religious issues have extended throughout history. The conflict in the Middle East between Israel and its Arab neighbors is well known. Less well known, however, is that in Israel there are regions that have been under formal or near martial law for hundreds of years (under rule by the Romans, Arabs, British, and the Israelis) because of constant conflict and local disputes.

Despite negotiations, there is apparently no end to the escalation of conflicts in such regions of Europe as the Basque areas of Spain or Northern Ireland. Belfast is an outstanding example of an ongoing conflict zone: people there are driven to violence by both political and religious fervor.

With the collapse of the communist regimes in Asia and Eastern Europe, political, ethnic, and other hidden conflicts have erupted into wars, as in Yugoslavia. Serious conflicts, such as the political tension between Azeris and Armenians or Russians and Moldovans, have become commonplace.

The recent escalation of these and other conflicts follows the emergence of an ineffective control mechanism, which almost invariably creates vicious cycles driven by a positive feedback loop.

Whenever one politically minded group gets ahead (politically, economically, or socially), it tries to establish and strengthen its advantage both symbolically and materially. This new position of relative power prompts that group to negotiate a compromise from its advantageous position. However, this inequitable relationship threatens the adversary group, leading that group, with a relative disadvantage, to respond aggressively as a demonstration of power. This show of power or muscle-flexing behavior is translated into war, acts of terrorism, uprisings, sanctions, and so on, all in an effort to reestablish a balance.

Cybernetic thinking sees many drawbacks to this kind of regulation: exchanges of information are limited and distorted, evaluation of the system's performance becomes colored by bias resulting in gross subjectivity, and simple systems designed to regulate the conflict come nowhere near matching the variety of the situation. The net results are predictable and repetitive. Each side considers its own aggressive behavior as a legitimate, warranted, and often necessary defensive action to restore a balance of power. However, the ever-accelerating positive feedback loop takes both parties further away from reaching any real balance. Even if the parties are able to understand the cybernetic logic in the escalation, the history of the conflict and lack of trust make any compromise more and more difficult to achieve.

impact of positive feedback loops is most evident in the tendency of tensions to escalate between nations when their intentions are in conflict. The global systems spotlight looks at the action of feedback loops in the context of the global political system.

Escalating conflict among competitors can also be costly for organizations vying head-to-head for increases in market share. All systems have limits on resources that limit the length to which conflicts can continue without causing internal decay and eventual decline of a company. Often the focus on an external threat causes managers to ignore the increasing inertia for disorder and chaos being created by internal feedback loops. Among the many prominent examples of such behavior is the demise of Eastern Airlines. Eastern's protracted conflict with unions drained its resources to the point where it could no longer withstand the pressures of competition in the airline industry.

FEEDBACK IN ORGANIZATIONS

A manager's ability to see various organizational activities in terms of a cybernetic framework is useful in a number of different ways. At the most general level, cybernetics is a tool to increase understanding of the causal relationships within a system. Specifically, using cybernetics helps managers see causes and effects as feedback loops. Feedback loops are information transmission structures that are self-reinforcing, meaning they exhibit circular causality. They are commonly known as vicious circles, virtuous cycles, snowball effects, and self-fulfilling prophecies. Richardson and Pugh (1981) have defined **feedback** as "the transmission and return of information" (p. 3). The concept of feedback assumes that information has the capability for recursion. **Recursion** is defined as "a next level that contains all the levels below it" (Beer, 1985, p. 17). Recursive activities usually take the form of cycles: an action is taken, and then information returns in the form of what is called a response.

Feedback involves causal relations. Every cause is an effect based on a preceding cause, and every effect is a cause of some future effect. Thus, influence flows in both directions (Senge, 1990). Therefore, feedback loops are self-reinforcing flows of reiterative causes and effects. For example, a drop in the price of a product usually triggers a response in the form of more orders for the product. The orders are the feedback. The cycle of cause and effect just described is a **causal loop.** Feedback occurs in the form of a loop of continuous causal relations.

Although it may not be immediately obvious, organizations are driven by such loops of cause and effect, known as **feedback loops.** Richardson and Pugh (1981) regard a feedback loop as a closed sequence of causes and effects, "a closed path of action and information" (p. 4). All human systems, such as organizations, act as feedback systems. They are wired like a telephone network with interconnected loops. Imagine a massive telephone conference call with people from around the world all participating in a single conversation. If the path of information flows were traced, various loops would become visible. Because feedback loops are recursive, whatever actions a manager takes change the state of the system. Consequently, it does not make much sense to create a static definition of a problem because the problem is continually changing in response to the managerial actions.

Feedback loops can be neatly divided into two basic types: positive and negative. Negative loops control systems by continually drawing the action of a system toward a goal or standard. A thermometer is a regulator based on a negative loop; it always tries to restore room temperature to the thermostat setting. In contrast, positive loops establish new goals or standards. For example, when a popular new product enters the market its sales are driven by the positive feedback loop of word of mouth advertising. Similarly, growing sales contribute to greater cash flow, which enables greater advertising, which drives sales up and on it goes.

Feedback loops can be powerful regulators of behavior or deregulators that drive systems toward chaos. As regulators, they are used to help achieve control. Feedback is based on data contained in a signal that is recognized by a regulatory system. The data becomes feedback when it is in a form comprehensible to the receiver—it now has meaning. In general, feedback takes the form of a pattern of information that signals the state of the system or one of its elements. The formation of a feedback loop also depends on the sensory and valuative capabilities of the system.

WHY REGULATE SYSTEMS?

All living and all social systems are purposeful. That is, systems exist for a particular reason, and much of their energy is devoted to attaining that purpose. An organization's purpose may be to provide high quality, affordable products and services or to offer rewarding rates of return on investment to stockholders. Regardless of the specific concern, the organization always needs to ensure that its performance meets the expectations of its stakeholders—for example, its customers, employees, and stockholders. Many forces, such as competition and economic cycles, may hinder the organization's ability to realize these purposes. As a result, various self-regulating processes must be in place to restore the organization to the direction in which it has the greatest likelihood of fulfilling its purposes. These processes may range from simple reporting procedures to strategic performance tracking systems. For example, GTE uses a formal strategic tracking system to monitor performance of the organization. The system's purpose is to ensure that areas where efforts were being focused (called "critical success factors") remain consistent with the strategic plan and to verify that actions taken are meeting the performance targets (Jones, 1986).

At one time management literature emphasized the importance of maintaining organizational equilibrium as a determinant for achieving success (Thompson, 1967). The general view today puts less emphasis on equilibrium in itself, recognizing that the value of balance lies in the many positive things that can happen for an organization when balance is present. Stability can encourage opportunities for investing in future capacity. However, it can also be a sign of a static organization suffering from dry rot. Nevertheless, for all the parts of an organization to interact together successfully, the system must have a measure of relative stability. Organizations characterized by continual chaos can neither plan effectively nor coordinate essential activities. Although planning to predict the future has dubious value, planning to continually learn about the future is necessary and requires both time and relatively stable conditions to be productive.

In dynamic, complex environments—such as many global business environments—managers are challenged to coordinate the activities of various subsystems that each interact with different dimensions of the larger environment. Managers' efforts to attain the goals of the whole system while still maintaining a sense of balance can be daunting without a conceptual framework that explains how the parts of a system are tied together and how they can be coordinated and controlled. Global organizations must interact with the world, often in a sophisticated manner, and still maintain a degree of internal order. The minicase describes the use of cybernetic thinking in the operation of the global retailer Benetton.

Uniting the Colors of Benetton with Cybernetics

From humble beginnings in a single shop in Venice, Italy, to over 5,000 shops in seventy-five countries by the 1990s, Benetton has become one of the premier global marketers. Bypassing wholesalers and retailers in favor of designing their own tightly controlled distribution system, Benetton has become the quintessential quick change artist. By integrating information processing capabilities with manufacturing technology, Benetton is able to respond in lightning fashion to changes in customer demand. The company's strategy of leaving sweaters undyed and waiting for "just-in-time" information on which colors are needed by retail stores allows it to rapidly adjust the product mix to fit the configuration of the market. This speed and coordination happen through an intricate communications system, high-speed computers, and a $20 million computerized warehouse that services the company's seven factories in Italy.

As Harvard's James Heskett says, Benetton has

> pioneered a retailing approach throughout Europe that promises to influence a number of other retailers worldwide: It substitutes information for assets. Because its first retail outlet offering knit outerwear in colorful fashions was very small, the Benetton family developed an approach to retailing that makes effective use of small spaces. Unlike its more traditional competitors with stores of perhaps 4,000 square feet, a typical Benetton outlet is not more than 600 square feet. Little space is wasted on floor selling or "back room" storage.
>
> An electronic communications system supported by a manufacturing process allows for dying to order and for rapid replenishment of the items in greatest demand during a fashion season. The result is a higher rate of inventory

turnover in the store and a level of sales-per-square foot that is often several times that of Benetton's competitors. Benetton's assets support many more sales because it has injected both communications and flexible technology into its service. In fact, Benetton can make changes in inventory in ten days that takes most retailers months.

Source: Based on T. Peters, *Thriving on Chaos* (New York: Harper and Row, 1987); L. Belkin, "Benetton's Cluster Strategy," *New York Times* (January 16, 1986), pp. D1, D5.

Benetton's strategy of rapid reaction to the market helps it stay in balance with the environment. By gathering virtually continuous feedback about the market through retail stores, the company is able to turn this information into a competitive tool. In some organizations, all this information and dynamism could mean utter chaos, but Benetton is built to take it. The organization is designed with integrated decision support systems tightly coupled with manufacturing technology. The effect is to give the organization more variety of actions than the market has. The result is that the company thrives on the market rather than being confounded by it.

REGULATING THROUGH COORDINATION AND CONTROL

Hage, Aiken, and Marrett (1971) have proposed that "All organizations require coordination and control." Several general approaches for achieving that coordination and control exist. The two most fundamental approaches are through feedback and programming (March and Simon, 1958). Attempts to regulate organizations through the use of programming may take such forms as establishing standard operating procedures. Such controls are designed to restrict behavior or performance by instituting proven solutions to repetitive activities before the fact. Control limits a recurring activity, keeping it within a predetermined standard range of behavior. Control through feedback works by monitoring these activities and responding to any deviations beyond the standard range.

Flamholtz (1979) has observed that "The ultimate criterion of an effective control system is the extent to which it increases the probability that people will behave in ways that lead to the attainment of organizational objectives" (p. 56). In such a control system an acceptable range of performance is established; any deviation from that range triggers the release of a signal alerting whoever is monitoring the system to the deviation. For example, if you attempt to use your credit card when you have exceeded your credit limit or when you have not paid your credit card balance on time, a feedback control will go into action, and your purchase won't be approved (see Figure 7-1).

With control through feedback the information used to make a decision is more likely to accurately reflect the current performance situation than it would under any other technique. Control through feedback allows managers

FIGURE 7-1

Source: Copyright © 1991. Reprinted courtesy of Bunny Hoest and *Parade* Magazine.

to develop customized actions on the spot to address the problem—within the limitations of the system's capacity to process and interpret information. This is particularly important in complex situations. High complexity often makes it extremely difficult to preplan; as a result, control through feedback becomes the most effective approach. There are a number of different types of complexity, with varying implications for the use of the cybernetic approach.

SYSTEM COMPLEXITY

The complexity of a system depends on both the characteristics of the system and the person viewing it. Complexity is a relative concept—it depends on the mental model of the person viewing the system. What may seem overwhelming to one person is simple to another. Finding his way around campus on the first day of class can be overwhelming to a college freshman while it may be a routine experience for the senior.

In general, complexity is determined by three main factors: perceptual factors, structural factors, and intrinsic factors. Perceptual factors have just been explained. Intrinsic factors are a function of the number of possible states that can be attained by a system. Structural factors are governed by a system's feedback relationships. For example, the structure of product markets is composed of both positive and negative feedback loops that produce a balancing effect. The forces of supply and demand, which control such markets, are driven by these feedback loops (Figure 7-2).

Beer (1959) identified four factors that contribute to **system complex-**

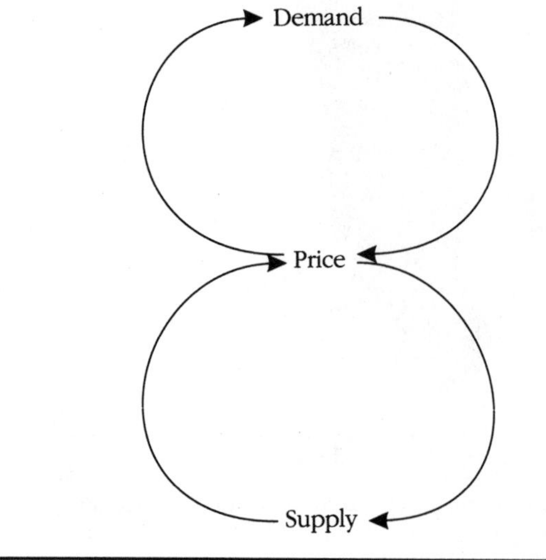

FIGURE 7-2 ***The Forces of Supply and Demand as Feedback Loops***

ity, three of which are intrinsic: (1) the number of elements comprising the system, (2) the attributes of each element within the system, (3) the number of interactions between the elements in the system, and (4) the degree of organization that characterizes the system. Campbell (1988) also identified four factors that create complexity: (1) multiple possible outcomes, (2) multiple paths to achieve outcomes, (3) interdependency, and (4) uncertainty.

The quantitative dimensions of complexity are more likely to cause physically differentiated states in a system. Normally, the greater the number of variables in a system, the greater the complexity of a system. The potential range of behaviors of each system element contributes to the array of possible states the system may attain. This is one of the reasons that human systems, such as business organizations, are so complex. Human systems differ from mechanistic rational systems because they have a survival orientation; they engage in various formal and informal activities to perpetuate their survival (Scott, 1987). The complexity of a system can also be influenced by the extent of interconnectedness among its elements. The degree of interdependency is a product of the structure of the system. You will recall that in systems with tight coupling, relatively small changes in one element cause relatively greater effects in the other elements of the system (Richmond, 1990). Thus a minor disturbance to the system's equilibrium may snowball to produce disproportionate effects throughout the system. Tightly coupled systems tend to amplify the effects of change throughout a system. This causes them to appear more complex, with greater potential for variety (Beer, 1985).

Uncertainty exists in organizations when there is a difference between the information needed to solve a problem and the information available (Galbraith, 1977). When uncertainty exists, it is impossible to determine clearly the effect that a given managerial intervention will have on a situation (March and Simon, 1958). Whenever the causal relations in a system are unclear, uncertainty will be pervasive.

System complexity is often increased by variety within the elements. **Variety** describes an element's ability to behave in a wide range of different ways while still maintaining its identity. Human systems normally exhibit a high degree of variability. In a system with high variety, the regulatory system must contain as much variety or more as the system as a whole to be effective. This idea, one of the central themes of cybernetic thinking, is the foundation for Ashby's Law of Requisite Variety. This law is mainly concerned with creating a strategy for controlling complex systems, a formidable task because most human systems have the potential for so much variety that all the possible problems cannot be recognized as they happen (Waelchli, 1989). The variety of many human systems is so great that attempts to control problems by focusing on each single one separately cannot keep pace with the rate at which new problems emerge. Consequently, a control system must have the capacity to recognize all states that a system is capable of creating. Such a control system must address the view that "only variety can absorb variety" (Beer, 1985, p. 35). To achieve effective control over another system, an organization must maintain relatively more variety than the system it wants to regulate. This can be accomplished in several ways:

1. Increase the organization's variety
2. Match the organization's variety exactly to the system to be controlled
3. Reduce the potential of the controlled system to generate variety (Waelchli, 1989)

How could this principle be applied to controlling the size of the national debt in the United States?

SECTION 2
CYBERNETIC THEORY

BASIC ASSUMPTIONS OF CYBERNETIC THEORY

Several basic principles govern the way that cybernetic systems operate. These principles are generalizations that collectively form the foundation for managerial cybernetic thinking. They can be potentially useful guides for the design and implementation of control systems.

THE NEED FOR REGULATION

All human systems are purposeful. As a result they periodically make internal adjustments in response to changes in external conditions. Systems increase the likelihood of attaining their purpose when they continuously evolve in a way aligned with their purposes. The most efficient way of aligning system changes with overarching purposes is to continuously monitor the extent to which current performance approaches desired future states (Hage, Aiken, and Marrett, 1971). This approach to regulating performance may be utilized at any level in the system. For example, the performance levels of individuals or work teams may be measured against goals to ensure unity of purpose. Therefore *the first basic principle of cybernetics is that all human systems require regulation.*

Related to this principle is the importance of compensation, which refers to efforts by the organization to counteract or neutralize any destabilizing effects resulting from responses to environmental pressures. The diverse elements that make up systems need to be coordinated. Subsystems may interact with the environment and behave in ways that will fundamentally conflict with the actions of other subsystems. To ensure that the overall system continues to function smoothly, a compensating mechanism must exist to offset the differing states of the various elements in the system. The human body is able to coordinate such activities very effectively. For example, digestion activities are normally reduced during sleep or rigorous physical activity.

THE NEED FOR INFORMATION

The regulation of systems relies on information gathered from monitoring both the system and its environment. Monitoring involves the collection, processing, and interpretation of information. Those systems that have access to mation that is timely, accurate, and relevant are more likely to remain vital. Wiener (1948) observed that control is primarily a process of sending messages that ultimately modify the performance of the recipient, emphasizing the interdependency between control and information. The mere act of receiving information about the state of the system changes the way the system is perceived. With this change in perception, behavior also changes.

In its basic form, cybernetics is a method for using information, in the form of feedback, to learn about both the state of a system and how it works. Such information becomes the basis for actions to correct any undesired states. Cybernetic systems collect data, regulate themselves, compensate for destabilizing effects, and coordinate activity within the system.

Coordination depends on information as it seeks to reconcile the differences between organizational subunits. Those differences could be in terms of time perspectives, location, goals, formalization, or perceptions (Lawrence and Lorsch, 1967). Coordination depends on the exercise of judgment to make decisions and the accurate communication of those decisions to the intended site. The effectiveness of coordination depends heavily on the capability of what are called **executive processes** within systems. Executive processes are used to plan, monitor, and evaluate how well a task has been accomplished (Baron and Sternberg, 1987). All cybernetic systems must have a built-in capacity to make judgments. This capacity can be simple, such as making a yes or no decision; in more sophisticated systems, the executive function may rely largely on human reason or on expert systems for decision making. (Expert systems are computerized systems that combine a data base with sets of algorithms and heuristics to act similarly to a human expert. For example, at General Electric, maintenance crews use an expert system named *DELTA* to identify malfunctions in diesel locomotives [Senn, 1990].)

When executive processes operate properly, information gathered from the system is interpreted and analyzed, and signals are sent to relevant subunits to either maintain the current status or change. When there is such a continuous cycle of data collection, analysis, and interpretation, learning opportunities are enhanced. Thus cybernetic systems have the potential to facilitate learning—although there is no guarantee that learning will occur. As the minicase about the NASA Hubble Telescope demonstrates, the corporate culture at work clearly had a linear project orientation that prized the achievement of outcomes rather than learning or quality.

NASA, Perkin-Elmer, and the Hubble Telescope

Part A

If there was a season of change for Perkin-Elmer and NASA, it came in 1969. For NASA the 1960's had been a space scientist's dream.... NASA's budget more than tripled in four years. Apollo 11 lifted off the moon in July 1969, on time and on budget. Yet soon after the first moon landing, the government cut NASA adrift. Its budget plunged to a third of its peak as the space agency searched for a new, defining mission.... Three months after Neil Armstrong stepped on the moon, NASA dropped plans for intermediate space telescopes intended to reduce the technical challenges. It was the first step in a decade-long effort to cut the telescope program by squeezing the budget, deferring spending, reducing staff, making design compromises, and eliminating tests and spare parts.

In 1969 Perkin-Elmer was celebrating more than 30 years of growing success as a maker of telescopes and other advanced optics and scientific instruments. But rapid growth was revolutionizing the corporate culture. Elmer had died in 1954. Perkin-Elmer sold stock to the public for the first time in 1960.... In 1964 the company got a new chief executive officer in Chester Nimitz Jr., a retired admiral like his famous father.... By May 1969, when Perkin fell ill on an airliner over Ireland and died in a Limerick hospital, the old days were really gone.

Thanks largely to Micralagin [a new product], sales and profits at Perkin-Elmer tripled between 1976 and 1980. But with all the growth and diversification, the management style of Perkin and Elmer seemed obsolete.... Nimitz began filling management positions with outsiders.... The company sent accountants and salesmen off to business schools for crash courses. When they returned they talked about a new way to organize a corporation. It was called "matrix management." ... Although the new system helped the company save money by making its work force more flexible, it also widened the rift between science and management. The scientists felt the managers no longer shared their goals. And many believed the system left people with less pride in their work and less responsibility for the product.

Part B

With its lofty goals and shrinking budget, NASA too was forced to relax the rigor with which it had done science programs since the space agency was founded in 1958. Kent H. Meserve, Perkin-Elmer telescopic project manager in 1980 and 1981, believed NASA had no choice. "The shuttle was in trouble and it was eating lots of money and that was the leading program for NASA. So a lot of science programs, the telescope being one of them, were being pushed to the right, or defunded or being adjusted to make room for everything." In the space agency's official view, computer simulations, good design, and careful supervision could overcome the need for prototypes or full-scale tests of the space telescope.... Many within Perkin-Elmer believed that abandoning the prototype hardware was "penny-wise and pound-foolish" (Paul E. Petty, V. P. Optical Technology).

Both the Marshall Space Flight Center in Alabama and the Goddard Space Flight Center in Maryland knew the space telescope program was important, and they competed hotly for the project. In May 1972, NASA headquarters split the project, giving Marshall responsibility for the telescope's optics and structure and Goddard responsibility for the science instruments. Although the solution kept both centers busy it resulted in friction that lingered throughout the program. . . . NASA's stripped-down posture shifted an unusual amount of responsibility to the people the agency sent to Danbury (CT) [Perkin-Elmer] to observe the work. Carl Fuller [is one example] . . . For the first time in his career as a NASA quality-assurance specialist, his bosses told him to ignore the quality of one job—the mirror.

Part C

During the making of the mirror, there were only three space agency engineers in Danbury, compared with more than twenty at a project NASA considers comparable—construction of the Saturn rocket boosters at Rockwell International Corp. in the 1960s. Money was one reason. Another was that, just down the corridor from the polishing room, Perkin-Elmer was still doing classified work for the Defense Department. High ranking NASA officials say the Pentagon was nervous about having outsiders running around the plant and talked the space agency into limiting the number of people it had there. Being host to a minimal staff from NASA suited Perkin-Elmer. Company officials worried that manufacturing secrets could fall into the hands of competitors. They sometimes discouraged their own workers from publishing scientific papers about polishing techniques. . . .

At a NASA conference a decade earlier, astronomers had identified the making and testing of the main mirror as the central challenge of the telescope project. But now, no one from the space agency was monitoring the quality of the mirror work. . . . Without a quality-assurance inspector on the mirror, [said Charles O. Brooks of NASA,] "You're momentarily out of control. One moment you have a complete grasp on what's going on, and then you move into shadow. It's like moving into an eclipse." . . . NASA required that one of its quality-assurance officials give final approval to each piece of equipment destined for space. Of all the approval documents on the thousands of parts that went into one of the most complicated scientific instruments ever assembled, only one is known to lack a signature from the quality office. Fuller just wouldn't sign off on the main mirror.

Source: R. Capers and E. Lipton, "The Looking Glass: How a Flaw Reflects Cracks in Space Science," *The Hartford Courant* (April 3, 1991), pp. 1, 4, 5.

SENSING SYSTEM STATES

Systems have limitations that narrow the range of states that can be exhibited. Performance standards, goals, cultural values, and beliefs all tend to set boundaries on the behaviors and types of performance that an organization can support. In effect, these factors also define an organization's understanding of the prerequisites for survival. Although there are limits to the variability of a system's behavior, it is normal for systems to behave dynamically over time. Cycles characterized by varying levels of performance are generated by the complexity of the system and its interaction with its environment.

All human systems have some capacity to internally generate random fluctuations in performance. They also are subject to the long-term, deteriorating effects of **entropy**, the powers of disorganization and disintegration present in the universe. When systems are not infused with energy and organization, they tend to decay. Two of the most fundamental polar forces that operate within all systems are order and chaos. When organizations are in equilibrium, there is something of a stalemate between these primal energies. When organizations are dominated by entropy, they are in the far-from-equilibrium state of chaos. Chaos may precede the demise of the system, or it may be a stage of evolution to a new level of equilibrium. For PeopleExpress, the chaos was final; for International Harvester (Navistar), it was instrumental to the redefinition of the organization.

In organizations, performance standards and goals trigger corrective actions, such as increasing quality standards. Although corrective actions do not always eliminate fluctuations in a system's state, they do generally dampen the range. For example, a thermostat setting of 72 degrees serves as a standard that will normally trigger a heater to start when the temperature drops below 68 degrees or shut off when it rises above 76 degrees (see Figure 7-3).

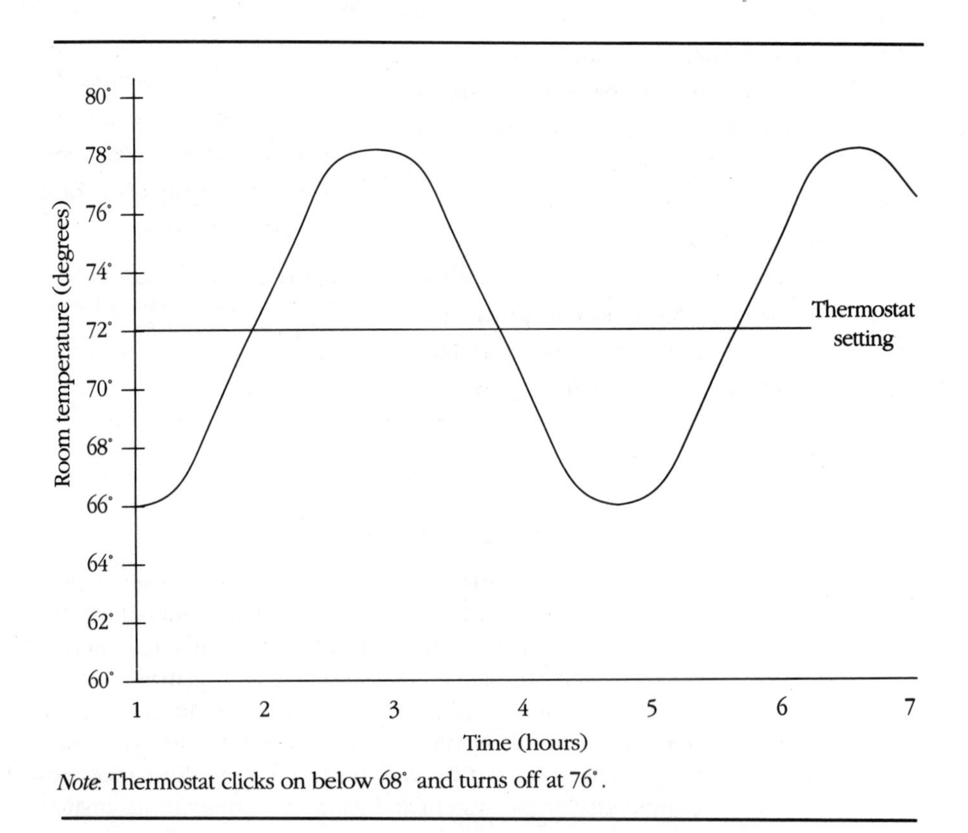

Note: Thermostat clicks on below 68° and turns off at 76°.

FIGURE 7-3 ***Thermostatic Setting as a Means to Reducing System Variability***

Because systems such as organizations are dynamic and do not maintain a single state indefinitely, multiple measures are necessary to determine the current state of the system (Beer, 1966). The frequency of applying such measures depends on the expected frequency or riskiness of the deviations. Passenger jets flying on transoceanic flights are often off course, but they maintain their desired direction by making frequent evaluations of their current state and making appropriate corrections. Some systems maintain continuous contact with their environment in order to limit variations in the system's output. The thermostat mentioned above maintains continuous contact with the living space to keep room temperature within a narrow range. The regulating device used in a thermostat is sensitive enough to detect significant variations in room temperature. This enables it to preserve a general zone of comfort for people in the room. A regulator that was oversensitive to variations in temperature would be economically inefficient. Indeed, because of time delays in the heating process, it could actually create more pronounced variations in output than would occur otherwise.

MAINTENANCE OF EQUILIBRIUM

The continuing goal of a regulatory mechanism is to ensure the survivability of the system. One way to accomplish this is by maintaining the system's equilibrium within a range that will support internal activities and enable the continuation of relations with the environment. When a system has achieved a balance, a level of **equilibrium** has been attained. Equilibrium is a relatively stable state where variations are relatively uncommon. In reality, it is unlikely that an organization would establish a single equilibrium point. It is more likely to perceive equilibrium as a dynamic fluctuation within a relatively restricted range of states. The concept of dynamic equilibrium is referred to as **homeostasis.** When wide variations are observed, a state of **disequilibrium** is said to exist.

Within the field of cybernetics there is a general assumption that it is desirable and even necessary to maintain equilibrium in order to promote organizational prosperity. Seeger (1992) notes that the widespread adoption of this belief by managers has prevented them from considering more dynamic models of how organizations function. For example, in the system dynamics approach the behavior of a system is seen as the product of many interactions that may produce a change in the system's equilibrium as a consequence of the normal dynamics of the system.

~ SYSTEM EQUILIBRIUM

When the term *equilibrium* is used in relation to an organization, the implication is that the whole organization is at a point of equilibrium. In relation to organizations, equilibrium is more a conceptual benchmark than a practical tool for understanding systems. In order for such a true equilibrium to occur, there must be a balancing of all parts within the system (Ashby, 1960). In complex, varied organizations, however, it is highly unlikely that all subsystems will simultaneously be balanced with one another. Especially for large organizations it is more useful to think in terms of relative degrees of equilibrium. Large organizations with many autonomous subunits, such as General Motors or Siemens, are less prone to need or achieve balance. Ashby and Gardner (1970) found that the stability of a system was directly related to the number of elements in the system and the number of connections between these elements. Clearly, then, more complex systems are less likely to sustain equilibrium.

In the past, equilibrium maintenance was widely seen as necessary for organizational survival. Today, the emergence of different system paradigms has raised the question of whether equilibrium maintenance is necessary or even desirable. Many analysts accept the argument that some organizations' efforts to maintain equilibrium may mask broader systemic changes that may require fundamental changes to the organizations' current patterns.

In fact, there may be many possible levels of equilibrium in which an organization may prosper. Cybernetic systems cannot adequately address the question of whether a given equilibrium level is the most supportive of an organization's chances to survive. The role of cybernetic thinking is limited to maintaining the necessary, relative degrees of stability within a system to permit critical functions to occur. The question of which is the most desirable level of equilibrium goes to a much more fundamental issue pertaining to the purposes and definition of the system.

There are clearly substantial reasons for valuing the maintenance of equilibrium within organizations. Efforts to maintain equilibrium support the creation of relative stability. Classical management views such an organization as able to plan effectively, based on the premise that the state of the system will not change dramatically. Stability is seen as increasing opportunities for organizational capacity building. Some theorists propose that this "predict and prepare" paradigm helps to perpetuate further stability for the system (Weinberg and Weinberg, 1979). Conversely, Ackoff (1974) argues that the predict and prepare paradigm is not generally workable because of the improbability that planners can clearly visualize the future. He asserts that organizations should continuously plan for the future rather than engineer the future course of the system.

Beyond helping an organization maintain equilibrium, cybernetic thinking and control systems are essential to organizations for closing themselves to unwanted environmental influences. They are also important mechanisms for controlling the growth processes that can dominate organizations at certain stages in their development. Just as there is conflict in systems between the polar opposites of order and chaos, there is also a similar relationship between control and growth. Both are necessary for survival, but a balance must be maintained between them. PeopleExpress Airlines is a classic example of the consequences of negative uncontrolled growth. This airline was vulnerable to competitor price cuts when it overextended itself and suffered bankruptcy in the late 1980s. Understanding that cybernetic systems are critical mechanisms for neutralizing the unwanted effects of growth and entropy, let us examine the basic elements and mechanics of a cybernetic control system.

SECTION 3
BASIC ELEMENTS OF A CYBERNETIC CONTROL SYSTEM

This section will present the primary components of a cybernetic system and explain how they work together. Many of the functions of a cybernetic control system can be performed quite mechanically. However, designing such a system involves recognizing which performances need to be controlled and judging the amounts and types of variety that should be built into the controller.

A SIMPLE CYBERNETIC CONTROL SYSTEM

In its simplest form, a cybernetic control system regulates by continually comparing current system states against an ideal. Discrepancies, once detected, are corrected by changing the inputs into the system. As Powers (1990) puts it: "Control systems control their inputs, not their outputs" (p. 7). The net effect, of course, is to control the outputs. This relationship can be simply represented in terms of an open-systems or productive-systems model, both of which transform inputs into outputs (Hage, 1974). The black box model clearly depicts the stabilizing effect that regulation of inputs can have on the state of a system. This model also illustrates how feedback loops in cybernetic systems will interactively compensate for variations in outputs to maintain equilibrium within the system.

The model is called the black box model because, as Ashby (1956) concluded, those systems with high levels of complexity are not easily understood. They must be regarded as opaque containers with unknowable contents. By noticing the consequences of varying actions within the system we may make an educated guess as to the contents of the box, yet ultimately its inner mechanisms remain shielded by the impenetrable clouds of complexity.

What is the purpose of the black box technique? The black box is a mechanism for lessening the complexity of information that has flowed from various locales within a system to managers. Information seen as relevant to maintaining the stability of the system becomes an ingredient in the organization's executive process. The goal of using the black box is to diminish undesired variation in the system while maintaining balance in other parts of the system. This is accomplished by a process of continual experimentation in which inputs are altered and their effects are noted. Eventually, after specific relations between these factors have been deduced, managers will be able to exert some level of control over the system without knowing the contents of the black box.

In organizations, the black box serves managers by acting as a tool for achieving control. The more complex the organization, the more sophisticated must be the black box. When framed in Ashby's (1956) terms, the black box must exhibit a level of variety that extends beyond that of the organization, in order to serve as an effective tool for control. The primary advantage of the black box technique is that it enables managers to bypass the complexities associated with understanding cause-and-effect relations throughout the system, and allows managers to simplify their focus to inputs and their attendant outputs (Richardson, 1991).

For example, managers often use compensation (salary, raises, bonuses, incentives) to improve the performance of organization members. Although money may have different meanings to different employees, the manager may be concerned only that it is an effective motivational tool, and not desire to know why.

The merits of this efficiency of action are a matter of debate since this approach reinforces the tendency of managers to avoid understanding the system and focus instead on outcomes. This is in direct contrast to the system dynamics approach, which seeks to explicitly identify and quantify causal relations within organizations (Richardson, 1991).

In organizations, the black box can be easily visualized in terms of a simple five-part model. The components of the **black box model** (see Figure 7-4) are:

1. Inputs—raw materials, human resources, energy, information, money
2. The black box—the site of all regulatory activity and transformational processes in the system
3. Outputs—products, services, energy, growth, and so on
4. Feedback—information relating to the extent to which outputs have been utilized or absorbed by the environment and to the current state of any aspect of the system's environment
5. Environment—other systems that interact with the system in focus.

Although the black box model provides a simplistic picture of the way cybernetic systems function, it does make a clear point that the outputs of such systems influence both the subsequent future inputs and the ongoing transformation processes in the system. The essence of this approach is that the inputs of the system regulate the future of the system through the feedback process; the system is self-regulatory.

The black box model is reiterative; that is, it is based on continuous cycles of processing to create output, measurement, adjustment, and subsequent processing. After the first cycle, inputs include both the raw materials needed for the production process and information in the form of feedback. Thus the figure shows two arrows as inputs; one represents raw materials, and the other represents feedback. The black box itself is the most complex part of the system, where all control originates.

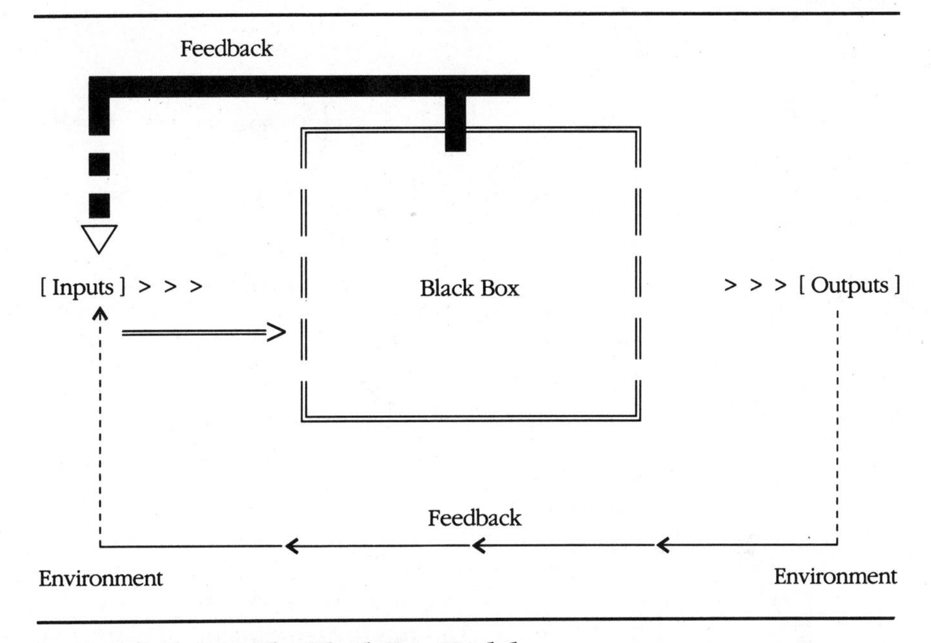

FIGURE 7-4 ***The Black Box Model***

To better understand the black box, it is necessary to describe its contents. They are the sensor, the comparator, the effector, and technology.

All cybernetic control systems have the capacity to detect changes in the state of the systems they control. This detection function is performed by the **sensor,** which is designed to recognize those changes that may prove capable of causing further disruption of the system's equilibrium. The sensor transmits its readings to the **comparator.** The comparator measures the feedback provided by the sensor against the relevant performance standards. The comparator often makes simple yes-and-no judgments as to whether the current system state is acceptable. Generally, the comparator makes assessments on the basis of a range of acceptable values rather than a single value.

The comparator sends a signal that carries its findings to the **effector.** Such findings indicate the extent of the variance from the standard (if any) and identify the type or class of problem. The effector is the executive component of the cybernetic control system. It interprets the information provided by the comparator and determines whether corrective action should be taken. It analyzes the extent of the variations from the standards and determines what set of actions are called for to neutralize them (Figure 7-5).

Another subsystem that appears in the black box—technology—is not often associated with cybernetics. **Technology** is the means employed to transform inputs into outputs. Technology may take the form of anything from machines or tools to forms of intelligence, such as knowledge and skills. Although it is subordinate in control systems, technology plays a central role in productive systems. Technology governs the amount of energy and resources

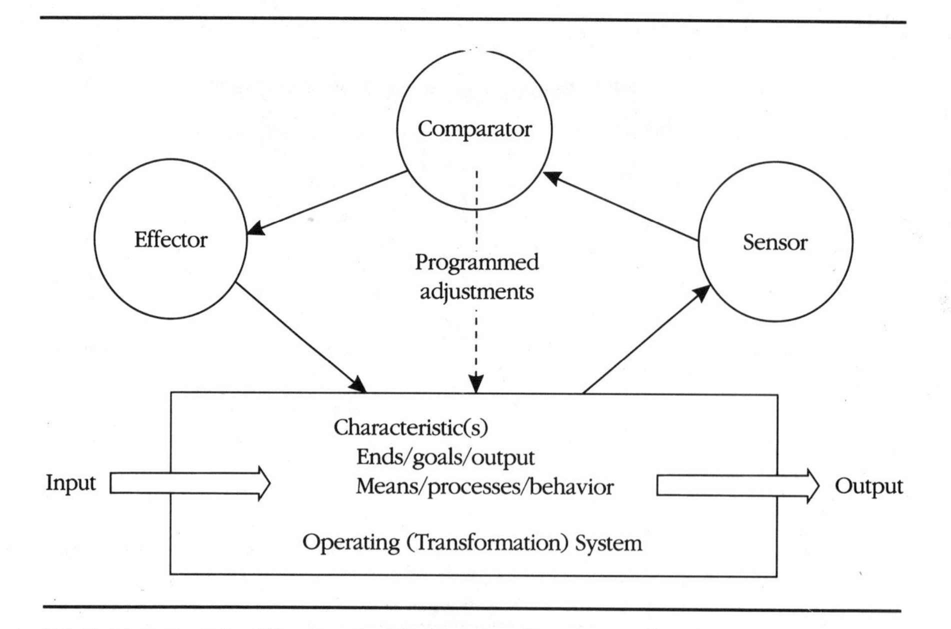

FIGURE 7-5 ***The Basic Elements of a Control System***

Source: John V. Baumler, "Defined Criteria of Performance in Organizational Control," *Administrative Science Quarterly* (September 1971), p. 340.

consumed in converting inputs into outputs. To the engineer, the cost of an electrical circuit is normally inconsequential, while technology is strategically important to the manager. In a similar fashion, the concept of quality may mean quality control to the engineer, but to the manager it means "total quality." A total quality perspective suggests thinking about managing all dimensions of an organization in a quality-conscious way.

APPLICATIONS OF CYBERNETIC THINKING: MANAGEMENT OF QUALITY

Cybernetic principles serve as the foundation for several noteworthy managerial practices, particularly the management of quality. The American Society of Quality Control has endorsed the term *management of quality* (MQ) to reflect the activities that are commonly associated with quality improvement. Some practitioners prefer to label their work as total quality management (TQM), implying that quality must exist in every business function from design to consumption. In any case, **quality management** is a continuous improvement process that relies on systems thinking, and specifically cybernetic principles, to achieve customer satisfaction. Quality has two basic dimensions: "product performance that results in customer satisfaction" and "freedom from product deficiencies, which avoids customer dissatisfaction" (Juran, 1988,

p. 332). In practice, TQM can—and must—be applied to the improvement of service quality as well as product quality.

Prior to the advent of the Japanese quality revolution, quality issues were normally regarded as pertaining to quality control. Yet, quality control has a limited scope; it is just one of the components of a program of total quality. Total quality management is a philosophy and practice based on the ideas developed by Armand V. Feigenbaum (1961). The thrust of his argument was that quality control should be the responsibility of everyone in an organization, not just managers. He also believed that quality defects should be identified and remedied at their point of origin as part of a continuous process. Today, there are a number of popular programs for quality management. The most widespread are the programs proposed by Philip Crosby, Edward Deming, Joseph Juran, and Genichi Taguchi. Although they all have their own slant, these philosophies all agree that quality begins at the design stage of product development and that management, not workers, is the prime cause of quality-related problems (Kathawala, 1989). A study of the quality improvement practices of over 200 U.S. companies by the consulting firm Zenger-Miller found the approaches used tended to argue for the same basic points and shared the same basic strengths and weaknesses (Dumas, Cushing, and Laughlin, 1987).

A number of elements in each of these perspectives illustrate cybernetic concepts particularly well. Figure 7-6 details how several key cybernetic concepts are used.

Cybernetic principles are emphasized in the management of quality for regulation and learning. The results often help to stabilize relationships with customers. Juran (1988) defines **customers** as "all who are impacted by our processes and products" (p. 11). The management of quality is also a methodology for maintaining equilibrium with stakeholders. Management of qual-

SCHOOL OF THOUGHT	CYBERNETIC CONCEPT
Crosby	Quality is defined in terms of requirements (standards)
Deming	Customer feedback loop Process feedback loop Continuous reduction in variation Building management's capacity to manage quality (requisite variety)
Juran	System-wide implementation Design systems with less variation Continuous raising of standards
Taguchi	Continuous improvement in product quality by reducing variations from product target values

FIGURE 7-6 ***Cybernetic Concepts in Leading Quality Philosophies***

ity helps organizations stay in tune with their environment by improving their products and services. Effective implementation of total quality management is fueled by continuous improvement based on cybernetic thinking.

SUMMARY

Cybernetics play a critical role in organizations as a result of the influence of cybernetic thinking. Such a frame of reference allows managers to orchestrate the flow of continuous improvement, to coordinate, to control, and to compensate for fluctuations within a system. These functions collectively help organizations to grow while still remaining balanced. While systems thinking provides a useful way of viewing organizations, its power is magnified when it is integrated with the use of specific tools. Managerial tools such as strategy, leadership, culture, structure, and procedures can enable managers to gain leverage to effect systemic change. Part III examines these systemic tools and discusses how they must be used together from an integrative perspective.

KEY TERMS AND CONCEPTS

cybernetics
feedback
recursion
causal loop
feedback loop
system complexity
variety
executive processes
entropy
equilibrium
homeostasis
disequilibrium
black box model
sensor
comparator
effector
technology
quality management
customers

QUESTIONS FOR DISCUSSION

1. Identify some of the most important areas in business organizations whose functioning is influenced by feedback loops. How could recognizing this influence change the way the organization is managed?
2. How can cybernetic thinking be used to help organizations become more responsive to the needs of their customers?
3. You have just been appointed the director of tax collection in your country. How can cybernetic thinking be used to ensure that everyone pays his or her fair share of taxes?
4. How can Ashby's Law of Requisite Variety be used to improve quality as part of a TQM program?
5. Because the demand for your company's new product has doubled over the past six months, your partner recommends that you double both advertising expenditures and plant capacity. Your plant is currently oper-

ating at 68 percent of total capacity. How do you use your knowledge of feedback loops to prepare a reply?

6. Your company, Acme Enterprises (based jointly in New York and London), has just merged with a Tokyo-based firm that specializes in manufacturing for niche markets. You have been asked to prepare a plan that will use information processing technology and manufacturing to achieve full customer satisfaction in the markets of the Global Triad. What are the key starting points of your plan?

REFERENCES

Ackoff, R. L. (1974). *Redesigning the Future.* New York: John Wiley & Sons.

Ashby, W. R. (1956). *Introduction to Cybernetics.* New York: John Wiley & Sons.

Ashby, W. R. (1960). *Design for a Brain.* London: Chapman and Hall.

Ashby, W. R., and Gardner, H. (1970). "Connectance of Large Dynamic Systems: Critical Values of Stability," *Nature,* p. 228.

Baron, J., and Sternberg, R. (1987). *Teaching Thinking Skills.* New York: W. H. Freeman.

Beer, S. (1959). *Cybernetics and Management.* London: English Universities Press.

Beer, S. (1966). *Decision and Control.* New York: John Wiley & Sons.

Beer, S. (1985). *Diagnosing the System.* New York: John Wiley & Sons.

Dumas, R., Cushing, N., and Laughlin, C. (1987). "Making Quality Control Theories Workable," *Training and Development Journal,* February, 41, pp. 30–33.

Feigenbaum, A. (1961). *Total Quality Control: Engineering and Management.* New York: McGraw-Hill.

Flamholtz, E. (1979). "Organizational Control Systems as a Managerial Tool," *California Management Review, 22,* no. 2, pp. 50–59.

Galbraith, J. (1977). *Organizational Design.* Reading, Mass.: Addison-Wesley.

Green, S., and Welsh, A. (1988). "Cybernetics and Dependence: Reframing the Control Concept," *The Academy of Management Review,* April, *13,* pp. 287–301.

Hage, H. (1974). *Communications and Organizational Control.* New York: John Wiley & Sons.

Hage, J., Aiken, M., and Marrett, C. B. (1971). "Organization Structure and Communications," *American Sociological Review, 36,* no. 1, pp. 860–71.

Jones, C. (1986). "GTE's Strategic Tracking System," *Planning Review, 14,* pp. 27–31.

Juran, J. (1988). *Planning for Quality.* New York: Free Press.

Kathawala, Y. (1989). "A Comparative Analysis of Selected Approaches to Quality," *International Journal of Quality and Reliability Management,* pp. 7–17.

Klir, J. (1965). *Cybernetic Modeling.* Princeton: D. Van Nostrand.

Lawrence, P., and Lorsch, J. (1967). *Organization and Environment.* Homewood, Ill.: Richard D. Irwin.

March, J., and Simon, H. (1958). *Organizations.* New York: John Wiley & Sons.

Powers, W. T. (1990). "Control Theory: A Model of Organisms," *Systems Dynamics Review, 6,* no. 1, pp. 1–20.

Richardson, G. (1991). *Feedback Thought in Social Science and Systems Theory.* Philadelphia: University of Pennsylvania Press.

Richardson, G., and Pugh, A. (1981). *Introduction to System Dynamics Modeling with DYNAMO.* Cambridge, Mass.: Productivity Press.

Richmond, B. (1990). "Systems Thinking: A Critical Set of Critical Thinking Skills for the 90's and Beyond." Lyme, N.H.: High Performance Systems.

Robb, F. F. (1985). "Cybernetics in Management Thinking," *Systems Research, 1,* no. 1, pp. 5–23.

Scott, R. W. (1987). *Organizations: Rational, Natural, and Open Systems,* 2nd ed. Englewood Cliffs, N.J.: Prentice-Hall.

Seeger, J. A. (1992). "Open Systems, Closed Minds," *Proceedings of the International System Dynamics Conference,* Utrecht University, Utrecht, the Netherlands, July 14–17.

Senge, P. (1990). *The Fifth Discipline: The Art and Practice of the Learning Organization.* New York: Doubleday.

Senn, J. (1990). *Information Systems in Management,* 4th ed. Belmont, Calif.: Wadsworth.

Thompson, J. (1967). *Organizations in Action.* New York: McGraw-Hill.

Waelchli, F. (1989). "The VSM and Ashby's Law as Illuminants of Historical Management Thought," in R. Espejo and R. Harnden (eds.), *The Viable System Model.* Chichester, England: John Wiley & Sons, pp. 51–75.

Weinberg, G., and Weinberg, D. (1979). *On the Design of Stable Systems.* New York: John Wiley & Sons.

Wiener, N. (1948). *Cybernetics.* Cambridge, Mass.: MIT Press.

~ ADDITIONAL READINGS

Baumler, J. (1971). "Defined Criteria of Performance in Organizational Control," *Administrative Science Quarterly,* September, p. 340.

Beer, S. (1968). *Management Science: The Business Use of Operations Research.* New York: Doubleday.

Carlzon, J. (1987). *Moments of Truth.* Cambridge, Mass.: Ballinger.

Checkland, P. (1981). *Systems Thinking, Systems Practice.* New York: John Wiley & Sons.

Gleick, J. (1987). *Chaos.* New York: Viking-Penguin.

Katz, D., and Kahn, R. (1966). *The Social Psychology of Organizations.* New York: John Wiley.

Lerner, A. Y. (1972). *Fundamentals of Cybernetics.* London: Chapman and Hall.

Senge, P. (1984). "Systems Thinking in Business," *ReVision,* 7, no. 2.

Shannon, C., and Weaver, W. (1949). *The Mathematical Theory of Communication.* Urbana: University of Illinois Press.

~ PART ~

III Systemic Tools

FREQUENTLY, MANAGERIAL FUNCTIONS are discussed as if they were linked together as part of a linear, sequential process. From a systemic perspective, this view of the relationship between managerial functions is seen as unrealistic. We propose that managerial functions are instead part of a dynamic system in which the way various managerial tools are used influences the individual effectiveness of each tool. Three basic relationships can emerge between the systemic tools: complementary-reinforcing, substitution, and antagonistic-neutralizing. Strategy and culture, for instance, may reinforce each other. Culture and procedures may serve as substitutes. Some tools—such as strategy and procedures—may be antagonistic to one another if they are combined inappropriately.

There are five major, systemic managerial tools that will be discussed in Part III: strategy, structure, procedures, culture, and leadership. Each chapter in this part will outline the key functions of one of these tools and then discuss how that tool interacts with the others. These tools are all presented in the Management Systems Model (MSM) (see Figure III-1).

The model is dynamic; the five tools can be combined in various configurations, based on the judgment of management. The MSM is intended to be used as a framework for viewing situations in organizations rather than as a prescriptive program. The specific use of any given tool depends on the organization's current needs, as well as how the other tools are being employed. By focusing on five major managerial tools, we do not mean to exclude other, lesser ones. Each systemic tool has numerous subordinate, supportive tools used in tandem with it.

The model represents how these tools play a critical role in managing the open-closed and social-technical balances in an organization. At the most fundamental level these tools have two main tasks: managing the equilibrium between the social and technical elements of the organization and controlling the extent to which an organization is open to interaction with its environment. The first of these considerations is clear; it involves the tension between technological and human concerns. The second may bear explanation. When an organization participates in a free exchange of resources with its environment, it is said to be relatively open. An organization that concentrates on preserving internal equilibrium or reduces environmental influences is relatively closed.

The management systems model works in the context of another dimension: that of

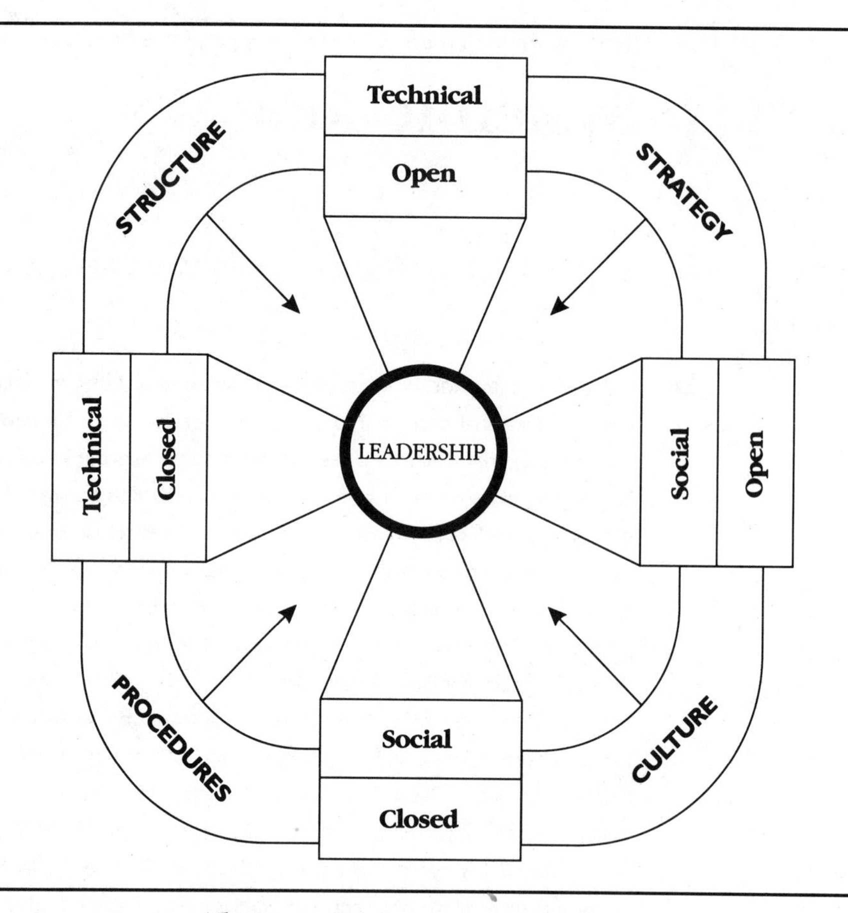

FIGURE III-1 ***The Management Systems Model***

dynamism and stability. All organizations require some level of adaptiveness in order to survive over the long term. The amount of adaptiveness desirable is a subject of debate. Some argue that adaptation is necessary to be able to respond to internal needs as well as to environmental threats and opportunities (Morgan, 1986). Organizations also need stability—that is, relative stability. Stability is not an end in itself, but rather a means to two ends, survival and prosperity. Stability provides the freedom to devote important resources to the work of adaptation and strengthening the organization's base. Organizations that are in crisis do not have this opportunity. On the other hand, organizations that are experiencing turbulence have many opportunities to learn presented to them. Stable organizations must pursue these learning opportunities aggressively to prevent organizational dry rot. There are myriad examples of organizations that were once stable and safe that suddenly found themselves rocked by dramatic decline in profits and loss of key markets: IBM, General Motors, TWA, and U.S. Steel are but a few. Simply put, organizations need to balance growth demands with the need to maintain their own vitality and repair and renew those areas that are not yet vital. Growth occurs when the productive centers of an organization are aligned with and linked to the environment. This occurs by increasing sensitivity to consumers, markets, and other environmental factors. However, this infusion of energy, money, and ideas must be balanced against the organization's current capacity to withstand the uncertainty associated with this process.

During these exchanges an organization must learn to balance the need for stability with the need for newness. Newness is often accompanied by confusion, awkwardness, and even chaos. These are necessary and tolerable in measured doses. To gain some degree of control over these and other forces, managers use the tools that are the components of the MSM. Let us now briefly consider each of these tools.

Strategy is often considered as a set of major goals and policies. In systems thinking, however, strategy relates an organization to its environment and is oriented toward preserving long-term stability and promoting growth. Therefore, it is a fundamental tool for dealing with the complex interface between organizations and their environments. Strategy determines an organization's opportunity for growth. Strategy is largely a social-oriented tool.

Structure allows for the division of tasks, authority, and information. It segments an organization into sections and departments with specialized tasks and responsibilities. Structure helps to regulate the organization's equilibrium by balancing unexpected disruptions mainly in technical operations. Structure is primarily a technical-oriented tool.

Procedures are rules and programs developed as a result of experience to deal with repetitive problems. Procedures are almost invisible to the casual observer as long as they are well designed and judiciously used. If procedures are abandoned or misapplied, order becomes disorder. In the dynamic business world, shortages arise, machines break down, inventory levels change unpredictably, and quality declines when there is nothing present to regulate them. To deal with these problems individually would be extremely costly and inefficient. The prime function of procedures is to reduce the complexity of internal operations into manageable programs and to prevent disorder. Therefore, procedures are best developed for the purposes of controlling the technical subsystem and for monitoring internal operations.

Culture is probably the oldest and most underestimated tool of management. It allows managers to exercise some influence over the complex social dimensions of an organization by aligning values and norms with the purposes of the system. The impact of culture is particularly visible in small organizations, where people know each other. It allows them to switch roles, replacing and substituting for each other. Most problems are solved through discussion and mutual alignment. It is even possible for such an organization to perform well without any formal structure or standard procedures, because every disturbance is balanced through adaptive behaviors. There are also organizations based upon ideology, such as churches, that by their very nature use culture as an fundamental tool.

Change can often be profound and disorienting for organizations. It may transform an entire system and break existing patterns of thinking and acting. To manage change in innovative ways, an additional systemic tool is needed: leadership. Leadership is a complex subject that is often difficult to define in relation to organizational performance. However, it is easier to grasp its essence in the context of the study of organizational dynamics. A leader's actions first destabilize an organization, then create temporary situations of nonequilibrium, and finally introduce a new level of equilibrium. Leadership is the core of the management systems model. It is the prime moving force that integrates all others. Leadership is the process that creatively combines the resources and tools of an organization into a performance that comes alive with the energy and dreams of the organization's leaders.

The five systemic tools play five main functions: regulating, compensating, buffering, opening, and driving. Regulatory tools, such as procedures and structure, are designed to reduce the amount of variety in a system. Doing so generally makes a system more stable and predictable.

Compensating functions seek to stabilize an organization by counteracting any effects that disturb the organization's equilibrium. Procedures can serve a compensating role, as can culture. In fact, these two tools may substitute for each other. In the military, for example, control is achieved by tightly enforcing strict codes, rules, and procedures. On the other hand, such organizations as Wal-Mart or Scandinavian Airlines Systems use few controls, but are still assured of good judgment from employees because they carefully select their members and have a rich culture with strong traditions.

Buffering tools protect the equilibrium of an organization or its subunits from destabilizing influences. Strategy can buffer an organization from environmental turbulence by detaching the organization from its relationship with that aspect of the environment that is unstable. Culture can also buffer an organization by requiring that organization members adjust their behavior to respond to unexpected changes.

Strategy is the primary tool for opening a system to its environment. The strategic actions of an organization establish or reinforce links with consumers, suppliers, distributors, and other environmental factors. These linkages act as conduits that channel the flow of resources, money, information, and ideas into an organization.

Culture is the major tool used to reinforce the driving forces in an organization. Driving forces are those that break old patterns through innovation and expansion. Culture may achieve this by centering the attention of organization members on a compelling new future that is an expression of its core values and beliefs. Alternatively, a culture may use the positive feedback that stems from expansion to build momentum and positive expectations for the future. It does this by amplifying small but significant changes and drawing attention to success.

Leadership is also a valuable tool for promoting driving forces. Leaders can stimulate organization members to break old patterns by persuading them of the attractiveness of an alternative future. Over the long run, however, leadership alone is usually insufficient to maintain the momentum of driving. For pattern-breaking forces to be sustained, there must be additional mechanisms such as culture, which can amplify the innovations.

▲

8 Strategy as a Systemic Tool

The function of strategy is not to "solve a problem," but to so structure a situation that the emergent problems are solvable.

R. P. Rumelt (1979)

1. To understand the role of strategy as a managerial tool for opening a system to its environment.
2. To clarify what strategy is and what it is not.
3. To recognize the difference between strategy content and strategy-making processes.
4. To visualize how the Miles and Snow model of strategy can be used as a systemic tool for strategic decision making.
5. To know the basic assumptions of the synoptic perspective on strategy.
6. To understand the incremental approach to strategy formulation.
7. To become familiar with the environmental assessment process.
8. To learn the role of strategy in relation to the other managerial tools in the management systems model.

SECTION I
UNDERSTANDING STRATEGY

INTRODUCTION

Strategy is a systemic tool used by organizations to open themselves to their environment. In contrast to procedures or culture, which are often used to close the system off from environmental influences, strategy supports exchanges with the environment. Typically these exchanges are in the form of information, materials, products, and energy. In practice, strategies are used to either increase an organization's involvement with and dependence on its environment or to limit that involvement and dependence. According to the management systems model, strategy ultimately should help to balance the relationship between the system and its environment (see Figure 8-1).

Opening to the environment means that an organization is placing a higher value on aligning itself in ever-increasing ways with its markets, customers, suppliers, and other external stakeholders. This process takes place simultaneously with efforts to juggle internal demands to meet operating requirements (such as efficiency). In practice, organizations achieve this balance

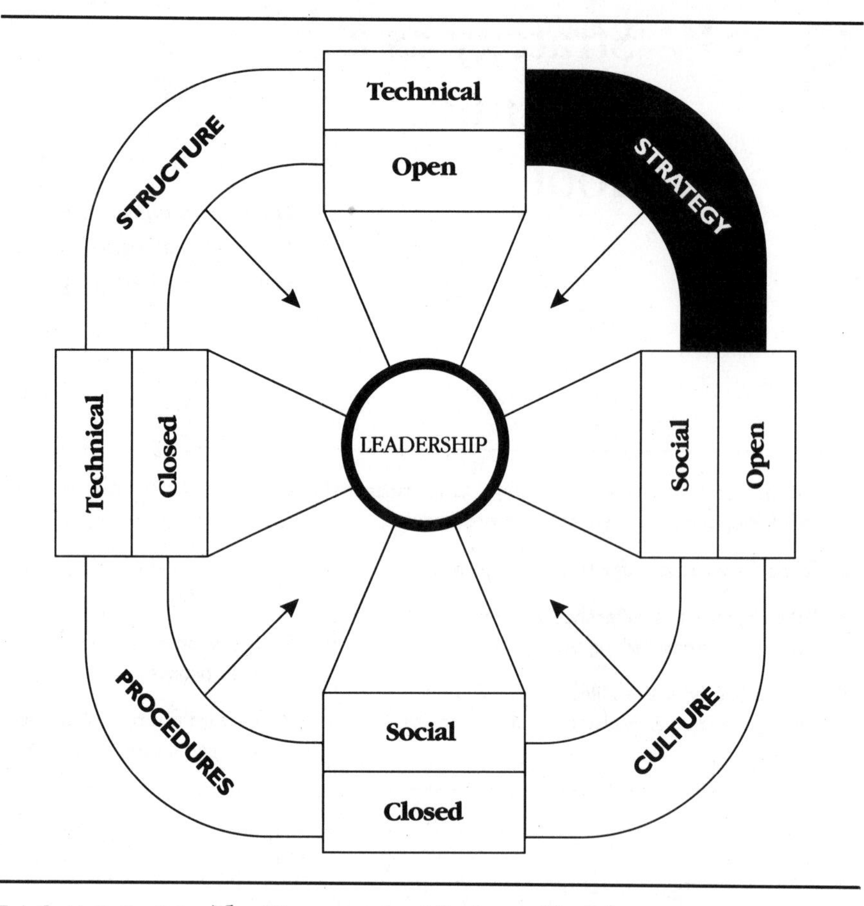

FIGURE 8-1 ***The Management Systems Model***

through product or market diversification, concentration on market segments, divestiture of business units, and other strategic alternatives.

Strategy can also be used to close off the organization from the environment. Such an approach may be used in response to an economic decline or changes in market evolution, such as the onset of maturity in the life cycles of products or markets. A visible example of this strategy was the move by insurance companies to withdraw from insuring customers in certain states during the early 1990s. They took this action based on a belief that the legal and economic environments made business in those states unprofitable.

To more fully appreciate the instrumental role of strategy in the dynamic process of balancing the relationship of an organization to its environment, it is necessary to consider how this perspective evolved. This chapter will examine and evaluate several of the systems-thinking models of strategy and then consider the role of strategy as a systemic tool.

Strategy not only interacts with the other functions in the MSM, it is also the link by which an organization interacts with the world. One of the finest examples of a firm that has been effective in interacting with the global business environment is the Japanese electronics firm Sony, as the global systems spotlight shows.

GLOBAL SYSTEMS SPOTLIGHT

The Systemic Strategy of Sony

Sony is a well-known Japanese firm: superbly managed, big, and profitable. In 1991, Sony employed 112,900 people worldwide, and had $27.5 billion in sales with a profit of $806.2 million. But there are many big, innovative, and well-managed companies in today's marketplace.

From a strategic point of view, Sony is different because it employs almost every possible model of strategy formation at the same time—and successfully. All the various types of strategy or strategy making are somehow present in the way Sony manages its strategy.

Sony is a typical prospector, pumping out almost 1,000 new products per year. Nearly 200 of those are products targeted to new markets. The company is also a defender because it constantly improves it product. In fact, about 800 out of 1,000 new products are improved, better-engineered versions of earlier ones that are intended to penetrate into markets even further. Sony pursues the differentiation strategy, probably better than any other Japanese company, and its image in the United States and Europe is better than that of Panasonic, JVC, or any other company. In a market full of almost indistinguishable products and copycat companies, Sony has successfully molded its image through timely innovations, consistent quality, and creative product design. The company also follows a cost-oriented strategy, as is true of most Japanese firms, and aims toward simultaneously marketing to broad markets and focusing on niche-building.

An example of strategy in action is the way Sony made the conversion from a broad-scope strategy to a focused one. This is illustrated by the Betamax videocassette standard. While the VHS standard became dominant in the industry, Sony's Betamax system became a limited professional standard, but was a profitable business all the same.

If we look into the strategy-formulating process, Sony has simultaneously used synoptic and incremental strategic assumptions. It constantly scans the environment searching for new ideas and for creative people and carefully observes its competition. Environmental scanning is also used to evaluate its position in relation to the environment, as well as to identify new long-term trends in the personal electronic business. Fueled by innovative strategies, Sony has leapfrogged past its competition and built markets from scratch. However, at the same time the company constantly improves its existing products and processes. Sony's flexibility in using many strategies contributes to its ability to thrive in many diverse environments.

THE DEVELOPMENT OF STRATEGIC THINKING

The literature on the theory and practice of strategy is rich with concepts that often conflict with each other (Bracker, 1980). Many definitions and models of strategy formulation and implementation have attained popularity for relatively short periods of time; few have endured for very long. In order to narrow the consideration of possible definitions of strategy, this chapter will start with three statements describing what strategy is *not*. These statements refer to and expand upon the work of Newman, Logan, and Hegarty (1989).

WHAT STRATEGY IS NOT

Strategy is *not* a vague statement describing the system's intentions, nor is it a set of precise quantitative targets. During the 1980s it became increasingly fashionable to state organizational strategy in terms of such pious intentions like "global leadership," "total quality," "total customer responsiveness," "integrity and respect in dealing with all stakeholders," and so on. These slogans were sometimes even printed on the back of the business cards of company managers and executives as a show of commitment to the theme. These truisms sound like great ideas, but without translation into feasible plans of action they remain useless. They might even become harmful if they communicate vague, overambitious, and unattainable intentions to customers or investors. Similarly, quantitative goals developed on the basis of forecasts of the company's balance sheet, statement of cash flow, or income statement (such as "a 15 percent increase in profits," "12 percent return on investment next year," and so on) do not grasp the essence of organizational strategy. Financial and production projections based on managers' perceptions cannot substitute for strategy. They do have a purpose, however. They function as useful short-term targets and performance standards that help employees concentrate on those valued activities within the system.

Strategy is *not* a set of short-term actions to put the organization back on course to recover from a period of substandard performance. Nor is it intended to enable the organization to avoid environmental turbulence. A company faced with stiff competition coupled with declining demand might decide to close down an obsolete plant or phase out a lackluster portion of its product line. Such balancing efforts are a necessary part of day-to-day management; in most cases, however, firms use procedures to lessen the effects of environmental fluctuations rather than strategic decisions. For example, faced with downturns in demand in the 1980s, American automakers relied on buyer rebates and layoffs to cope with the environmental fluctuations rather than examining whether their fundamental strategy was still in alignment with the market. Rather than encompassing corrective actions made by firms, strategy provides a general direction and systemic position for the organization vis a vis its environment. An analogy is the activity of sailing a ship on the sea. The compensating reactions to winds and waves required to maintain a ship on course cannot be considered strategy; a general course directing a ship to a desired destination is much closer in concept to strategy.

Finally, strategy is *not* a tool meant to stand alone. It should not be formulated and implemented separately from organizational structure, procedures, or culture. The management systems model stresses the fact that strategy is a systemic tool of management, not the only tool. This statement does not diminish the importance of strategy as a tool for managing complex systems. Strategy can be envisioned as a bridge that connects a system with its environment. As a bridge, strategy must take into account the limitations and numerous possible contingent situations created by structure, internal operations, and culture. Nevertheless, the importance of the interdependence among strategy and the other elements of the MSM cannot be overstated. Vital, high-performing structures, strong culture, and finely tuned procedures can multiply the effects of a well-conceived strategy. Similarly, a carefully crafted strategy will enhance the performance of these elements. The need for alignment among these elements is critical for the success of the entire system. Many firms have built elaborate or innovative strategies only to have their plan fail as a result of faulty implementation or follow-through. A classic example is Texas Instruments' lightning-like entry and exit from the digital watch market.

This discussion has helped us zero in on what **strategy** is—a systemic concept of how an organization defines the relationship between its internal operations and its environment. This concept is defined for each organization through a series of processes that are based on deliberate actions and spontaneous reactions. In general, strategy either follows the systemic logic of interconnectedness with the other tools in the MSM, or it creates that interconnectedness. In any case, it cannot be effective if separated from its synergetic connections with those other tools.

A SYSTEMS PERSPECTIVE ON STRATEGY

A systems approach to strategy must integrate two divergent modes of thought that have evolved in the literature and practice of management and that have had virtually no relation to each other until recently.

The first mode is oriented toward exploring **strategy content,** an issue best illustrated by the question: "What strategies are successful and unsuccessful in a given type of environment?" The second mode is oriented toward exploring the **strategy-making process.** It deals more directly with the question: "How do you formulate and implement a successful strategy?" These two approaches are divergent, creating conflict and misunderstanding despite the fact that they have common origins. Their central point of agreement is the recognition of the importance of strategy as a managerial tool (see Figure 8-2).

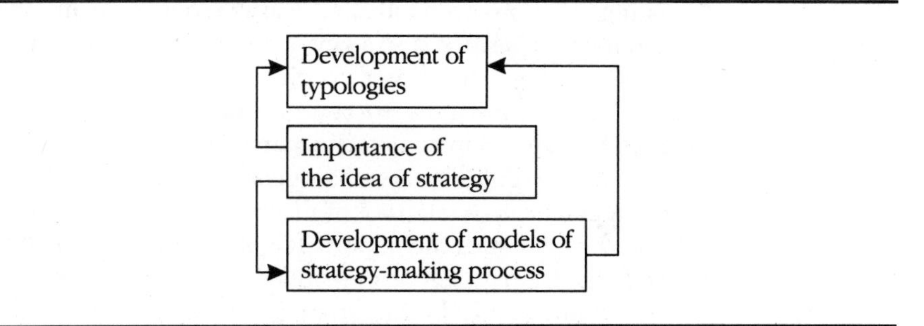

FIGURE 8-2 ***Two Modes of Strategic Thinking***

THE IMPORTANCE OF STRATEGY AS A MANAGERIAL TOOL

Strategy is extremely important in both theory and practice. Its importance is underscored by the fact that it is one of the few variables that is consistently employed in research on organizations. Moreover, such research suggests that strategy is usually a prime variable. Further, all research results, even when inconclusive, have been incorporated into a number of widely used theoretical frameworks. The majority of these frameworks are oriented toward describing the process of how organizations relate to their environment. Two of the most influential frameworks that address strategy content were developed by Miles and Snow (1978) and Porter (1980). These models will be discussed in the next section.

The practical importance of strategy is evident in two practical criteria. First, managers find strategy to be an indispensable tool for changing organizational performance. It is the prime vehicle to move a company from here to there. Indeed, global strategies and competition are considered the most important and relevant issues for managers of the 1990s (Lyles, 1990). Today, more than ever before, practicing managers must develop meaningful strategies in order to set directions, focus efforts, and coordinate actions in global environments.

Second, strategy is a pragmatic concept with symbolic value. The existence of a formal strategy in an organization is a signal to stakeholders that top management is responsible and works in tune with accepted standards of rationality (Bryman, 1984).

SECTION 2
THE CONTENT OF STRATEGY

INTRODUCTION

In the study of strategy content, scholars have tried to classify various types of strategies. The two best known and influential typologies or classifications were developed by Miles and Snow (1978) and Porter (1980).

THE MILES AND SNOW TYPOLOGY OF STRATEGIES

Miles and Snow (1978) offer a powerful framework of four generic strategies: defender, prospector, analyzer, and reactor. This typology is based on an interpretation of existing findings in the literature and an in-depth study of four different industries. The typology was subjected to intense scrutiny through further research and remains highly regarded (Smith, Guthrie, and Chen, 1989).

Miles and Snow start with an assumption that every firm must make strategic choices based on three broad problems. First, a firm must deal with the entrepreneurial problem of deciding on market definition, customer base, product mix, marketing approach, and means for attaining a competitive edge. Second, managers have to deal with engineering issues like technological and production questions, numbers of products to make, outlook toward costs, investments, and innovations. Third, every firm must solve its administrative problems by deciding on the type of structure and management process it wishes to use. By dealing with these three broad problems, managers try to develop a successful strategy in the process of solving recurring problems created by dynamic relations among the environment, structure, and managerial processes. Miles and Snow identified four potential strategies: defender, prospector, analyzer, and reactor (see Figure 8-3).

The defender strategy addresses how an organization defines its entrepreneurial problems, such as "how do we create a narrow product-market domain"? In this particular type of strategy, little environmental scanning takes place, and new opportunities are very carefully analyzed. The defending organization grows by concentrating its efforts on achieving market penetration, limited product development, and superb engineering. For example, Rolls-Royce, the British luxury automobile producer, relies on this generalized approach.

The prospector strategy involves scanning the environment to identify new opportunities and challenges and to monitor trends that are taking shape. Prospectors define the market broadly in order to facilitate new product development, to promote innovation, and to redefine the ways they seek to achieve a competitive edge. A firm following this strategy usually offers a wide

DEFENDER Attempts to "seal off" and exploit a portion of the total market to create a stable set of products and customers.	*PROSPECTOR* Seeks to locate and exploit new products and opportunities.
ANALYZER Seeks to locate and exploit new product and market opportunities while simultaneously maintaining a firm base of traditional products and customers.	*REACTOR* Exhibits pattern of adjustment to environment that is inconsistent and unstable. No consistent response mechanisms available.

FIGURE 8-3 ***Miles and Snow Model of Organizational Adaptation***

Source: Adapted from R. Miles, C. Snow, A. Meyer, and H. Coleman, "Organizational Strategy, Structure, and Process," *Academy of Management Review* (July 1978), pp. 546–62.

range of products and tries to fit production capabilities to product design. General Electric or 3M are examples of prospector firms.

The analyzer falls between the defender and prospector, being a hybrid strategy. Analyzers consider their competitors as major frames of reference for environmental scanning efforts. They normally observe particular traditional markets, but they sometimes monitor new market opportunities and products as well. They maintain multiple product lines and try to balance traditional markets and customers, where they behave like defenders, with new product entries into new markets, where they act like prospectors. To achieve this aim, analyzers combine two technological cores. One set of technologies is stable and enables the organization to achieve low-cost production advantages. The second technology, primarily exploratory, facilitates the creation of new generations of products or completely new designs. IBM represents a prototype of an analyzer strategy. The computer giant was able to maintain a dominant position in the mainframe computer market business and still venture into the newly established markets for PCs, desktop computers, computer networks, and software.

The fourth strategic type is the reactor strategy, in which a firm responds to environmental changes in a random and arbitrary fashion. Such a company may maintain a wide or narrow range of products and attempt to switch from defender to prospector modes without proper support in place. In short, reactors do not have a clear and cohesive alignment of framework, structure, and technology. Chrysler Corporation's inability to concentrate on a single market, its ill-defined acquisitions, and its erratic product development and marketing policies in the late 1980s culminated in the creation of an unstable financial position in the 1990s.

The Miles and Snow typology is useful in analyzing systemic patterns in a firm's behavior. It relates the environment, technology, and structure together into a single perspective. It focuses on the content of strategy, not on the process of its formulation or implementation (Segew and Gray, 1990). In general, research has confirmed the value of this typology and underscores two major points about strategy content:

- There is no one best strategy in a given market or situation. Successful strategies might have different logics, as demonstrated by the diversity of the defender, prospector, and analyzer strategies.
- Each strategy enables the organization to achieve a desired equilibrium between the system and its environment.

However, this state of equilibrium can only be attained when the organization remains systemically consistent. It must align its strategy with internal management processes and structures. A lack of fit can result in the erratic behavior of the reactor organization.

PORTER'S TYPOLOGY OF STRATEGIES

Michael Porter's typology of strategies (1980, 1985) is widely recognized and probably the most widely referred to in the American business community. The model includes four successful strategies for business organizations: cost leadership, differentiation, focus-cost, and focus-differentiation. The typology is similar to that of Miles and Snow in that it recognizes an inconsistent, failure strategy called ***stuck in the middle.***

The four successful strategies are positioned along two dimensions as shown in Figure 8-4. The first dimension distinguishes between a broad or narrow strategic target. A company can decide to establish an industrywide position or focus on a narrow segment of the market. The second dimension is related to the strategic advantage a company seeks, either through low cost or perceived uniqueness.

Similar to the ideas of Peters and Waterman, Porter's ideas have become popularized through several books with widespread appeal, especially *Competitive Strategy* (1980) and *Competitive Advantage* (1985). Each of Porter's four generic strategies will be discussed briefly. In the cost leadership strategy, the firm pursues aggressive policies of efficient-scale facilities, production cost reductions from experience, and tight control of overhead cost. It avoids marginal customer accounts and aims toward the mass consumer. Low-cost positions facilitate maintaining or expanding market share and favorable relations with vendors; they thus replicate a favorable strategic position of the firm. Black and Decker (tools), du Pont (chemicals), and a number of the Japanese car producers are exemplary cost leaders.

		COMPETITIVE ADVANTAGE	
		Lower Cost	Differentiation
COMPETITIVE SCOPE	Broad Target	1. Cost Leadership	2. Differentiation
	Narrow Target	3A. Cost Focus	3B. Differentiation Focus

FIGURE 8-4 ***Porter's Typology of Generic Strategies***
Source: M. Porter, *Competitive Advantage* (New York: Free Press), p. 12.

The differentiation strategy or brand strategy aims at creating a perception among customers of the uniqueness and exclusivity in the products or services offered by the firm. An organization may try to achieve this position through product design or brand image. Examples of high-end producers who have successfully employed this strategy include Mercedes or Ferrari in automobiles, Corning in the glass industry, Bang & Olufsen in audio equipment, and Federal Express or Caterpillar (given its ability to deliver any spare parts around the world in twenty-four hours) in customer service and distribution. Differentiation produces above-average returns and insulates a brand producer from competitors by generating brand loyalty.

The focus-cost strategy is a version of cost leadership applied only to a particular consumer group, a segment of a product line, or a geographic market. The strategy rests on the premise that limiting the market scope combined with vigorous cost reductions will create a viable, albeit narrowly defined, strategic niche. The American machine-tool company, Fadal Engineering, which holds 20 percent of the market for inexpensive, reliable metalworking machines tailored to the needs of small manufacturers, pursues such a strategy. Its operational strategy is to design products with a minimum of parts and frills, to tightly control costs, and to service its products through a network of reliable distributors.

The focus-differentiation strategy is a small-scale version of the differentiation strategy. Organizations employing this strategy try to do an exceptional job of serving a narrowly defined segment of the market while developing a brand name to create a sense of uniqueness about the product. Swiss firms like Nagra (professional tape recorders), Rolex, or Philipe Patek (watches); the Italian car producer Lamborghini; the American producer of super-computers Cray; and many other niche players insulate their limited markets by offering superb products and cultivating intense loyalty from customers.

According to Porter, a firm must follow one of these strategies to succeed. Those firms that pursue diffuse or undefined strategies end up being stuck in the middle: they lack the market share, capital investment, or differentiation of products and services needed to succeed. Such firms will experience mediocre performance and eventually decline into a vicious cycle of low profit margins, low investment, and an absence of uniqueness.

EVALUATING TYPOLOGIES OF STRATEGIES

The Miles and Snow and Porter models do not exhaust all of the theoretical or practical perspectives. Nevertheless, both approaches vividly show the content orientation of the first approach to strategy, as well as its practical and theoretical importance. The prime value of these typologies is that they offer general frameworks managers might use to judge the strategies of their own organizations. For example, during strategy reformulation, executives may ask the basic questions: "Are we defenders or reactors?" "Are we focusing, differentiating, or maybe stuck in the middle?" These and similar questions are powerful guidelines for managers because they orient their strategic perspective and allow them to choose different types of strategies (Ansoff, 1981). Despite the fact that these types of strategies might be considered very general, they provide managers with both direction and enhanced rationality in their strategic thinking.

The major strength of the content approaches to organization strategy is their practicality. Managers may not be interested in theoretical differences or the subtleties of different typologies, but they are interested in flexible and practical mental frames of reference into which they can put facts, expectations, and emotions. These models supply convenient frames on which to build mental models of reality. The weakness of such an approach is that imitation of successful strategies does not necessarily produce results. Imitating the cost-leadership or focus-differentiation strategy of an industry leader might not be a good prescription for the future because, in a changing environment, the best strategies may have changed. Hamel and Prahalad (1989) noted this point, crediting much of the success of the best Japanese companies to the fact that they do not follow common prescriptions but always develop new and imaginative strategies.

SECTION 3
THE STRATEGY-MAKING PROCESS

INTRODUCTION

A second mode of research on strategy revolves around the problems connected with the strategy-making process. It, too, sees strategy as practically and theoretically very important, but this orientation explores how strategies are made.

Two almost contradictory perspectives on the process of strategy formulation and implementation have emerged. The first perspective can be termed a synoptic (deliberate) strategy-making process while the second is an incremental (emergent) strategy-making process (Mintzberg and Waters, 1985). We will examine each of these two approaches in detail.

THE SYNOPTIC PERSPECTIVE

The **synoptic perspective** postulates that organizations follow a planned, rational, hard systems-like pattern of strategy development. This pattern consists of four stages: formulation of the strategic objectives and goals, assessment of the environment, development of an organizational profile, and choice and implementation of a strategy.

The starting point of the synoptic strategy-making process is to formulate strategic objectives for the organization as a whole. Strategic objectives represent the basic purposes of an organization; they are statements of an organization's desired state. For example, General Electric's strategic objective is to be number one or two globally in each product line. Strategic objectives provide broad direction in qualitative terms; these objectives are then defined as goals, which are usually medium- or short-term, precise, quantitative, and broken down to the level of individual business units and departments. The function of goals is simple, and powerful: to provide an organization with performance standards that enable continuous monitoring of processes and outputs related to the system's strategic objectives.

The second stage of the synoptic strategy-making process consists of building an organizational profile of capabilities. Such a profile is normally framed in terms of questions aimed to identify strategic strengths and weaknesses. An organization's advantages, or strengths, are features that distinguish it from other organizations in such areas as finance, human resources, operations, marketing, research and development, and management innovativeness. The analysis of organizational disadvantages, or weaknesses, aims to establish an inventory of organizational resources and skills that may be exploited. A common rule of thumb is that organizations will seek to position themselves in markets where they can build on their strengths in a way that will take advantage of competitor weaknesses.

The third stage of the synoptic strategy-making process is the environmental assessment. This phase is usually viewed in terms of possible external threats and opportunities. Such an assessment normally begins with an examination of the general social, political, economic, and technological shifts or trends that can be perceived in the environment. Such scanning efforts are focused on searching for early warning signals of major changes or emerging issues that may potentially impact the interests of stakeholders. The aim of this step is to establish a broad sketch of the future environment that the organization is most likely to encounter while pursuing its strategy.

This broad view should be followed by a more precise analysis of industry, competitors, markets, and products aimed to build a map of the environmental "territory" (Oliva, Day, and DeSabro, 1987). Such a map will address such issues as an industry's structure; its patterns of growth; the identity of major competitors; the significance of future threats of new competitive entrants or substitutes; and the effect of relations among producers, buyers, and suppliers on competition. These questions are often instrumental in conducting an environmental scan. The main functions found in an environmental scanning system are in Figure 8-5.

Some organizations perform very elaborate environmental assessments. They may use a battery of sophisticated methods and tools; but others use relatively simple and qualitative techniques. The most common methods are trend extrapolation, statistical modeling, scenario building, strategic maps, and product-market matrices.

A qualitative approach may be combined with quantitative approaches, as General Electric does. At GE managers prepare one-page answers to five fundamental questions:

1. What are your market dynamics globally today, and where are they going over the next several years?
2. What actions have your competitors taken in the last three years to upset those global dynamics?
3. What have you done in the last three years to affect those dynamics?
4. What are the most dangerous things your competitor could do in the next three years to upset those dynamics?
5. What are the most effective things you could do to bring your desired impact on those dynamics?

The answers to these questions are then combined and regularly updated to provide GE with an overall picture of environmental dynamics (Koerner, 1989; Tichy and Charan, 1989).

MAIN FUNCTIONS
Support choice of strategic objective
Provide early warning of threats and discovery of opportunities
Assess competitors and trace their strategies
MAIN PROCEDURES
Collect data
Process information, analyze it, and make forecasts
Disseminate results
Provide feedback for further scanning of environment

FIGURE 8-5 ***Main Functions and Procedures of an Environmental Scanning System***

The fourth stage of the strategy-making process deals with the selection of an effective strategy that ensures attainment of objectives while still capitalizing on internal strengths and environmental opportunities. Implementation—from top executives down through middle and lower levels of management to all staff—follows. Consistent goals and courses of action are delegated to various subsystems, and processes for monitoring performance are simultaneously implemented. This sequence of strategy formulation is known by the acronym **OSWOTS** (objectives-strengths-weaknesses-opportunities-threats-strategy); it is depicted in Figure 8-6.

The OSWOTS process described above is partially consistent with the management systems model (MSM). This model recognizes explicitly the external-internal dimension incorporated in the OSWOTS process through the analysis of external factors (environmental scanning) and internal factors (the company profile). The social-technical dimension of the MSM appears in the OSWOTS framework's recognition of technical-economical contingencies on one hand and managers' values and social expectations on another. The balance of stability and dynamism in the MSM can be compared to the OSWOTS process recognizing stable trends and possible discontinuities and events, as the five questions asked at GE show.

The strength of the synoptic perspective is its comprehensive search for a fit between internal resources, external contingencies, and strategic objectives. Its central weakness is its reliance on planned strategy and the outdated belief that managers operate in purely rational fashion (Stopford and Baden-Fuller, 1990).

THE INCREMENTAL PERSPECTIVE

The synoptic view purports to describe how managers should approach the strategy-making process; the incremental perspective looks at how they actually engage in this process. The **incremental perspective** suggests that managers often wait to get results on how effective a strategy is in the initial stages before proceeding with full commitment. This gives them a greater opportunity to make adjustments, if necessary. Although organizations may go to great lengths to establish precise, well-analyzed plans, they do not exert the same

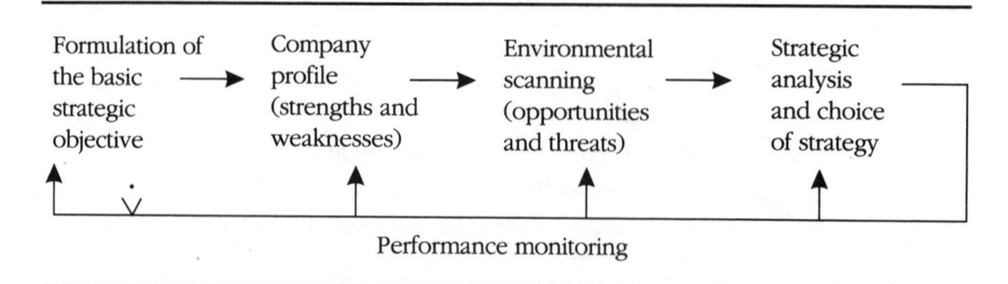

FIGURE 8-6 ***The Synoptic Process of Strategy Formulation (OSWOTS)***

effort in adhering to them. In many cases plans are rendered useless by changing action, and a quick, spontaneous response becomes more beneficial than following a predetermined route. Quinn (1980) has proposed that companies often follow a strategic pattern of **logical incrementalism.** He found that when well-managed organizations adjusted their strategies, their methods break away from using the rational, analytical systems traditionally prescribed in the strategic planning community.

Although the synoptic perspective dominates the process-oriented approach to strategy, the incremental approach, which evolved from research on decision making in organizations, is also influential (Fredrickson, 1983). Lindblom's (1983) analysis of patterns of strategic decisions revealed that decision makers often mix objectives and means. They try to make "small steps," introduce limited improvements, and intuitively react to situations in order to find satisfying solutions in difficult situations. They use rules of thumb (heuristics) instead of precise rules (algorithms). Their behavior is more dynamic than the rationalist perspective acknowledges. They make real time adjustments, taking into account interests of different stakeholders in the organization as well as possible environmental contingencies. If they discover that their intentions were wrong or if the course of action yielded limited success, they change actions accordingly. In short, as Lindblom (1983) aptly described it, managers often "muddle through."

Two premises of the incremental approach are especially noteworthy. First, decision makers are not perfectly rational. Their rationality is limited because their ability to process information is finite, and their values may be in conflict.

Second, the differences between statements and actions in organizations can be very revealing. The prime function of the synoptic process may be to serve as a vehicle to appease stakeholders rather than a genuine game plan that has the commitment of managers. Over time, managerial actions usually establish an almost invisible pattern of momentum. This inertia perpetuates the status quo, regardless of whether the strategy continues to be appropriate. This pattern, through which organizations adapt, becomes the actual strategy. The best illustration of such a process is found in the case of Honda.

Honda

Honda made a number of marketing errors in the way it approached the American motorcycle market. In late 1958, Honda sent two executives to establish an American subsidiary and evaluate the American market. They were incompetent in their command of the English language and had no idea how large, seasonal, and demanding the market was. They did not have any strategy or money to back up their vague idea of a challenge. Honda tried to sell directly to retailers, but did not know how to approach them. It ran advertising in the motorcycle trade magazines for dealers, and few responded.

They started to sell large bikes, but the bikes proved to be of poor quality. During the first six months of its efforts, Honda made virtually all the possible marketing mistakes. Then luck struck.

Honda executives in Los Angeles were riding a small, lightweight 50cc SuperCub moped on an errand. They attracted lots of attention—and offers (even from a Sears buyer) to sell the bikes. Honda did not want to sell the small bikes, afraid that they would hurt the Honda image! Finally facing disaster in large bike sales, Honda, almost out of despair, decided to sell the 50cc bikes in sporting goods stores. The sales rise was as unexpected as it was phenomenal; sales grew from $500,000 in 1960 to $77 million in 1965. In 1961–1963 Honda followed the lucky strike by lining up distributors, increasing regional advertising, and capitalizing on economies of scale in production and sales.

In 1963, a UCLA undergraduate student prepared an ad campaign for Honda to fulfill a routine course assignment. Its theme: "You meet the nicest people on a Honda." The professor encouraged him to sell his work to Grey Advertising, which in turn offered it to Honda. Honda management split on the idea. The president and treasurer favored another proposal, but the director of sales endorsed the student's campaign promoting lightweight bikes. Finally Honda adopted Grey Advertising's proposal, which proved to be extremely successful. Pascale concludes: "Thus, in 1963, through an inadvertent sequence of events, Honda came to adopt a strategy that directly identified and targeted that large untapped segment of the marketplace that has since become inseparable from the Honda legend" (1989, p. 76).

Source: Based on R. Pascale, "Perspectives on Strategy: The Real Story Behind Honda's Success," in M. Tushman, C. O'Reilly, and D. Nadler (eds.), *Management of Organizations* (New York: Harper and Row, 1989), pp. 68–90.

Such incremental approaches to organizational strategy are rather different from the synoptic perspective. The major differences are portrayed in Figure 8-7.

THE POPULATION ECOLOGY PERSPECTIVE

Another view of strategy formulation—the **population ecology perspective**—views strategy in broader terms. It looks at industries as a dynamic collection of organizations continually reacting to each other's strategic moves. Strategies are seen to reflect not only perceptions of threats and opportunities, but also expectations of competitor moves and countermoves. Unlike traditional perspectives on strategy, the ecological paradigm views competition as being largely dynamic, rather than static. There is no assumption that the system is in equilibrium. To the contrary, it suggests that, because of tight coupling between competitors, industries are likely to exhibit near-continuous flux. This flux is generated not only by sequences of action and reaction, but by anticipatory actions. Action in response to expected competitor moves is known as *teleological causation*. The consequence of adopting this view is that strategic effectiveness cannot be evaluated with traditional procedures.

FEATURE	SYNOPTIC PROCESS	INCREMENTAL PROCESS
Strategic intent	The process begins with identifying objectives in response to an environmental assessment.	The process is initiated in response to current problems.
Relations between objectives and means (alternatives of action)	The objectives are identified independent of the analysis of environment, organizational profile, and analysis of feasible alternatives. The process is an "ends-means" process.	The remedial change outcome is considered at the same time that means are analyzed. The processes are intertwined or independent.
Environmental assessment	Plays an important role and is vigorously pursued using formal and informal methods and techniques.	Plays a limited role; managers concentrate on a limited number of issues and themes.
Organizational profile development	The process of organizational evaluation tends to be exhaustive and comprehensive.	The process is limited and randomly pursued.
The choice of strategy	Follows as a choice of the best action after identification of feasible alternatives in terms of the objectives, environmental assessment, and organizational profile.	The strategy emerges as a pattern of loosely linked decisions handled individually by managers solving sequentially formulated problems.

FIGURE 8-7 ***Comparing the Synoptic and Incremental Approaches***

The strategy evaluation process becomes increasingly difficult, yet more anchored to the systems perspective.

The ecological view also emphasizes the role of resources in limiting or enabling the growth of certain organizations within an industry (Hannan and Freeman, 1977). Changes in the availability of resources may, in theory, alter the competitive dynamics and patterns of competitive interaction in the industry. As a result, those strategies that had demonstrated prior success are less likely to continue being successful. According to this perspective, when resources are abundant, firms compete most directly for control of resources with firms that are their own size. In an industry of declining resources, large organizations are more likely to compete with each other or with medium-sized organizations for resources. At the same time, relatively smaller organizations will be forced to attempt competing with ones of greater size for market share. The effect is that medium-sized firms often get squeezed by firms on both sides seeking new sources of business. Small firms are comfortable in niches and encroach selectively in markets where firms of greater size are weaker. Boeker (1991) investigated the interactive effects of organization size and competitive strategies in the brewing industry. He found general support for the population ecology view in his study. He found that when resources declined, regional brewers (medium-sized firms) had trouble competing against both national brewers as well as smaller local brewers.

STRATEGIC IMPERATIVES

One of the most important contributions of the systems approach to strategy is the insight that strategy is part of an overall dynamic system within an organization. This perspective departs from traditional postures on strategy. Early theorists viewed strategy as one of the duties of executives: to provide future direction for the organization. Strategy was seen as the managerial tool that could translate a firm's mission into a concrete action plan. More recent efforts to view strategy from an open systems perspective generally regard it as a single component in a linear flow model (see Figure 8-8). The entire process that encompasses strategy development is known as *strategic management*. Strategic management approaches are valuable because they see strategy in system terms, but they generally minimize the importance of interaction between the components of the model.

Churchman (1968) proposed that five key considerations—called **strategic imperatives**—provide a general framework for strategic decision making. These five factors are interdependent, dynamic, and operate as a system. They are:

1. Identify the basic values of the organization and the goals that develop as an extension of them.
2. Appraise the organization's environment to learn about critical forces, opportunities, and threats.
3. Evaluate the resources and capabilities of the organization.

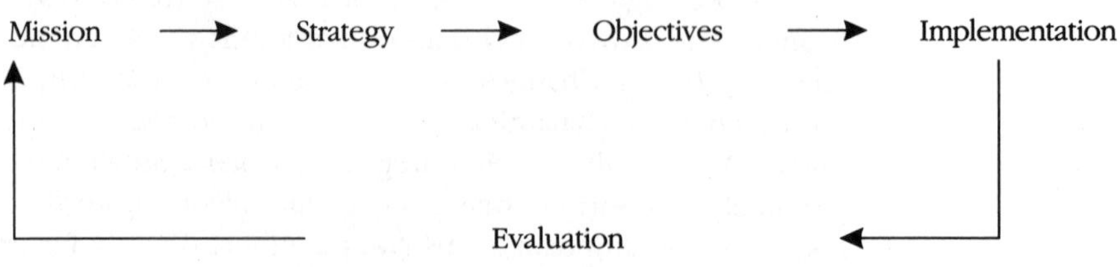

FIGURE 8-8 ***Strategic Management Flow Model***

4. Design the structure of the system, including organization structure, information flows, creation of subunits, and so on.
5. Create the management decision-making process.

According to Churchman, the organization's values form the basis of all strategic decisions. Values are expressed as priorities of what is most important and relevant to the desired future of the organization. Value statements serve as the common reference point for all future decisions.

SECTION 4
SYSTEMS THINKING ON STRATEGY

STRATEGY AS A SYSTEMIC MANAGERIAL TOOL

This investigation of organizational strategy began with a review of two major trends in the literature of last three decades. We noted that a major source of confusion and disagreement is the lack of a system approach to strategy. Typically, scholars studied relationships between two metavariables, like strategy and structure, environment and strategy, or strategy and performance (Segew and Gray, 1990). Similarly, the separation of strategy formulation and strategy content approaches results from adopting different, but not contradictory, views on strategy.

The strategy-making perspective focuses on the actions of strategy makers. Its strength lies in its comprehensive coverage of the internal processes and events that interact to shape the process of strategy formulation. In essence, its central concern is to understand how managers make strategic choices among alternatives. In one view, managers act in a very orderly or rational way; this view reflects the hard systems perspective. Another perspective suggests that they behave in a way that emphasizes a qualitative, invent-as-you-go, more incremental approach that reflects soft systems thinking.

The major thrust of this line of research is a search for goodness of fit among internal contingencies: structure, organizational culture, and strategy. Although there is an attempt to connect the organization to the environment via environmental scanning processes such as identifying threats and opportunities, this is quite limited. The reasons for the limitation are many. First, the identification of threats and opportunities, as a part of OSWOTS, is a static rather than dynamic process. The image formed of the market and competitive environment is more like a snapshot than a movie. Second, because these attempts are usually static, there are few opportunities to actually learn. When continuous learning takes place in the process of reaching out to the environment, a relationship is built with important sectors that becomes the basis for establishing linkages.

An example of the two different approaches can be seen in the ways Japanese and American automobile manufacturers market their products. In Japan, attempts are made to establish life-long relationships with consumers. This is done through a sales force that visits the family's home and gets to know them. This continuing relationship helps Japanese companies to learn in detail about how the family's transportation needs change over time.

In contrast, American firms typically prefer to take periodic macro-level snapshots of the overall marketplace. This approach supplies a different view of consumer needs and profiles—one that contributes less to these firms learning about consumers. More importantly, this approach restricts their ability to establish linkages.

Stoner (1989) has developed a conceptual model that articulates how the interests of customers can be more effectively integrated into an organization's processes. This approach, known as integrated process management (IPM), provides a theoretical framework that brings together the concepts of statistical process control, continuous improvement, and project teams in systemic interaction. The systemic function of strategy is to reach out to the environment, but this is not emphasized in the strategy-making perspective. The environment is of greater concern in the strategy content school of thought, however.

The strategy content view assumes an external perspective. It inquires into the success or failure of a particular strategy in the context of an environment. If an organization is examined from the vantage point of its environment, it would not be possible to determine how the organization arrived at a particular strategy. What is more likely to be perceived from this vista is the organizational outcome of prior strategic commitments. The panorama seen through the strategic telescope is likely to reveal the position of the organization in terms of brand image and product positioning, of how the firm stands vis a vis competitors, products, customers, and markets. The external perspective ties together three important systemic factors: a strategy, an environment, and organizational performance.

Integration of these two major perspectives is offered in Figure 8-9. The figure permits us to reach two major conclusions about the role of strategy in the process of balancing an organization's interface with an environment.

First, the strategy-making process results in strategy formulation; then the effects of strategy formulation and implementation are fed back to the strategy-formulation process. Therefore, from a systemic point of view it is more appropriate to speak of **strategic management,** an ongoing process of strategy formulation, implementation, and reformulation. The organization's strategic heartbeat, the root of the company's core ideas about direction, might not change over time, but the constant evaluation of the degree to which that strategy matches the environment and internal factors will lead to modifications and continuous improvement. In this way, the traditional opposing forces of the synoptic and incremental approaches to strategy are reconciled. Briefly put, the synoptic formulation of strategy leads to the implementation of incremental improvements. The way that these two processes intertwine depends on managerial perceptions of the systemic nature of an organization.

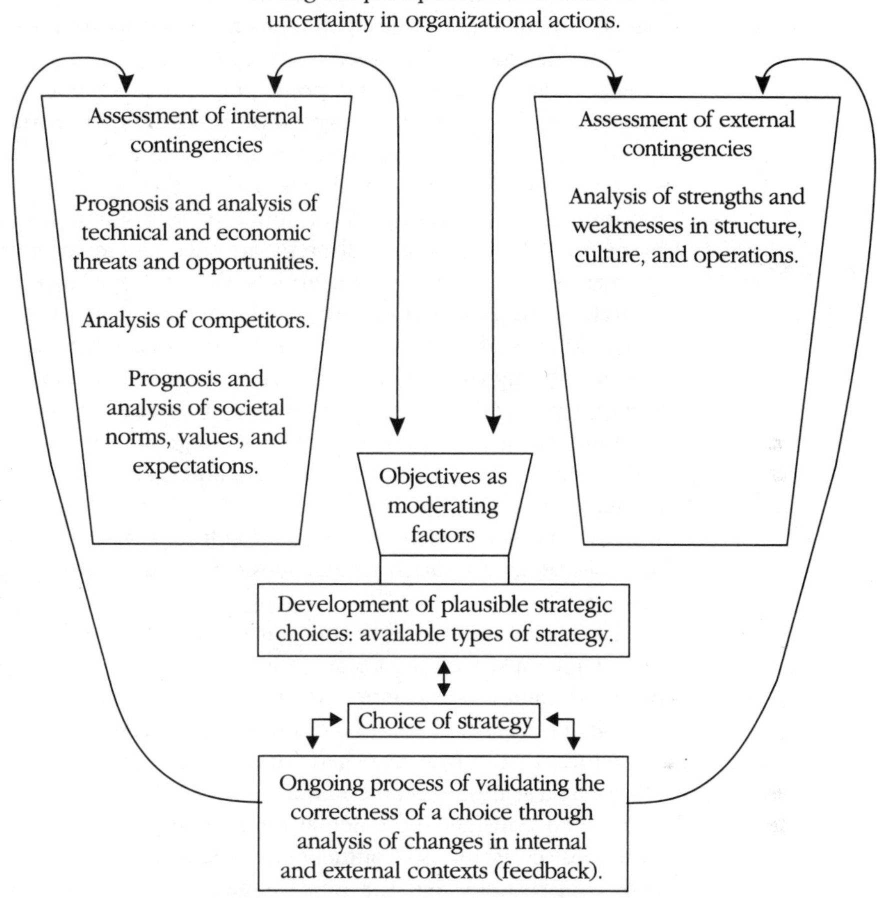

FIGURE 8-9 ***Systemic Approach to Strategy Development***

The management systems model (MSM) specifies three main dimensions of systemic behavior: closed versus open, technical versus social, and stable versus dynamic. An organization that efficiently performs in a relatively stable and controlled environment is partially isolated from both technical and social disturbances. The mode of strategic management for such an organization will be different than that for an organization that must perform as a very open, dynamic system. Simply stated, managers reduce the perceived uncertainty of actions by increasing their knowledge about environmental and organizational contingencies. Hence, the balance between environmental and organizational assessment depends on the extent of managerial perceptions of uncertainty. If managers perceive the environment as a major source of uncertainty, they will perform environmental scanning comprehensively and continuously. If they perceive the environment as placid or stable—or if they believe they are powerless to influence it—they will concentrate on internal issues.

The discrepancy between strategic perspectives of some British and American firms illustrate this point perfectly. A recent survey of 150 major British companies shows that about 60 percent of companies do not use information about competitors or marketplace when setting their strategies (Peat Marwick Management Consultants, 1990). This study suggests that in Britain the intuitive approach based on hunch or feel dominates, and the traditional model of annual revision of strategy prevails. The major source of information for managers is internal financial data and previous track record. Such **introcentric strategy development** can be explained by the fact that British (and many other European) companies defend themselves from competitive turbulence through the practice of selective isolation. This is accomplished by erecting protectionist barriers, which we found throughout the European Community, against foreign competitors, especially Japanese and South Korean traders. In this way, companies achieve a degree of stabilization over environmental turbulence and consequently smooth the interface between the organization and the environment. The conditions that favor the implementation of this strategy are much greater in the EC than in other regions of the global marketplace.

American companies do not currently enjoy the benefits of such protection. They must face the combination of direct, fierce competition with Japanese companies; deregulation; declining social trust; and other environmental turbulence in the process of strategy development and implementation. Therefore, in the best-managed American companies, strategic management efforts are a continuous and mixed process of synoptic formulation, implementation, and constant incremental improvement of a chosen strategy. In such situations, constant environmental assessment plays a major role in both the choice of strategy and the improvements made to it (Klein and Linneman, 1986; Robert, 1990).

The second conclusion to draw from Figure 8-9 is that the use of strategy as a management tool depends on using other tools such as culture, structure, and procedures. Culture, as a tool for regulating the social domain of a company, strongly influences the process of strategic management: formulation, choice, and implementation. This impact is most visible if we look at the norms and values of top managers. Thus a changing strategy for Apple Computer (attempting greater penetration of the low-priced consumer and corporate markets) required changing cultural values and norms which resulted from the change of top management in 1990.

Traditional management viewed these tools in an isolated way; strategy was a tool for balancing the organizational interface with a perceived environment while procedures regulated internal operations. From the MSM perspective, this distinction is artificial, based on the culture-laden choices of managers and on the way internal operations and environmental interface are related.

A brief comparison of typical Japanese and American companies and their modes of performance will help to clarify this point. Japanese corporations often vigorously pursue cost-leadership strategies based on control over operations in terms of costs, quality, and cycle times. To the Japanese this is a strategic matter; to many American companies this has been historically viewed as a technical or operational matter.

The need to abandon this separation of strategy and procedure is especially important given international or global expansion by many firms. The distinction between strategies and procedures becomes blurred for global companies operating in a highly competitive framework. At Benetton, GE, Asea Brown Boveri, or Toyota the drive to become number one exerts constant pressure to perform faster and more efficiently. For global companies, methods such as TQM, high-speed information processing, or team-based approaches cease to be limited as operational experiments; they become instrumental parts of the global strategy, to be implemented at every level of the organization.

From a systems perspective, the relation between strategy and procedures can be visualized in terms of a feedback loop. The choice of strategy determines the configuration of technology, operations, and procedures that will be adopted. Further, the degree of routinization and standardization built into the system will partially define the state of internal equilibrium within the system. Consequently, the technical or operations areas of the firm come to influence strategy. For example, complex procedures limit strategic flexibility. An organization that relies very heavily on standard procedures to regulate the system of activities can easily become bureaucratized and inflexible. An organization that does not develop such procedures and relies primarily on improvisation gains flexibility but loses the cost and time benefits of standardization, specialization, and formalization, which can reduce many recurrent uncertainties. The need for a balance within the system is crucial in relation to strategy. The desire to move forward in a straight line toward goals must be tempered by the realization that moving a large organization is more like moving a caravan of wagons than propelling a high-performing sports car.

SUMMARY

Organizational strategies have traditionally been regarded as linear mechanisms to link the desired ends of a business with the means to achieve them. When seen from a systems perspective, strategy is a managerial tool to influence a business's interaction with its environment. It is also the force that serves to open a system to external influences through the integration of customer needs with internal operations. This cybernetic link is critical in the global business environment because the most effective firms can derive significant competitive advantage by building competitive strategies on internal structures and operating efficiencies.

KEY TERMS AND CONCEPTS

strategy
strategy content
strategy-making process
synoptic perspective
OSWOTS
incremental perspective
logical incrementalism
population ecology perspective
strategic imperatives
strategic management
introcentric strategy development

QUESTIONS FOR DISCUSSION

1. Should strategy be a definitive plan that is precisely executed or a general vision modified to fit the demands of the circumstances?
2. Miles and Snow have identified four basic strategies: the defender, prospector, analyzer, and reactor. Select at least one global organization that fits each type and explain its performance in terms of the strategy.
3. Some people argue that market dynamics are created mainly by economic factors. An alternative explanation for such dynamics is the population ecology model. Which model provides the most robust explanation of the entry and exit of firms to and from an industry?
4. The strategic management process seems to have much less power as a conceptual tool when it is viewed as a linear process without feedback. Using the basic process outlined in Figure 8-8, add the feedback loops to the model that would make strategic management a more valuable process. How does the addition of the loops change the way you view this process?

REFERENCES

Ansoff, H. I. (1981). *Strategic Management.* London: Macmillan Press.

Baron, J., and Sternberg, R. (1987). *Teaching Thinking Skills.* New York: W. H. Freeman.

Boeker, W. (1991). "Organizational Strategy: An Ecological Perspective," *Academy of Management Journal, 34,* no. 3, pp. 613–35.

Bracker, J. (1980). "The Historical Development of the Strategic Management Concept," *Academy of Management Review, 5,* pp. 219–24.

Bryman, A. (1984). "Organization Studies and the Concept of Rationality," *Journal of Management Studies, 21,* pp. 391–408.

Churchman, C. W. (1968). *The Systems Approach.* New York: Dell.

Fredrickson, J. W. (1983). "Strategic Process Research: Questions and Recommendations," *Academy of Management Review, 8,* no. 4, pp. 565–75.

Hamel, G., and Prahalad, C. K. (1989). "Strategic Intent," *Harvard Business Review,* May–June, pp. 63–76.

Hannan, M., and Freeman, J. (1977). "The Population Ecology of Organizations," *American Journal of Sociology, 82,* pp. 929–64.

Hannan, M., and Freeman, J. (1989). *Organizational Ecology.* Cambridge, Mass.: Harvard University Press.

Klein, H. E., and Linneman, R. E. (1986). "Environmental Assessment: An International Study of Corporate Practice," *Journal of Business Strategy, 5,* no. 1, pp. 66–75.

Koerner, E. (1989). "GE's High-Tech Strategy," *Long Range Planning, 22,* no. 4, pp. 11–19.

Lindblom, C. (1983). "The Science of 'Muddling-Through,'" in D. S. Pugh (ed.), *Organization Theory.* London: Penguin.

Lyles, M. A. (1990). "A Research Agenda for Strategic Management in the 1990s," *Journal of Management Studies, 27,* no. 4, pp. 362–75.

Miles, R., and Snow, C. (1978). *Organization Strategy, Structure, and Process.* New York: McGraw-Hill.

Mintzberg, H., and Waters, J. A. (1985). "On Strategies, Deliberate and Emergent," *Strategic Management Journal, 6,* pp. 267–73.

Newman, W. H., Logan, J. P., and Hegarty, W. H. (1989). *Strategy: A Multi-Level, Integrative Approach.* Cincinnati: South-Western.

Obloj, K., and Kasperson, C. (1989). "Strategic Management as Paradox," working paper, School of Management, Warsaw University.

Oliva, T. A., Day, D. A., and DeSabro, W. S. (1987). "Selecting Competitive Tactics: Try a Strategy Map," *Sloan Management Review, 5,* pp. 5–15.

Peat Marwick Management Consultants (1990). "Introvert Strategies," *Financial Times,* October 31.

Porter, M. (1980). *Competitive Strategy.* New York: Free Press.

Porter, M. (1985). *Competitive Advantage.* New York: Free Press.

Quinn, J. B. (1980). *Strategies for Change: Logical Incrementalism.* Homewood, Ill.: Richard D. Irwin.

Robert, M. M. (1990). "Managing Your Competitor Strategy," *The Journal of Business Strategy,* March–April, pp. 24–28.

Segew, E., and Gray, P. (1990). *Business Success.* Englewood Cliffs, N.J.: Prentice-Hall.

Smith, K. G., Guthrie, J. P., and Chen, M. (1989). "Strategy, Size and Performance," *Organization Studies, 10,* no. 1, pp. 63–80.

Stoner, J. A. F. (1989). "What Is Integrated Process Management?" working paper #89–103–1. New York: Fordham University, Graduate School of Business, Lincoln Center.

Stopford, J. M., and Baden-Fuller, C. (1990). "Corporate Rejuvenation," *Journal of Management Studies, 27,* no. 4, pp. 399–415.

Tichy, N., and Charan, R. (1989). "Speed, Simplicity, Self-Confidence: An Interview with Jack Welch," *Harvard Business Review,* September–October, pp. 112–20.

~ ADDITIONAL READINGS

Astley, W. G. (1987). "Administrative Science as Socially Constructed Truth," *Administrative Science Quarterly, 36,* pp. 497–513.

Cameron, K. (1986). "Effectiveness as Paradox," *Management Science, 32,* pp. 539–53.

Chandler, A. D. (1962). *Strategy and Structure: Chapters in the History of the Industrial Enterprise.* Cambridge, Mass.: MIT Press.

Dess, G., and Davis, P. (1984). "Porter's Generic Strategies as Determinants of Strategic Group Membership and Organizational Performance," *Academy of Management Journal, 27,* pp. 467–88.

Fredrickson, J. W. (1986). "The Strategic Decision Process and Organizational Structure," *Academy of Management Review, 11,* no. 2, pp. 280–98.

Hambrick, D. C., and D'Aveni, R. A. (1988). "Large Corporate Failures as Downward Spirals," *Administrative Science Quarterly, 33,* pp. 1–23.

Hrebiniak, L. G., and Joyce, W. F. (1985). "Organizational Adaptation: Strategic Choice and Environmental Determinism," *Administrative Science Quarterly, 30,* pp. 336–49.

Kozielecki, A. (1987). *Koncepcja transgresyjna czlowieka.* Warsaw: PWN.

Lawrence, P. R., and Lorsch, J. W. (1967). *Organization and Environment.* Homewood, Ill.: Richard D. Irwin.

Lewin, K. (1951). *Field Theory in Social Science.* New York: Harper & Row.

MacGregor, D. (1960). *The Human Side of Enterprise.* New York: McGraw-Hill.

March, J. G., and Olsen, J. P. (1975). "The Uncertainty of the Past: Organizational Learning Under Ambiguity," *European Journal of Political Research, 3,* pp. 147–71.

Masuch, M. (1985). "Vicious Circles in Organizations," *Administrative Science Quarterly, 30,* pp. 14–33.

Meyer, M. W., and Zucker, L. G. (1989). *Permanently Failing Organizations.* London: Sage Publications.

Miller, D., and Friesen, P. H. (1986). "Porter's Generic Strategies and Performance: An Empirical Examination with American Data," *Organization Studies, 7,* pp. 37–56.

Miller, D. (1989). "Matching Strategies and Strategy Making: Process, Content and Performance," *Human Relations, 42,* pp. 241–60.

Mintzberg, H. (1987). "The Strategy Concept I: Five P's for Strategy," *California Management Review, 6,* pp. 11–32.

Mintzberg, H. (1988). "The Strategy Concept II: Another Look at Why Organizations Need Strategies," *California Management Review, 30,* no. 1, pp. 25–31.

Obloj, K., and Davis, A. S. (1989). "Innovation Without Change: Case Study of One Industry in Poland," *Proceed-*

ings of the XVth Annual Conference of the European International Business Association, December 17–19, Helsinki, Finland.

Pascale, R. T. (1989). "Perspectives on Strategy: The Real Story Behind Honda's Success," in M. L. Tushman, C. O'Reilly, and D. A. Nadler (eds), *The Management of Organizations: Strategies, Tactics, Analyses.* New York: Harper and Row.

Pascale, R. T., and Athos, A. G. (1982). *The Art of Japanese Management.* New York: Penguin Books.

Pearce, A. J., and Robinson, R. B. (1988). *Strategic Management.* Homewood, Ill.: Richard D. Irwin.

Schein, E. (1985). *Organizational Culture and Leadership.* San Francisco: Jossey-Bass.

Schendel, D., and Hofer, C. W. (1979). *Strategic Management: A New View of Business Policy and Planning.* Boston: Little, Brown.

Tilles, S. (1963). "How to Evaluate Corporate Strategy," *Harvard Business Review,* July—August, pp. 111–21.

Wiener, N. (1948). *Cybernetics or Control and Communication in the Animal and Machine.* New York: John Wiley.

9 Structure as a Systemic Tool

Managing an organization is like building with Legos, and the best structure is the one that balances forces most gracefully.

Henry Mintzberg (1991)

1. To understand how organization structure functions as a regulatory tool to balance an organization's technical equilibrium.
2. To distinguish technical equilibrium from social equilibrium.
3. To comprehend how structure functions in the context of the management systems model.
4. To become familiar with the basic processes required for designing an organization structure: designing roles, designing information flows, and segmenting into subunits.
5. To recognize the primary considerations in the organizational design process.
6. To explain the premises underlying the contingency approach to organization structure.
7. To appreciate the implications of Mintzberg's effective organization model.

SECTION I
UNDERSTANDING STRUCTURE

INTRODUCTION

An organization's design, its structure, is the product of dividing work into distinct tasks and responsibilities. There is also a corresponding division of information and authority that is established to ensure coordination among these tasks (Mintzberg, 1983; Hall, 1982). Structuring is the major tool for balancing the technical system of the organization. The **technical system** is the part of an organization primarily concerned with tasks rather than people. It is the counterpart of the social system.

Although the prime function of structure is to balance the technical system, it may also serve as an important managerial tool in other capacities.

Among the most prominent of these are: as a mechanism to reduce uncertainty (Galbraith, 1977), as a vehicle for increasing strategic vision (Greiner, 1983), and as a means to control behavior (Litterer, 1980). In general, an organization's structure is the product of purposeful efforts to engineer sets of tasks and relations to balance major subsystems. All such efforts to rationalize the system must be viewed against the background of a social system of values and beliefs that will influence which engineered approaches are acceptable. These two opposing forces have been called "visible and invisible hands." The rational design aspects of structure are the visible hands giving shape to the system, yet the softer elements of culture and leadership have a more subtle but no less important effect. Simply stated, the hard and soft elements of structure co-create each other. Structure influences the type of culture that emerges, and culture shapes the structure. This chapter describes the various ways in which the hard and soft systems come together when structure is used as a tool for balancing the system.

Though structure is a systemic managerial tool in the technical dimension of the organization, it is not a device intended to regulate technology. The statement simply reflects the fact that structure is oriented primarily around task requirements—the needs of the job to be done—not with people. Structure is designed to ensure that people engage each other and the other parts of the system in a way that meets the requirements of tasks. The effect of structure is to ensure that the various elements of the system interact in a coordinated way that permits the system to reach its goals. In essence, **structure** is a managerial tool that directs the flow of inputs toward the attainment of systemic purposes through the regulation of power, information, behavior, and ultimately organizational states of balance. Figure 9-1 shows the relationship of structure to the social and technical dimensions of organizations in the context of the MSM.

Structure is a powerful tool for promoting stability in systems, but it cannot accomplish this goal without synergistic support from the other elements—strategy, procedures, culture, and leadership. The global systems spotlight that follows—on the European Community airline system—presents an example of how a system can quickly move away from equilibrium when the structure is not coordinated with the rest of the management system.

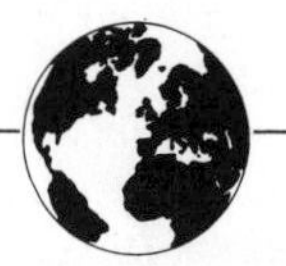

GLOBAL SYSTEMS SPOTLIGHT

The European Community Airline System

Deregulation of the U.S. airline industry in 1978 changed the structure of the industry. First, it considerably lowered the barriers to entry into the industry, and 120 new airlines were launched over the next few years. Second, it created an increasingly competitive environment that caused air fares to drop to bargain levels. The price wars exacted a large toll; causing the demise of many smaller firms and resulting in a market dominated by a few big carriers. A new structure eventually evolved to replace the old, quasi-monopoly that had existed for decades and was seen as antithetical to a free market. This is not a scenario European legislators wish to see repeated.

They are working to liberalize, not deregulate the airline system within the countries of the European Community (EC). The structure of the European airline system has traditionally been decentralized and lacking in coordination mechanisms. The carriers, such as Alitalia, Air France, KLM, and Lufthansa, are national flagships. They are highly subsidized and politicized institutions that enjoy a virtual monopoly in their own countries. Further, the infrastructure of the EC is highly fragmented, with twenty-two separate air-traffic control systems spread across forty-four centers, each with its own operating standards. The new structure of the transportation industry became an important element in the drive to integrate the EC. A plan for the liberalization of air transport in the EC and the European Free Trade Association (EFTA) countries aims at creating a cohesive and coordinated industry structure, harmonization of rules and standards, increased competition between carriers, improved services, and lower fares.

The EC approached the problem using a hard systems approach by developing two packages of liberalization measures. However, the structure that started to evolve within the regulatory framework is somehow different from the intentions of the regulatory bodies. Instead of choosing from a larger number of airlines offering competing services, European travelers are finding major routes dominated by mega-carriers or strategic alliances of medium-sized carriers. Some smaller airlines have already gone bust, and more will follow. The economic climate does not encourage startup airlines. Bigger carriers are too short on cash to be tempted into lengthy price wars. Airline costs are spiraling upwards. Recession and the Gulf War depressed 1991 air passenger numbers worldwide by 1.7 percent and in Europe by 7 percent. Annual growth projections in passenger volume have been revised downward, to an average of 3.9 percent per year from 5–6 percent.

Thus an attempt to create a more effective and efficient organizational design by removing the fragmentation of national structures resulted in the emergence of alliances built not on adaptation, but on survival of the fittest in a competitive environment.

If the designers of the EC airline system had the opportunity to redesign the system, what should they consider? The next section discusses what factors should be addressed in designing a structure.

STRUCTURE: A TOOL FOR REGULATING EQUILIBRIUM

All elements of an organization need to be balanced at various times to increase the opportunities for them to interact in ways that will aid the attainment of goals. The system's efforts to create and maintain equilibrium in the

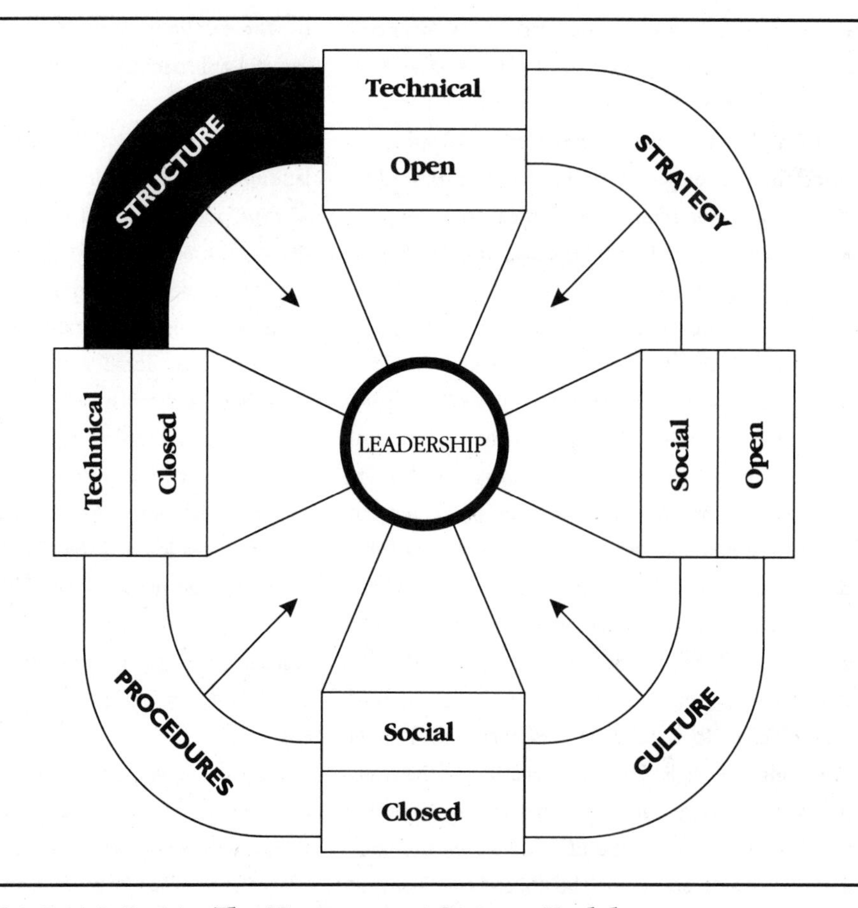

FIGURE 9-1 ***The Management Systems Model***

technical dimension are evident in the strategies used to address three main problems: designing organizational roles, designing information flows, and segmenting equilibrium into subsystems.

DESIGNING ORGANIZATIONAL ROLES

The first problem addressed in structuring an organization is the design of organizational **roles.** Roles are the expected behaviors that specific individuals or groups of workers will play in the technical system. Roles represent the prescribed activities and responsibilities of each position as defined in the system. Roles are necessary in order to perform the specialized tasks that create continuity of work throughout an organization. For example, salespeople, engineers, machine operators, and corporate attorneys all operate within the scope of a set of guidelines that place boundaries on what they will do and how they will perform the task.

In traditional hierarchical organizations, these boundaries may be very clearly defined and strictly enforced. In highly interactive organizations, such as Quad Graphics or some divisions of Ford, the distinctions are blurred. For example, at Quad Graphics, a printing press operator can be found trying to make a sale. At Ford, the design team for the Taurus brought together designers, engineers, marketers, and financial people into a single team. In such **concurrent forms of structure,** a salesperson participates in the design process to offer the consumer's view, indicating how the product might be perceived in the marketplace. The interaction among various task performers that takes place in teams is changing the way responsibilities are defined in organizations.

DESIGNING INFORMATION FLOWS

The question of designing the flow of information and division of authority and responsibility must be addressed by managers. An effective structure brings a continuity of decisions and actions throughout the organization's various subsystems. The *raison d'etre* of such a design is to engineer decision-making patterns along a continuum ranging from centralization to decentralization. The product of this design process is to differentiate the organizational structure vertically.

SEGMENTING EQUILIBRIUM INTO SUBSYSTEMS

The third problem involved in structuring a system deals with segmenting the overall equilibrium of the organization into a set of individually balanced subsystems. This is accomplished by grouping similar organizational roles into larger subsystems. Subunits such as sections, departments, and divisions are established. This grouping allows for continuous coordination and control of performance among similar tasks and roles. The structural solution to the segmenting problem, then, differentiates the structure horizontally.

Let us examine in depth how each of these three problems are approached.

~ CREATING ORGANIZATIONAL ROLES

Every organized activity starts with a division of tasks that need to be performed. The process of establishing such distinctions is part of **task design.** Griffin (1982) views task design as "specification of an employee's task-related activities, including both structural and interpersonal aspects of the job, with considerations for the needs and the requirements of both the organization and the individual" (p. 4). Class lectures in a university are typically structured into two roles: one is performed by the professor (talking, instructing, answering questions), and the other is performed by students (listening, taking notes, asking questions, learning). Assembly-line workers perform highly specialized tasks organized on the basis of a task design.

Regardless of the setting, organized activities start with a design plan that specifies the degree of specialization and formalization of jobs. The plan must identify what must be performed to achieve a desired end (design a book, purchase merchandise for a retail outlet, or manufacture a headlight).

TASK SPECIALIZATION

One of the primary decisions to be made in task design is how much specialization will characterize each job. **Task specialization**—which is involved with the variety and repetitiveness of tasks—can be viewed as existing on a continuum ranging from low to high. There are tradeoffs associated with each position on that spectrum. Narrowly defined jobs with little task variety (high specialization) permit the rapid mastery of necessary skills. Jobs low in specialization are less likely to promote boredom and its accompanying stresses.

The advantages of task specialization are simple and well known. Specialization creates a pervasive logic based on a perceived degree of order and built on repetition. Most organizations are created to operate in cycles for efficient performance over a long time. Therefore, logic suggests that it is reasonable to decide what tasks should be repeated and assign specialists to perform them. Specialization typically increases productivity because it allows people to learn through repetition. Repetition leads to the discovery of optimal ways to accomplish tasks. It also facilitates standardization, which eventually provides an organization-wide basis for tasks to be performed uniformly and efficiently (Mintzberg, 1983). Companies such as Texas Instruments are widely known for their attempts to build competitive strategies based on the learning curve that accompanies the increased cycles of repetition that accrue over time in relation to both specialization and volume.

FORMALIZATION

Closely related to specialization is the extent to which tasks are formalized. **Formalization** refers to the level of discretion given to a job holder in deciding how a task is to be performed. The greater the formalization, the lesser the variability of possible behaviors. The structure of a class offers an example of the tradeoffs that apply to this dimension. The structure of a class can be viewed along the continuum ranging from informal to formal. A highly formal class limits the range of possible behaviors by students. In informally structured classes, control ultimately shifts in the direction of students and the range of their possible behaviors is expanded. In both designs, the extent of job specialization (what is to be done) is very much alike; the basic difference lies in the level of formalization (how tasks are performed).

The same principles apply to the design of organizational roles. The formalization of roles is usually achieved in one of two possible ways. The first approach is through a job description that specifies the role. The second approach is through the use of procedure manuals, which spell out how and when every part of a task is to be done. Formalization is further ensured by the extent of routinization in the technology used to accomplish the task. For example, highly mechanized technology normally places many limitations on the degree of freedom that operators have in choosing alternative courses of action.

The fundamentals of the relationship between specialization and formalization are outlined in Figure 9-2. This figure performs two functions. The obvious one is to exemplify four types of roles present in most organizations. The second is to indicate two major tendencies—toward hard and soft modes in the design of roles.

A SYSTEMS VIEW OF ROLES

As mentioned previously, the presence of these two polar modes of systems thinking can be recognized in Figure 9-2. Hard systems thinking influences task design by maximizing the number of highly specialized and formalized tasks. Such a design strategy tends to exert a stabilizing effect on the equilibrium of the technical system of the organization. It does so by greatly reducing the unpredictability and variability of human behavior and by supporting efforts to optimize economic and technical operations. An example of such a situation can be found in the design strategy used to engineer assembly lines. This tendency is also apparent in the manner in which bureaucratic institutions control behavior by establishing precise specifications for a narrowly designed role.

However, Morgan (1986) suggests that designs that overemphasize specialization or formalization eventually create internal disturbances of the social equilibrium. While some control-oriented structures may appear to be perfect

		FORMALIZATION High	FORMALIZATION Low
SPECIALIZATION	High	Low-skilled, highly specialized jobs	Professional jobs, specially enriched jobs
SPECIALIZATION	Low	Special skilled jobs, low-level managerial jobs	Managerial roles

FIGURE 9-2 ***Relationship of Two Main Parameters of Task Design***

Source: Adapted from H. Mintzberg, *Structure in Fives: Designing Effective Organizations* (Englewood Cliffs, N.J.: Prentice-Hall, 1983), p. 33.

from the technical viewpoint, they often produce unexpected social consequences. Eventually, such disturbances in the social equilibrium may have an impact on the technical equilibrium. The **technical equilibrium** of an organization is the relative balance between its tasks and technology and the rest of the system. For example, if customers demand relatively large quantities of high-quality products but the technology is not capable of providing either the quantity or the quality, the technical system is in disequilibrium. Technical disequilibrium also exists if organization members need to be free to act creatively to respond quickly to customers or to increase innovation in new product development, and the way their job is designed makes this impossible. The technical equilibrium of an organization must always remain in a relative long-term state of balance with the organization's social system and its environment. This is true as long as the organization's purpose remains the same or as long as it is not engaged in a process of transformation.

Soft systems thinking also appears in the design of structure. This view advocates relaxing the rigidity in design and using methods such as job enrichment, job enlargement, autonomous groups, and quality circles. It advocates extensive training of organizational members and encourages them to exchange roles in order to avoid the dysfunctional effects of routine tasks. Such an approach reinforces the **social equilibrium** of the system: the relative balance of human factors. However, if used too intensively, this reduction of the formal structure might decrease technical efficiency and increase costs. Consequently, a tradeoff in these two tendencies emerges. Ideally, the most effective structure establishes a balanced system, where gains in the technical equilibrium compensate for a lower level of social balance or vice versa.

~ INFORMATION MANAGEMENT

Information management is largely a perspective of dealing with environmental uncertainty. Environmental heterogeneity arises from such factors as the number of variables influencing the task environment, the patterns of change or dynamism in the environment, and the extent of time delays between environmental causes and effects. The task of understanding environmental complexity can be daunting.

Jay Galbraith (1977) developed a model of organizational design that deals generally with uncertainty without requiring that the specific causes of the uncertainty be known. Galbraith's model assumes that, at some time, relevant uncertainties will be transformed into discontinuities in work flow. Ultimately, this creates exceptional circumstances, which demand resolution through managerial action. Galbraith believes that organizations are hindered because the information needed to resolve such exceptional problems is not always readily available. This restricted information flow may result from rigidities in the structure, adherence to bureaucratic procedures, or the use of an inappropriate structure.

To optimize information flow throughout the organization, Galbraith recommends a number of different structural variations—alternative organizing modes—that can increase the efficiency with which information is managed (Figure 9-3).

The Galbraith strategy has advanced the systems perspective by illustrating the importance of information flows in systems. In some regards, this approach integrates rational decision making and organizational design. It marries the work of Herbert Simon in administrative theory with the ideas of James D. Thompson. This is an example of a primarily hard systems approach to engineering organizational structure. The result is a strategy for reducing uncertainty by contingent use of rules, procedures, and frameworks. Unfortunately, this model has not been elaborated on sufficiently to link the alternative design modes with other system elements such as goals and feedback, or with softer system elements such as culture.

BUILDING AGGREGATE UNITS

THE BENEFITS OF GROUPING TASKS

Grouping roles into larger aggregate units is necessary for all but the smallest organizations. There are five reasons why creating such aggregates is important. First, the hierarchical arrangement of most organizations requires that their goals be subdivided into increasingly specialized locations within the system. This process is intended to raise competency levels of those performing tasks and ensure accountability for results.

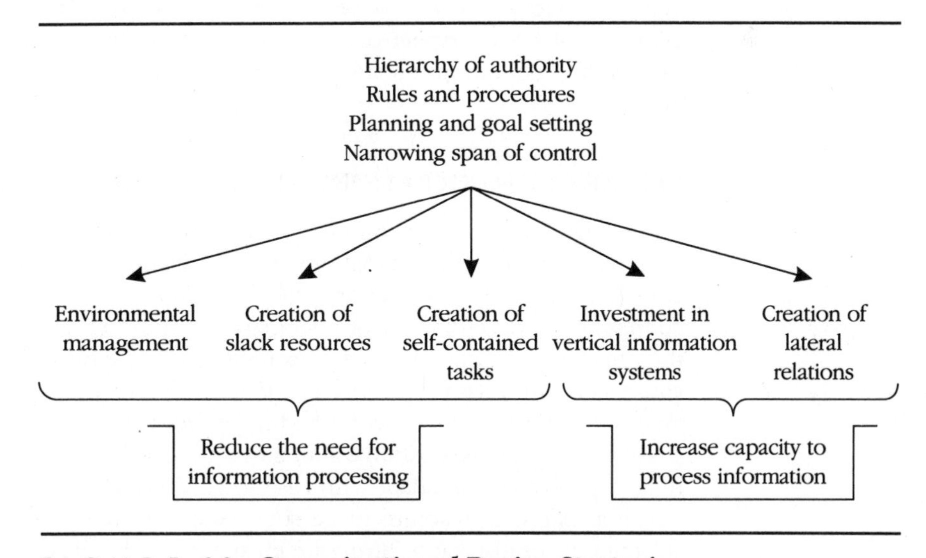

FIGURE 9-3 ***Organizational Design Strategies***

Source: J. Galbraith, *Organization Design* (Reading, Mass.: Addison-Wesley, 1977), p. 49.

Second, grouping similar roles makes it possible to establish common supervision. While this technique has less value in loosely structured, flat organizations, it is common in hierarchical systems. In hierarchies, the technique is used to increase control, improve coordination, and reduce costs. Cost reductions result from having the managerial responsibilities spread over a broader scope.

Third, grouping enhances and improves communication. It creates clear lines for the exchange of information, both within the unit and with other units. Without grouping, each organizational member would have to communicate with other members of the organization on a case-by-case basis.

Fourth, grouping allows organizations to increase their efficiency in using resources. A typical aggregate unit shares common resources such as facilities, equipment, and support staff.

Finally, from a systems perspective, unit grouping allows organizations to simplify the problem of managing the system's equilibrium. By grouping roles into large units, the complexity of the equilibrium problem can be segmented into the need to achieve equilibrium for each of the units. In each unit different indicators of balance are developed. For example, the strategies and procedures used to manage the organization's technical core must be measured and controlled differently than administrative units. Under these circumstances, actions to restore partial imbalances may be undertaken at the local level on a subsystem and need not have undue effects on total operations.

The process of engineering the framework of the organization to match its needs plays a critical role in balancing the system. The structural division of tasks and the creation of coordination mechanisms reduce the complexity of managing the equilibrium of the total system. The ability to visualize the system as subsets of related, yet discrete, partial equilibria is a powerful conceptual tool. This perspective enables managers to tailor actions to the uniqueness of local conditions created by such factors as size, environment, and technology.

PRIMARY CONSIDERATIONS IN ORGANIZATIONAL DESIGN

The configuration of units into larger aggregates, such as departments or divisions, is a standard process that typifies organizational design efforts. The more complex aspects of this design process center on two key issues. The first addresses the development of a strategy to guide the actual construction of the organization's framework. For example, does the business want to be lean and mean, with relatively few layers of management and the ability to react to environmental challenges with blinding speed? Or is the company pursuing a strategy of global expansion, wanting to saturate the market by opening many retail outlets that can adapt with ease to local conditions? Such a strategy requires a different structure to be successful. The link between strategy and structure is unmistakable. Structure must accurately reflect both the organization's intentions and its purpose. A firm that places a premium on customer service must design a unique structure to enable its members to freely interact with customers in the most beneficial way possible.

The second key issue of organizational design involves top management's mental model of what an effective structure looks like. Certain fundamental assumptions are the building blocks of any model of a structure that will work for a given organization. Some firms place the highest value on teamwork as a way to increase responsiveness to customer needs. Such firms may tend to prefer a structure that increases the interaction among its members, such as a matrix form.

Steers, Mowday, and Ungson (1985) have identified four major perspectives that have guided organizational design efforts in the past (Figure 9-4). Organizations are likely to be significantly influenced by these perspectives and incorporate their elements into what Argyris (1990) calls their "theory-in-use." (Theories-in-use are master programs built on sets of values and beliefs that guide the people's actions.)

Four major themes, or theories-in-use, are revealed in the four perspectives shown in Figure 9-4:

Perspective	*Theme*
1. Bureaucratic model	Manager-centered, emphasizing control and uniformity.
2. Human relations model	Employee-centered, emphasizing job satisfaction and motivation.
3. Contingency model	Configuration-oriented; design depends on key factors.
4. Neocontingency model	Structure is tool for enacting main competitive strategies.

SECTION 2
THE CONTINGENCY PERSPECTIVE

INTRODUCTION

The **contingency perspective** proposes that the best way to organize is tailored to match the characteristics of the situation. Organizational design efforts are generally thought to be based on the successful configuration of a framework that aligns critical internal and environmental subsystems. Such design decisions are typically based on judgments revolving around three main factors: size of the organization, type of technology, and type of environment (Hall, 1982). While these are the three main factors, a number of others also influence design. These other factors include the nature of the task, the clarity of goals, and the capabilities of employees.

With the advent of the contingency approach, organizational design was

	1890	1910	1930–1950	1965	1965–1970	1975–Present
Prevailing Work and Theory	Weber's work on bureaucracy	Taylor's work on scientific management	Hawthorne experiment: Birth of human relations movement	Harvard's school of contingency design	Examination of contingency variables: Environment, technology, size, power, control	Strategic choices
Perspectives on the Organizational Design problem	I. Universalistic theories with emphasis on managerial prerogatives.		II. Universalistic theories with an emphasis on employee prerogative	III. Contingency theory with an emphasis on contingency design variables.		IV. Neocontingency theory with an emphasis on strategic choices or managerial discretion.

FIGURE 9-4 ***Historical Approaches to Organizational Design***
Source: R. Steers, R. Mowday, and G. Ungson, *Managing Effective Organizations* (Boston: PWS-KENT, 1985), p. 143.

conceived as being dynamic for the first time in modern history. The goal of organizational design became not only to find the right match to balance the various subsystems but also to create an organization capable of adapting to its environment. The breakthrough reflected in the contingency perspective was that both the structure and the design process could be adaptive and organic.

The influence of the biological elements of systems thinking in this perspective is evident. In departing from the view that organizations behaved mechanically or as a simple social system, the contingency model favors the principle that organizations exhibit properties similar to those of living organisms. The essential systems concepts of openness, boundaries, equilibrium and interdependence were translated from the works of Bertalanffy (1950), Miller (1978), and Parsons (1960) into the contingent organizational theories of Emery (1969), Kast and Rosenzweig (1979), and Katz and Kahn (1978).

The contingency approach contributed to managerial thinking by recognizing the potential for the environment to behave dynamically. This dynamic potential was viewed as an uncertain force that would disrupt the normal equilibrium of organizations and interfere with the rational systems that had been constructed. These developments were thought to create a need to align critical subsystems of the organization with the corresponding environmental changes. Much of the interest in the contingency perspective was reflected in research on the relations between environment and structure (Burns and Stalker, 1961; Lawrence and Lorsch, 1967). The common theme that characterized the contingency approach was an interest in understanding how environmental dynamism generated uncertainty throughout the subsystems of an organization.

FACTORS THAT INFLUENCE DESIGN IN THE CONTINGENCY APPROACH

SIZE

Conventional wisdom says that large organizations have more elaborate structures than smaller firms. The growth of organizations normally argues for the expansion of physical capacities (for example, the number of production lines in factories or retail outlets of a fast-food chain), personnel, and the variety of inputs and outputs. Along with increasing growth, tasks become more specialized and numerous. Consequently units become larger and the hierarchy more complex. The belief that large organizations must have relatively complex structures has been virtually unquestioned for many years. Today, however, examples abound of large organizations with relatively simple structures. GE's dramatic downsizing efforts over the past decade is one instance.

General Electric

When the revolution hits GE, it's time to sit up and take notice. Concepts such as strategic planning, decentralization, and market research—indeed, the very notion that management is a discipline that can be taught and applied across a broad spectrum of business—were all either invented at GE or first used systematically there. . . .

GE was a different company from the one [CEO Jack] Welch had taken over in 1981. By 1989 he had squeezed 350 product lines and business units into 13 big businesses, each first or second in its industry. He had shed $9 billion of assets and spent $18 billion on acquisitions. He collapsed GE's management structure, a wedding cake that had towered up to nine layers high, and scraped off its ornate frosting of corporate staff; 29 pay levels became five broad bands. Victims dubbed Welch "Neutron Jack" after the neutron bomb, a Pentagon idea for a weapon that would kill people but leave buildings standing. That was a misnomer: Welch eliminated 100,000 jobs and flattened buildings too. Those tough actions beefed up GE's total stock market value from $12 billion in 1980 (eleventh among U.S. corporations) to $65 billion today (second only to Exxon).

Source: "GE Keeps Those Ideas Coming," *Fortune*, (August 12, 1991), p. 42.

ENVIRONMENT

The nature of the organization's environment also vitally affects organizational structure. In a classic study, Burns and Stalker (1961) matched two types of environments and organizational structures. They observed that organizations that performed in simple, stable environments, tended to exploit elaborate, centralized, formalized, and standardized structures. The lack of distinctive, changing environmental threats and opportunities allowed organizations to use efficient, rigid, and mechanistic structures. Conversely, when an environment behaved in a dynamic fashion with no obvious pattern (a turbulent environment), an organic structure became appropriate.

In organic structures, systems are designed to reflect the functional requirements of the task in the organization's form. As a result of this strategy of form follows function, roles tend to be less specialized and standardized, unit grouping more flexible, and coordination less formal. In organic structures, division of labor more naturally reflects the requirements of work processes and the needs of organization members. As a result, work is accomplished by whomever has the greatest expertise in a given area. Patterns of information flow and communication are oriented to support specific task needs rather than to adhere to bureaucratic reporting procedures.

TECHNOLOGY

Contingency theorists understood the practical ramifications of changing demand for products, both in terms of demand levels and innovations. The research in this area centered on the relationship between technology, structure, and uncertainty. The search for a link between technological uncertainty and structure found its broadest application in the work of Charles Perrow. Many recent studies done on this relationship have considered possible links between specific types of technologies and certain dimensions of structure. The most common elements of structure to be examined are: specialization, formalization, centralization, and autonomy (Fry, 1978).

It is unclear to what extent technology is important in organizational design. The entire question of determining technology's role in an organization is muddied by the fact that over the years no single operational definition of technology has ever been accepted. Furthermore, technology has been measured at various levels in organizations, ranging from individuals and small groups to the entire organization. The net effect is that it becomes difficult to reach a consensus.

Writers such as Mintzberg (1979) have argued that when the effects of different ways of measuring technology are accounted for, there is significantly more evidence to suggest the importance of technology. More specifically, there is evidence that technology, formalization, and centralization are all related (Hage and Aiken, 1967). In general, though, it is reasonable to argue that there is a lack of evidence to support the idea that technology and structure are strongly related.

CONTINGENCIES, STRUCTURES, AND SYSTEMS

Early contingency theorists viewed structure as depending on technology, environment, and size. Based on this belief, they conceived the prime role of management as adapting the organization's framework to be in harmony with these factors. While this approach represented a movement toward a more open and systemic way of visualizing structure, it still was narrowly concentrated on understanding a limited set of relationships. The perspective viewed managers as primarily reactive, adapting to conditions outside their direct control.

Where technological routines and environmental stability prevail, this view allows for precise determination of the prerequisites to achieve organizational equilibrium. Subsequently, reliable methods and efficient techniques can be established to maintain organizational balance. The problem of maintaining a system's technical equilibrium is relatively well defined under these conditions; consequently, so is its solution.

This approach becomes progressively less viable as the technology and environment become relatively more complex and dynamic. In order to maintain systemic balance under these circumstances, the organization can no longer rely on routine operations. The recognition of situational threats and opportunities demands timely, fluid reactions from organization members. This response becomes impossible from an organization with a highly standardized, formalized structure with clearcut specialization.

The difference between these two approaches can be easily seen in the different ways armies structure artillery units and commando groups. An artillery unit performs under relatively stable, predictable conditions. Therefore, they take full advantage of the benefits of formalization such as strict role specialization, role standardization, stable unit groupings, and clear line of command. A commando unit usually performs in a hostile, unpredictable environment; it must have a loosely coupled structure. Such a framework contains ill-defined individual roles, minimal standardization, and lack formalization (Obloj and Joynt, 1986). The same principles apply to business organizations.

CORE DESIGN STRATEGIES OF THE CONTINGENCY VIEW

A number of core design strategies are regarded as representative of the contingency approach. Each exemplifies a variation on the basic theme of adapting the organization's framework to match the uncertainty created primarily by the environment or by technology. These strategies generally assume that the efficiency of organizations can be optimized through the reduction of uncertainty. Contingency strategies advocate reducing uncertainty through structural avenues rather than by using softer approaches such as leadership or culture.

Lawrence and Lorsch (1967) first acknowledged the inadequacy of the prevailing organization theories of their time and asked, "What kind of organization does it take to deal with different environmental conditions?" (p. 3). After conducting a comprehensive study, they concluded that different units in organizations interact with various subsystems of the total environment. In environments with many, varied subsystems, the effect is as if each unit dealt with discrete environments. They then proposed that firms must distinguish among the various environmental subsystems and treat them individually rather than seeing the environment as a single homogeneous system. This process is known as **differentiation.** Differentiation promotes a divergence of interests among the subunits of an organization. Lawrence and Lorsch saw a need to compensate for these effects. To restore the commonality among the units, they developed a process that meshed the subunits. Known as **integration,** it could be expressed as the creation of multi-unit teams or groupings.

The concepts of differentiation and integration provide managers with mechanisms to translate their efforts of interacting with the environment into a practical strategy. However, from a systems perspective, this explanation fails to capture the complexity found in organization-environment relations. In particular, it neglects to acknowledge the role of perception in distinguishing the significant from irrelevant aspects of the environment.

APPRAISING THE CONTINGENCY APPROACH

The general picture painted thus far of the relationship among technology, size, environment, and organizational structure is somewhat misleading. First, the tacit assumption was made that technology and the environment are homogeneous, that the organization essentially uses the same technology and faces one type of environment. While this might be true of small businesses, it is not usually as relevant to larger organizations like IBM, Ford, Bayer, Toyota, or Olivetti. Different phases of production might demand different technology;

while R&D operations are customized and nonroutine, assembly lines utilize mass production and routine technology. In this situation, major subsystems like R&D, production, and marketing are facing different environments (Lawrence and Lorsch, 1967). The differences apply both in terms of complexity and uncertainty. The lack of homogeneity characterizing the environment and technology is reflected in different demands and conditions for the maintenance of equilibrium within the organization's internal subsystems.

A second lack in the contingency approach results from the assumption that contingencies influence organizational design per se. It is not necessarily true that managers perceive these contingencies in a broad spectrum of ways. The meaning attributed to each contingency depends on a variety of individual and organizational frames of reference. Organizational culture plays an important role as an information filter, shaping the criteria used to make choices relating to structure. This is particularly true in regard to the rationality of the design process. Contingency advocates admit that the environment often changes the inner workings of organizations in unpredictable ways, often creating unintended consequences. Yet, in the main, contingency theorists such as Galbraith (1973), Perrow (1967), and Thompson (1967) have maintained a deterministic perspective suggesting that desired information about the state of each element in the system will be known to managers. This assumes that somehow managers will have an accurate mental picture of how all the variables match each other and fit together.

Karl Weick (1979) has even been more ardent in his criticism of the rational perspective of organizational design. He advocates that the primary concern of managers in regard to structure should be the process of organizing rather than the form that the organization takes. Weick bases this position on the belief that an organization does not exist as a freestanding entity separable from the people who constitute it. As Weick (1974) proposes,

> The word organization is a noun, and it is also a myth. If one looks for an organization one will not find it. What will be found is that there are events linked together, that transpire within concrete walls and these sequences, their pathways, their timing, are the forms we erroneously make into substances when we talk about an organization. (p. 358)

The case below discusses one organization where managers failed to follow this principle.

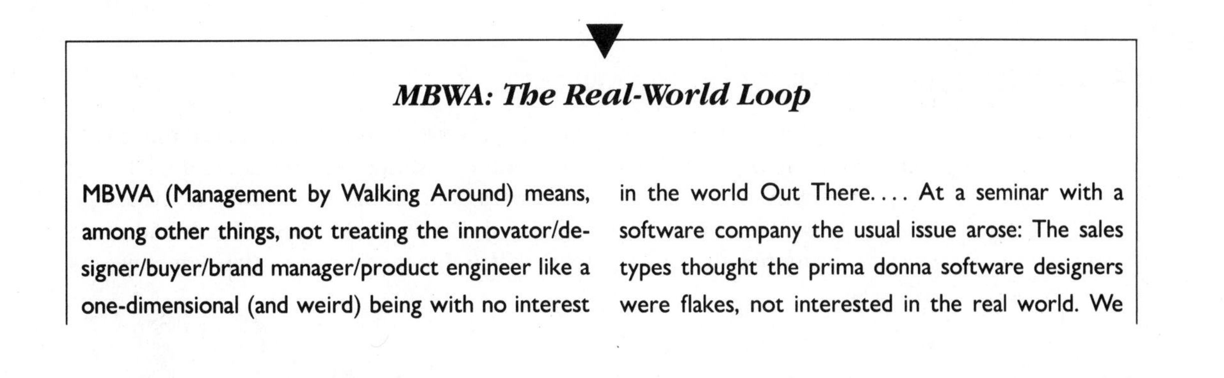

MBWA: The Real-World Loop

MBWA (Management by Walking Around) means, among other things, not treating the innovator/designer/buyer/brand manager/product engineer like a one-dimensional (and weird) being with no interest in the world Out There.... At a seminar with a software company the usual issue arose: The sales types thought the prima donna software designers were flakes, not interested in the real world. We probed a bit deeper, and discovered that given the salespersons' *assumption* that "all software designers are flakes," they did in fact treat the software designers like flakes. A designer pleaded: "How do you expect me to be excited about the sales and marketing end? You never give us any information on it—no sales data at all, no customer information except major glitches, in conjunction with which you simply run around calling us a bunch of 'impractical nerd ________.'" It turns out that this sad tale is an all too common one. The sales types . . . act out their negative assumptions; thus the "far out" types are not brought into the real-world loop, and they become farther out as a result, with predictable effects on innovation.

Source: T. Peters and N. Austin, *A Passion for Excellence* (New York: Random House, 1985), p. 28.

Clearly, an organization's structure exists in the minds of its members in ways that differ from the official diagrams. Ransom, Hinings, and Greenwood (1980) have also recognized the process elements of organizing. They view organizing as a continuous interactive process in which the structure is both the outcome as well as an input. This view is consistent with the systems perspective. Like culture, structure can create self-reinforcing patterns. For example, existing structures can create unintended consequences, which managers can compensate for by designing new structures.

Traditional approaches to organizational design have generally ignored the psychological and process dimensions of building structure. A systems perspective must try to account for as many of the variables that influence the process as is realistically possible. Of course, it is impractical for managers to attempt to understand *all* the potential influences on a system. But through continuous investment in organizational learning and empowering assignments, managers will develop over time a much clearer sense of the key issues and how they relate to the purpose of the organization. This accumulation of information and knowledge will enable them to be selective in the way they establish priorities for each element in the design process.

▶ SECTION 3 SYSTEMS MODELS OF ORGANIZATION DESIGN

MINTZBERG'S EFFECTIVE ORGANIZATION

Mintzberg (1991) developed one of the most well-conceived systemic explanations of the organizational design process. Mintzberg's model of the effective organization (see Figure 9-5) integrates factors that were never convincingly brought together by the contingency theorists. Earlier researchers had established a number of contingencies that influenced organization effectiveness.

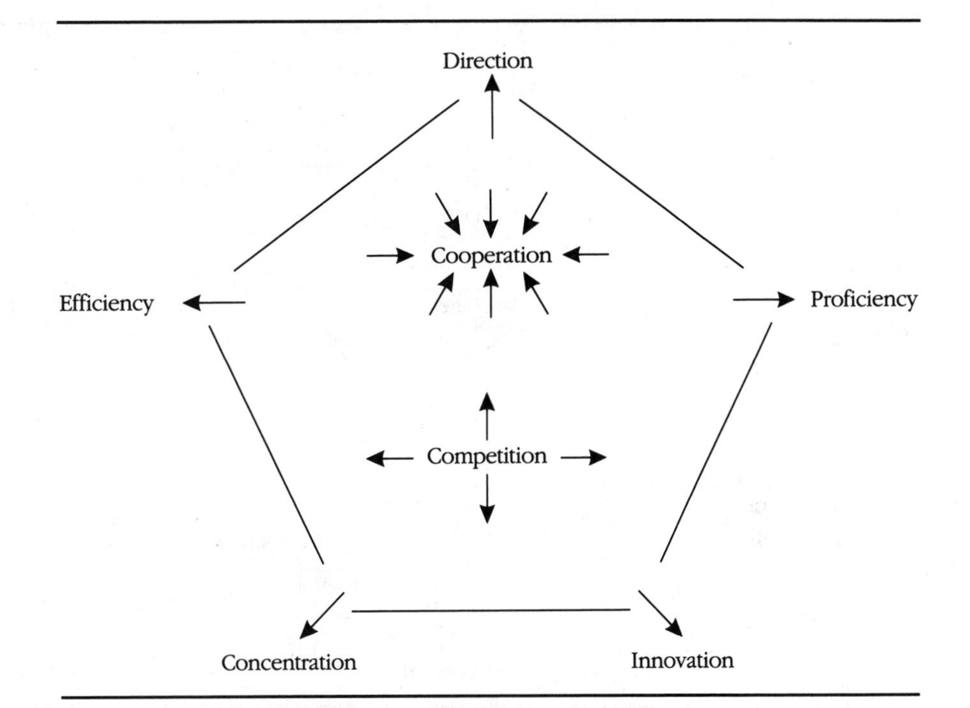

FIGURE 9-5 ***Mintzberg's System of Forces in Organizations***

Source: H. Mintzberg, "The Effective Organization: Forces and Forms," *Sloan Management Review* (Winter 1991), p. 55.

Mintzberg has identified a set of broader factors, which he refers to as *forces*. Most importantly, these forces interact together in the systemic framework.

Mintzberg proposes that the forces maintain a natural equilibrium that will be disturbed if any single force dominates the organization. But there are also problems when there is no clearly commanding force. Specifically, such arrangements promote dysfunctional patterns resulting from contamination and cleavage. Cleavage is best expressed by the adage "neither fish, nor fowl." **Cleavage** occurs when no force is strong enough to set the tone for the direction of the system. As a result, a number of forces offset each other and come to loggerheads; the organization shows no movement. **Contamination** occurs when a single force is so powerful that it begins to change the fundamental character of the other forces in the system. These two patterns represent polar opposites of system behavior.

To manage these opposing patterns, there must be a method to resolve the incompatibilities. Resolution relies on the skillful manipulation of the forces of competition and cooperation. Because these forces are also polar opposites, they require continuous effort to balance them simultaneously with the other forces in the organization.

At this point Mintzberg integrates function and form by proposing that

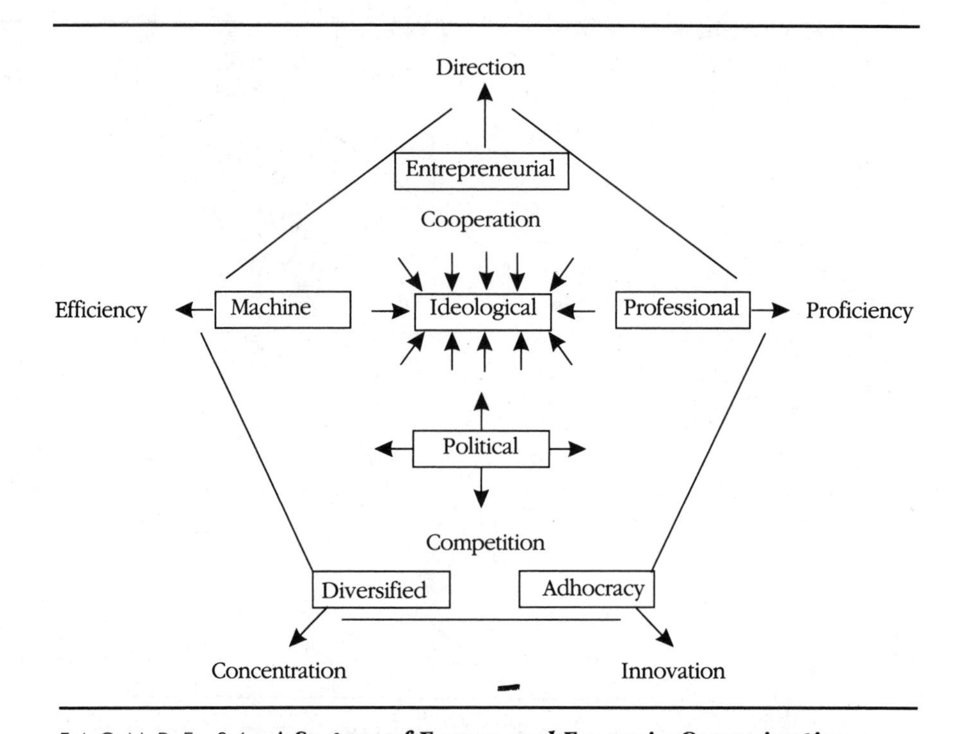

FIGURE 9-6 ***A System of Forces and Forms in Organizations***
Source: H. Mintzberg, "The Effective Organization: Forces and Forms," *Sloan Management Review* (Winter 1991), p. 57.

each unique configuration of forces will yield a unique structural arrangement (Figure 9-6). This kaleidoscopic view of organizations assumes that the specific arrangement of form depends on a highly interactive system of dynamic forces. Even when the forces themselves are not particularly dynamic, they are continually agitated by the patterns of cleavage and contamination.

Mintzberg explores all the hybrid forms produced by all the possible permutations and combinations of forces as well as their implications. More importantly, he eloquently explores the potential dilemmas associated with balancing the contradictory forces operating within systems. For example, the presence of a clear ideology to which organization members are committed can promote cooperation. On the other hand, dominant ideologies can become rigid dogmas enforced by peer pressure in oppressive cultures. The net effect can be competition and rebellion against that dominant ideology.

The Mintzberg model effectively presents many of the subtle elements that color organizations and does so from a view that is relatively more systemic than the contingency models. However, some serious questions remain regarding the method used to select the seven forces. Also, the model appears more explanatory than prescriptive. Practicing managers will be challenged to determine with any precision where their organization falls within Mintzberg's model or how to move it from one form to another.

A SYSTEMS VIEW OF UNIT GROUPING

Systems thinking about organizational design indicates that there are two fundamental criteria for unit grouping, which reflect the potential sources of technical disequilibrium, and that these forces originate both externally and internally. When managers perceive a potential source of turbulence inside the organization, they often select a structural resolution that is internally oriented. The implicit goal is usually to efficiently eliminate or smooth over any significant imbalances within the organization. In this perspective, promoting internal efficiency is the main frame of reference for evaluating the desired level of organizational equilibrium.

When managers perceive the environment as the major source of disturbances to systemic equilibrium, they implement externally oriented structures to maintain systemic balance. In this case, they consider external effectiveness as a major indicator of equilibrium. In internally oriented structures, unit groupings are designed on the basis of function, work process, and human resource skills. Externally oriented structures use products and markets as the benchmark for grouping or aggregating units into larger, more complex, forms such as that shown in Figure 9-7.

In the global economy, organizations have grown in size and entered more markets with more product lines. Consequently, they face extensive and differentiated competition and greater variety in consumer behavior. These and other sources of uncertainty—largely external—have caused a transition from a mindset oriented to internal production processes to one pointed toward the environment. As the internal production orientation was replaced by a more external marketing orientation, structure followed this pattern. However, in true systemic fashion, the internal equilibrium became problematic again, challenged by the growing sophistication of skills and knowledge of employees and by fast-paced technological change. At the same time, the external equilibrium was further disturbed by increased globalization of financial and commodities markets. Additionally, the increased rate of change in technological, economic, social, and political domains exerted increased pressure on structures.

Responding to these pressures, many organizations have moved to the third stage of structural solutions: the hybrid structure. **Hybrid structures** combine the strengths of internally and externally oriented designs. They allow organizations to solve problems related to flexibility in internal and external technical equilibria simultaneously.

The hybrid structure is organized in the following fashion:

1. Top management is externally oriented and grouped usually by market criteria. The prime responsibility of top managers is to manage environmental interfaces and ensure the maintenance of global equilibrium by tuning up the whole system to new contingencies, threats, and opportunities in the environment.
2. Intermediate-level units are grouped according to products and are self-contained. They are responsible for balancing particular environmental in-

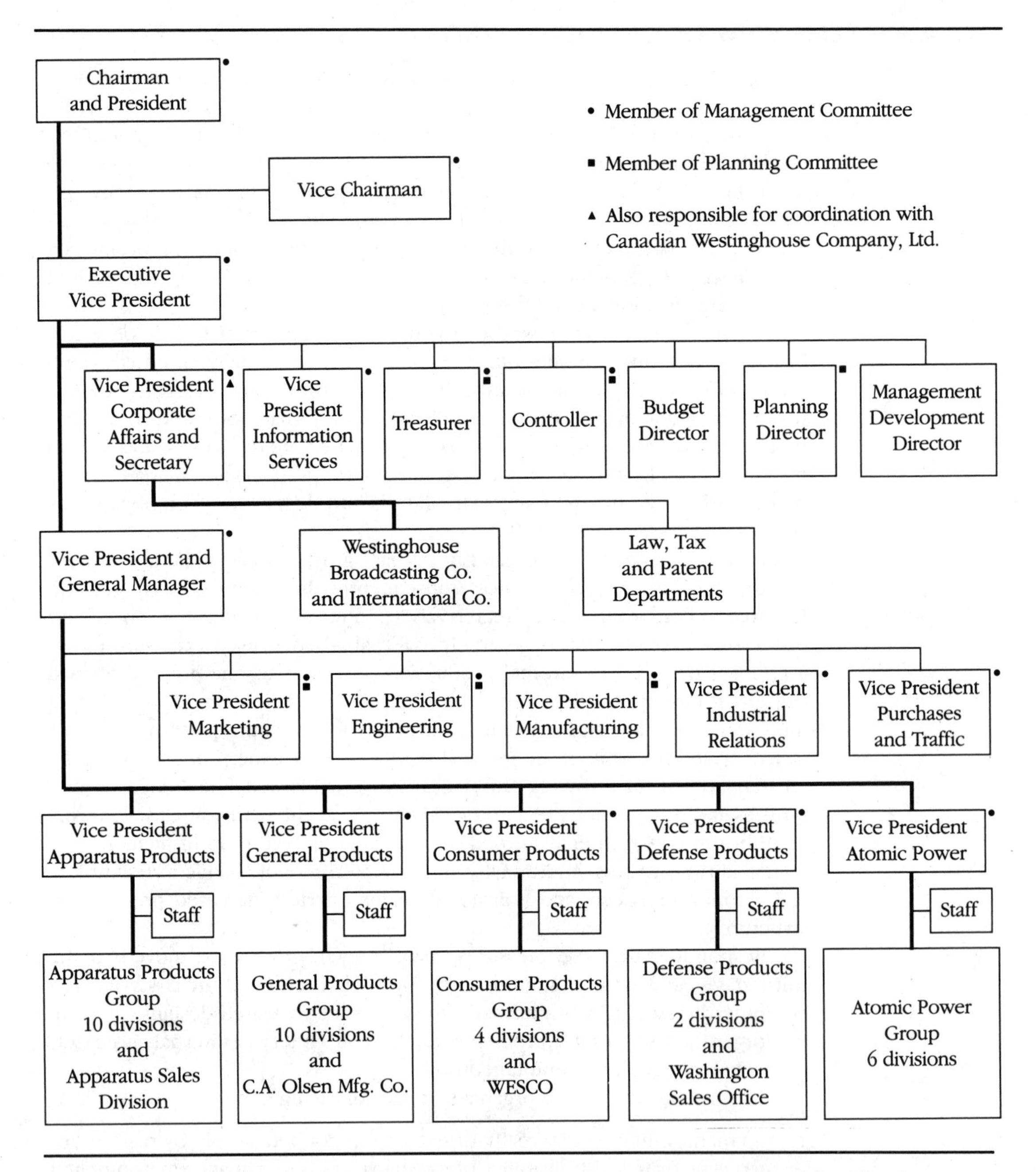

FIGURE 9-7 ***Product and Market Structure: Organization Chart for Westinghouse Corp.***

Source: J. Montanari, C. Morgan, and J. Braker, *Strategic Management* (Hinsdale, Ill.: Dryden Press, 1990) terfaces—such as selected groups of consumers—and maintaining internal balances within local self-contained units. This level also translates the decisions of top management into directives for lower levels. Therefore, it is a crucial link between external and internal dimensions of technical balances.

3. Lower levels are organized according to functions or workflow processes. The system's frame of reference for equilibria is shifted from external to internal problems at this level. Managers at this level are responsible for maintaining internal balances; they are relatively protected from the problems of turbulent environments. They are also responsible for the implementation of changes initiated by top management.

The hybrid structure segments the problem of technical equilibrium into three separate levels: the internal balance of self-contained units, the balance-environment interfaces, and the balance of the system as a whole.

The split of responsibilities for technical balance achieved in the hybrid structure leads to the last issue of system design: decisions involving centralization and decentralization.

ACHIEVING BALANCE USING CENTRALIZATION AND DECENTRALIZATION

The relationships between systemic balance and dispersal of formal power in the organization are complex. To explain them, it is useful to consider three major systemic options: centralization, selective decentralization, and decentralization.

CENTRALIZATION

Centralization means that formal decision-making power, information processing, and control of performance is concentrated in the hands of top managers. The problem of organizational equilibrium is solved in three steps:

1. What constitutes technical equilibrium is decided upon by top managers and implemented.
2. Roles are specialized and formalized so that total balance is achieved. Of course, the achievement of balance depends on people following the demands imposed by the role design.
3. Internally oriented unit grouping is adopted. Balance within units is achieved through the standardization of functions, and coordination and control of performance are managed through hierarchical information flows.

Three conditions are required to effectively use such a centralized approach. First, information must be highly reliable. Second, decision-making processes must be effective. Finally, the organization must avoid unnecessary time lags in the problem-solving cycles.

Despite the limitations in efficiency of this approach, it has proven to be workable in both smaller and larger organizations. The minicase describes how ITT's CEO, Harold Geneen, effectively managed using this structure.

Centralization in ITT

International Telephone and Telegraph (ITT), under the rule of legendary president and CEO Harold Geneen, used a centralized approach to achieve equilibrium in a complex and diversified corporation by combining both social and technical solutions. Technical balance was achieved by establishing precise targets for the corporation as a whole and for each division. Once yearly profit indicators were established for each division, Geneen's constant, obsessive, and personal control ensured that the targets were met. Only Geneen's extraordinary abilities to process vast amounts of information and the tension and fear he engendered in all managers helped to attain goals and maintain technical balance of the complex and diversified corporation in spite of far-reaching centralization. The total dominance of technical equilibrium could create social disequilibrium; so social peace had to be somehow bought, as Sampson (1973, p. 132) explains: "Geneen provides his managers with enough incentives to make them tolerate the system. Salaries all the way through ITT are higher than average—Geneen reckons 10 percent higher—so that few people can leave without taking a drop. . . . At the very top, where demands are greatest, the salaries and stock options are sufficient to compensate for the rigors. As someone said, 'he's got them by their limousines.' Having bound his men to him with chains of gold, Geneen can induce the tension that drives the machine. . . . The tension goes through the company, inducing ambition, perhaps exhilaration, but always with some sense of fear: what happens if the target is missed?"

Source: Based on A. Sampson, *The Sovereign State of ITT* (New York: Stein and Day, 1973).

SELECTIVE DECENTRALIZATION

Designs that utilize **selective decentralization** delegate power for different types of decisions down the hierarchy to unit levels. This approach can occur both in internally and externally oriented structures. The rationale of selective decentralization is twofold. First, the complexity of total systemic equilibrium is reduced by focusing on the equilibria of units. Each unit manager is delegated responsibility for maintaining the local balance. Many American and European firms use such a form, creating market- or product-oriented divisions. Second, upper managers reserve the right to intercede in local-level events.

DECENTRALIZATION

With **decentralization,** most of the power lies at the bottom of the hierarchy, either because of the particular nature of the industry or as a consequence of a decision by top management. Decentralized designs accomplish technical equilibrium through the broad dispersion of roles at the level of the individual or small subunits. The maintenance of technical balance is problematic, however. This difficulty results from the lack of coordination between the variety occurring in local decisions, and the organization-wide actions intended to support local balances. In the long run, these incongruent forces will disrupt the equilibrium of the system as a whole if they are not reconciled.

KIEDEL'S TRIANGULAR DESIGN

Robert Kiedel (1990) argues that the organizational design process has historically been preoccupied with the issue of centralization versus decentralization at the expense of considering cooperation and teamwork. In much the same vein as Mintzberg, Kiedel proposes that there is an equilibrium that is central to all designs and that all design choices involve tradeoffs. Kiedel has reduced the centralization-decentralization issue to several fundamental geometric forms (see Figure 9-8):

1. *The point design.* In the ideological organization the decision regarding the issue of centralization or decentralization is really no issue at all. You do one or the other. It is a simple, mutually exclusive choice.
2. *The linear design.* In this design, the alternatives of centralization and decentralization exist on a continuum. Companies that choose this way of viewing the issue continually adjust their position on the continuum to match their current needs.
3. *The angular design.* This design is characterized by the interaction of centralization and decentralization. The two are not viewed as mutually exclusive. Rather, the issue focuses on deciding which combination of the two will be most effective. The two forces can be visualized as existing together in a 2×2 matrix.
4. *The triangular design.* This strategy views the organizing process as the result of three considerations: centralization (called *control*); decentralization (called *autonomy*), and teamwork (called *cooperation*). The **triangular design** is based on three fundamental premises. First, every organization and organizational unit must blend these three variables. Second, autonomy, control, and cooperation represent tradeoffs. Third, not all organizational blends are viable. The triangular strategy allows for three pure forms of organization and three hybrid forms: autonomy/control design, control/cooperation design, and autonomy/cooperation design. The six possible designs and their corresponding advantages and tradeoffs are shown in Figure 9-9.

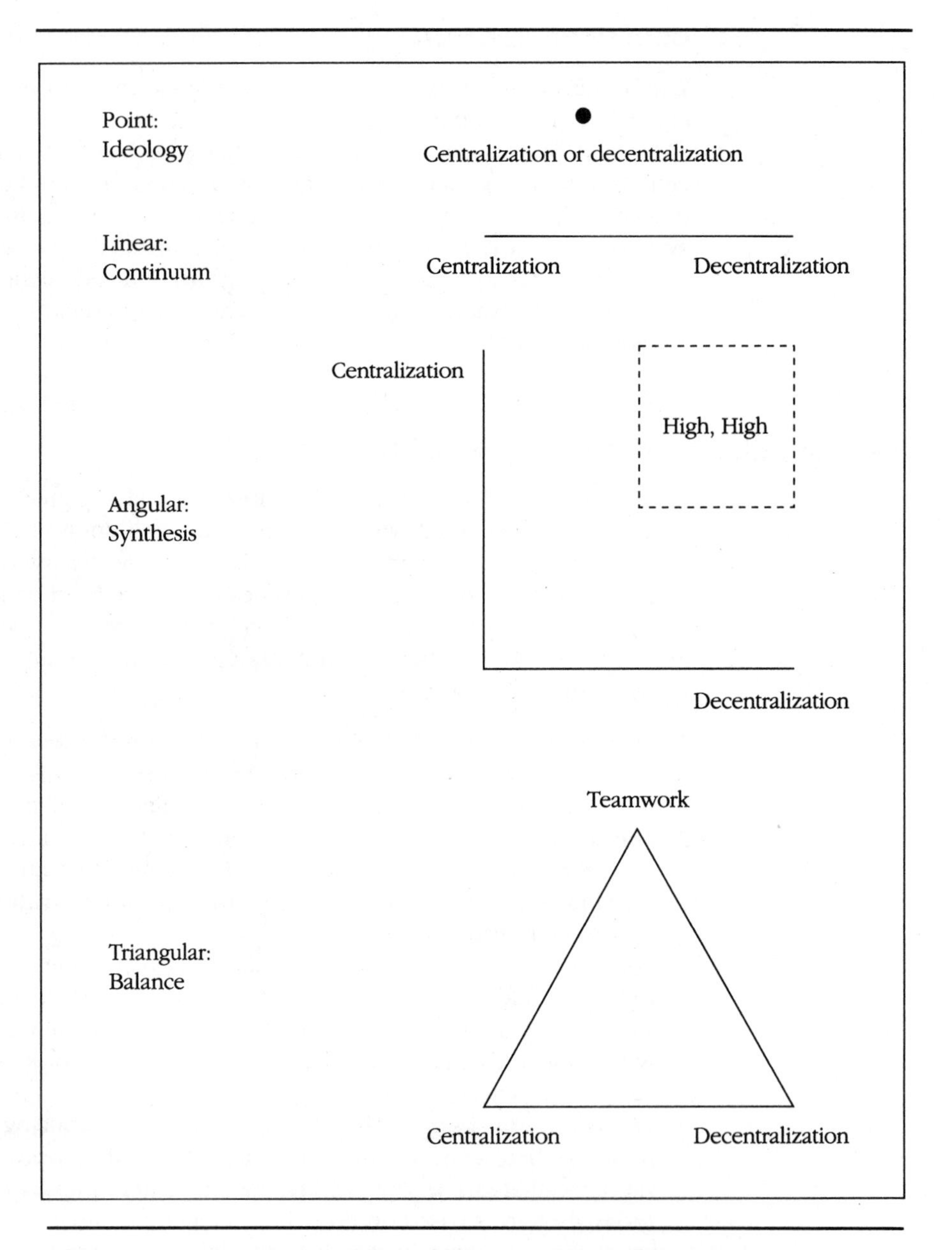

FIGURE 9-8 ***Geometries of Organizational Design***

Source: R. Kiedel, "Triangular Design: A New Organizational Geometry," ***The Executive*** (November 1990), p. 22.

TYPE OF DESIGN	MODE	ADVANTAGE	EXAMPLE
Autonomy	Decentralized	Solo performers	Cannondale bikes
Control	Centralized	Reduced costs	McDonald's
Cooperation	Teamwork	Innovation	Nucor Steel
Control/ cooperation	Humanistic hierarchy	Group interaction	Volvo
Autonomy/ control	Shared power	Accountability	Long's Drugs
Autonomy/ cooperation	Lateral relations	Cooperation	3 M

FIGURE 9-9 ***Six Organizational Forms According to Triangular Design***

As Mintzberg has recognized the pitfalls of an unbalanced system, Kiedel proposes that both pure forms and undifferentiated types are vulnerable to failure. He states, "A firm that tries to realize all three variables—autonomy, control, and cooperation—in equal measure [the undifferentiated form] will likely exhibit no organizational priorities. It will become stuck in the middle of the Organizational Design Triangle" (p. 28). The various types are depicted in Figure 9-10 (see p. 238).

SUMMARY

Structure is a managerial tool that has roots primarily in the hard systems approach. Its primary functions are to provide stability, differentiate task responsibilities, and coordinate the relationships between various subgroups. From a systems perspective, structure is the primary tool for maintaining the long-term equilibrium of a system. This equilibrium is constantly disrupted by strategies that seek to form new linkages with the environment. Structure is a more effective tool for preserving an organization's technical equilibrium than its social balance. Mintzberg's model elegantly captures how structure attempts to balance these competing forces. The next chapter explains how another managerial tool—procedures—is used to maintain an organization's technical equilibrium.

KEY TERMS AND CONCEPTS

technical system
structure
roles
concurrent forms of structure
task design
task specialization
formalization
technical equilibrium

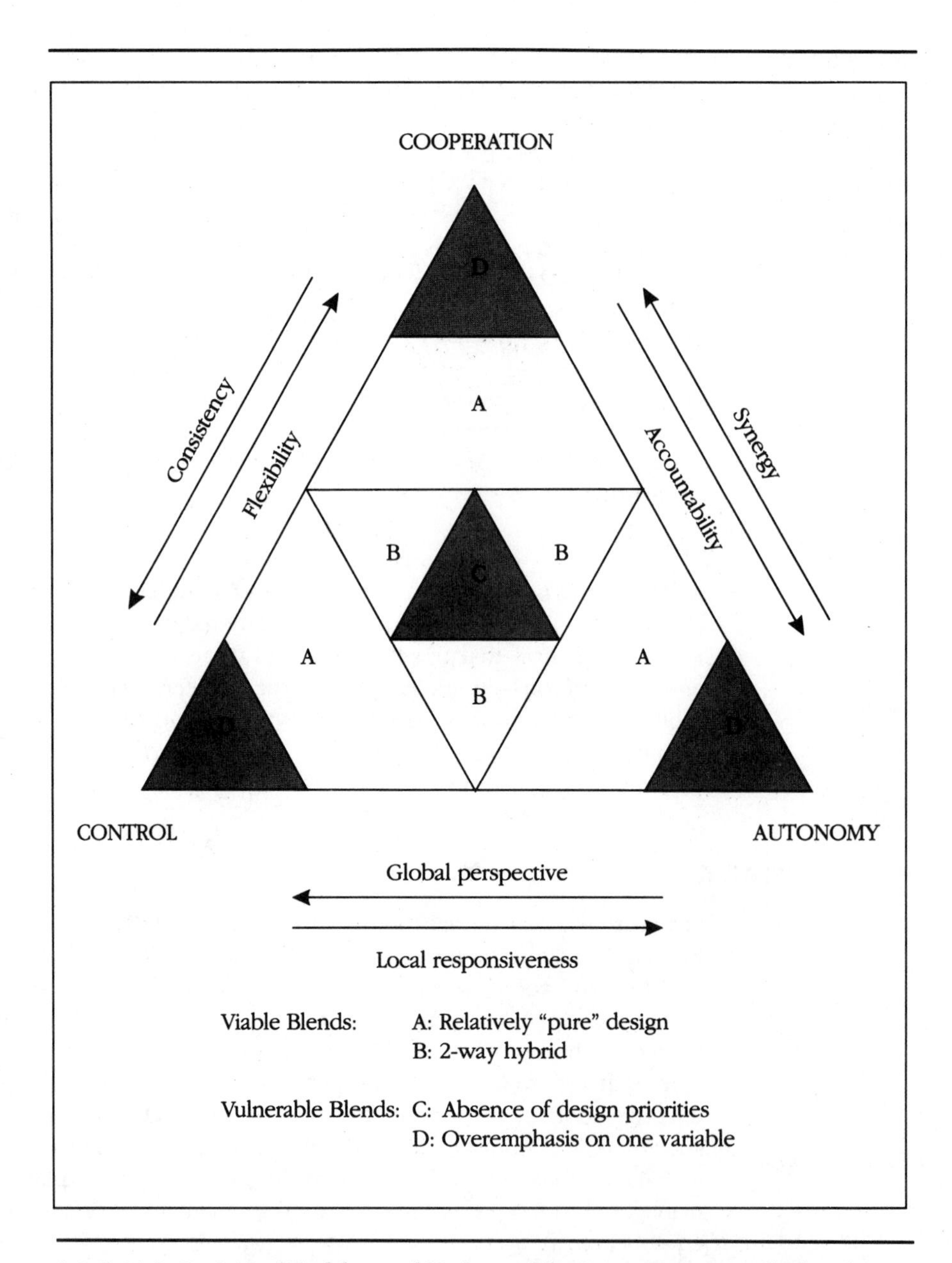

FIGURE 9-10 ***Viable and Vulnerable Organizational Blends***

Source: R. Kiedel, "Triangular Design: A New Organizational Geometry," *The Executive* (November 1990), p. 22.

social equilibrium
contingency perspective
differentiation
integration
cleavage
contamination
hybrid structure
centralization
selective decentralization
decentralization
triangular design

QUESTIONS FOR DISCUSSION

1. In the 1980s and 1990s many large organizations reduced the number of layers in their structures through a process known as *downsizing*. How will downsizing affect a large firm's ability to maintain its technical equilibrium?
2. What can a company do to ensure that its managers are able to differentiate the various segments of its environment?
3. You are the division manager of the property and casualty division of a large insurance company. The company developed a reputation over sixty years of business for mediocre performance, poor service, and lackluster quality. You have been assigned the task of transforming the company in light of an increasingly turbulent environment. Profits are down as a result of increased litigation, poor investments in real estate, and increased regulation by individual states. Which of Kiedel's six systems would you try to install? Why?
4. Which of Mintzberg's forces do you believe is the most important for managers to attend to, and why?

REFERENCES

Argyris, C. (1990). *Overcoming Organizational Defenses.* Needham Heights, Mass.: Allyn & Bacon.

Bertalanffy, L. (1950). "The Theory of Open-Systems in Physics and Biology," *Science, 3,* pp. 23–29.

Burns, T., and Stalker, G. M. (1961). *The Management of Innovation.* London: Tavistock Institute.

Emery, F. E. (ed.) (1969). *Systems Thinking.* Harmondsworth, England: Penguin.

Fry, L. (1978). "A Rationalization and Test of the Technological Imperative Through a Typology of Technology-Structure Relationships," unpublished doctoral thesis, Ohio State University, Columbus, Ohio.

Galbraith, J. R. (1973). *Designing Complex Organizations.* Reading, Mass.: Addison-Wesley.

Galbraith, J. (1977). *Organization Design.* Reading, Mass.: Addison-Wesley.

Greiner, L. (1983). "Senior Executives as Strategic Actors," *New Management, 1,* no. 2, p. 13.

Griffin, R. (1982). *Task Design: An Integrative Approach.* Glenview, Ill.: Scott, Foresman.

Hage, J., and Aiken, M. (1967). "Relationship of Centralization to Other Structural Properties," *Administrative Science Quarterly, 12,* no. 1, pp. 72–91.

Hall, R. H. (1982). *Organizations: Structure and Process,* 3rd ed. Englewood Cliffs, N.J.: Prentice-Hall.

Kast, F., and Rosenzweig, J. (1979). *Organization and Management: A Systems and Contingency Approach.* New York: McGraw-Hill.

Katz, D., and Kahn, R. (1978). *The Social Psychology of Organizations,* rev. ed. New York: John Wiley & Sons.

Kiedel, R. W. (1990). "Triangular Design: A New Organizational Geometry," *The Executive, 4,* no. 4, pp. 21–37.

Lawrence, P. R., and Lorsch, J. W. (1967). *Organization and Environment.* Homewood, Ill.: Richard D. Irwin.

Litterer, J. (1980). "Elements of Control in Organizations," in M. Jelinek, J. Litterer, and R. Miles, (eds.), *Organizations by Design.* Plano, Texas: Business Publications Inc., pp. 29–40.

Miller, J. G. (1978). *Living Systems.* New York: McGraw-Hill.

Mintzberg, H. (1979). *The Structuring of Organizations.* Englewood Cliffs, N.J.: Prentice-Hall.

Mintzberg, H. (1983). *Structure in Fives: Designing Effective Organizations.* Englewood Cliffs, N.J.: Prentice-Hall.

Mintzberg, H. (1991). "The Effective Organization: Forces and Forms," *Sloan Management Review,* Winter, pp. 54–67.

Morgan, G. (1986). *Images of Organization.* Newbury Park, Calif.: Sage Publications.

Obloj, K., and Joynt, P. (1986). "Strategic Reserve of the Firm," working paper, Norwegian School of Management.

Parsons, T. (1960). *Structure and Process in Modern Society.* New York: Free Press.

Perrow, C. (1967). "Framework for the Comparative Analysis of Organizations," *American Sociological Review, 32,* no. 2, pp. 194–208.

Ransom, S., Hinings, R., and Greenwood, R. (1980). "The Structuring of Organizational Structures," *Administrative Science Quarterly, 25,* no. 2, pp. 1–17.

Steers, R., Mowday, R., and Ungson, G. (1985). *Managing Effective Organizations.* Boston: PWS-KENT Publishing.

Thompson, J. (1967). *Organizations in Action.* New York: McGraw-Hill.

Weick, K. (1974). "Middle Range Theories of Social Systems," *Behavioral Science, 19,* pp. 357–67.

Weick, K. (1979). *The Social Psychology of Organizing,* 2nd ed. Reading, Mass.: Addison-Wesley.

~ ADDITIONAL READINGS

Ansoff, I. H. (1981). *Strategic Management.* London: Macmillan.

Chandler, A. D. (1962). *Strategy and Structure.* Cambridge, Mass.: The MIT Press.

Hofer, C., and Schendel, D. (1978). *Strategy Formulation.* St. Paul, Minn.: West.

Nadler, D. A., and Tushman, M. L. (1988). *Strategic Management Design.* Glenview, Ill.: Scott, Foresman.

Peters, T. (1984). "Strategy Follows Structure: Developing Distinctive Skills," *California Management Review, 26,* no. 3, pp. 111–25.

Peters, T., and Austin, N. (1985). *A Passion for Excellence.* New York: Random House.

Peters, T., and Waterman, R. (1982). *In Search of Excellence.* New York: Harper & Row.

Sampson, A. (1973). *The Sovereign State of ITT.* New York: Stein and Day.

10 Procedures as a Systemic Tool

The peak of Utopia is steep; the serpentine-road which leads up to it has many tortuous curves. While you are moving up the road you never face the peak, your direction is the tangent, leading nowhere. If a great mass of people are pushing forward along the serpentine they will, according to the fatal laws of inertia, push their leader off the road, and then follow him, the whole movement flying off at a tangent into nowhere.

Edward Koestler (1945)

1. To realize how procedures are used as a managerial tool to regulate the technical system of organizations.
2. To understand the role of procedures within the broader context of the management systems model.
3. To appreciate the function of the technical system of organizations.
4. To realize the relationship between technology and the technical system of organizations.
5. To visualize the relationship between technology and uncertainty in organizations.
6. To distinguish among compensating, buffering, and regulating procedures.
7. To recognize when it is appropriate to use procedures versus rules and algorithms to stabilize the technical system of organizations.

SECTION I
UNDERSTANDING PROCEDURES

THE ESSENCE OF PROCEDURES

This chapter explores the various ways that procedures are used as a systemic tool to regulate and stabilize the technical system of organizations. Since procedures are regulatory devices, they work in a complementary fashion with structure, which also regulates the technical system. The chapter will examine in detail how procedures control performance. There will also be an extended discussion of the core function of any technical subsystem, technology.

Fundamentally, **procedures** are routinized programs of action. They serve as useful tools for controlling basic repetitive operations, primarily in the technical and administrative systems of organizations. Koontz, O'Donnell, and Weinrich (1980) have noted that "Procedures are plans in that they establish a customary method of handling future activities. They are truly guides to action rather than to thinking, and they detail the exact manner in which a certain activity must be accomplished. Their essence is chronological sequence of required actions" (p. 615).

Organizations are productive systems. Because they generally transform a set of inputs into an output that has value to other people, they are called **input-transformation-output systems.** This simple model includes three features: (1) inputs, such as raw materials, are the focus of the transformation efforts; (2) operations and technology drive transformation processes; and (3) the final result, or output, is a value-added product or service. These transformative processes occur within the broader framework of an organization's **technical system.** The technical system is one of the two major systems in organizations; the social system is the other (Figure 10-1). The technical system includes structure, procedures, technology, and resources that are not human resources.

From the perspective of the management systems model, procedures are a hard systems tool used primarily for regulating the system rather than for increasing its generative capacities (see Figure 10-2). Generative systemic tools such as strategy, leadership, and culture extend and balance relations between the environment and the organization. They are open to the environment and tend to drive the creative functions that are necessary for growth.

Regulatory or stabilizing systemic tools are internally oriented, closed to the environment, and directed toward reducing uncertainty and increasing stability within the organization. Procedures are most directly aligned with the technical system and function primarily as controls. Procedures reinforce existing patterns within the system. As a consequence, they buffer the system from the effects of change, and they are incapable of initiating transformations. Procedures work synergistically with structure to limit the range of freedom of decision makers. These tools operate together especially well in such areas as finance, information systems, and manufacturing.

PROCEDURES AND CONTROL

All systems have a need to regulate their own activity to remain guided toward achieving their purposes. In simple systems, such as a thermostat, the ability to self-regulate is well ensured because the outputs also become the inputs. (The output is heat and the input is the level of heat.) When the room temperature increases beyond a set limit, the thermostat automatically turns off the heat source, in effect lowering the temperature until the heater needs to be switched on again. In more complex systems, such as organizations or cities, it is more difficult to regulate all aspects of the system because the systems are less stable and the variables more complex. For example, in large cities transportation systems, social service programs, or violence may be out of control. In some large cities, such as New York City, postal service may be

TECHNICAL SYSTEM	SOCIAL SYSTEM
Technology	Culture
Procedures	Leadership
Structure	Strategy
Resources	

FIGURE 10-1 ***The Technical and Social Systems of an Organization***

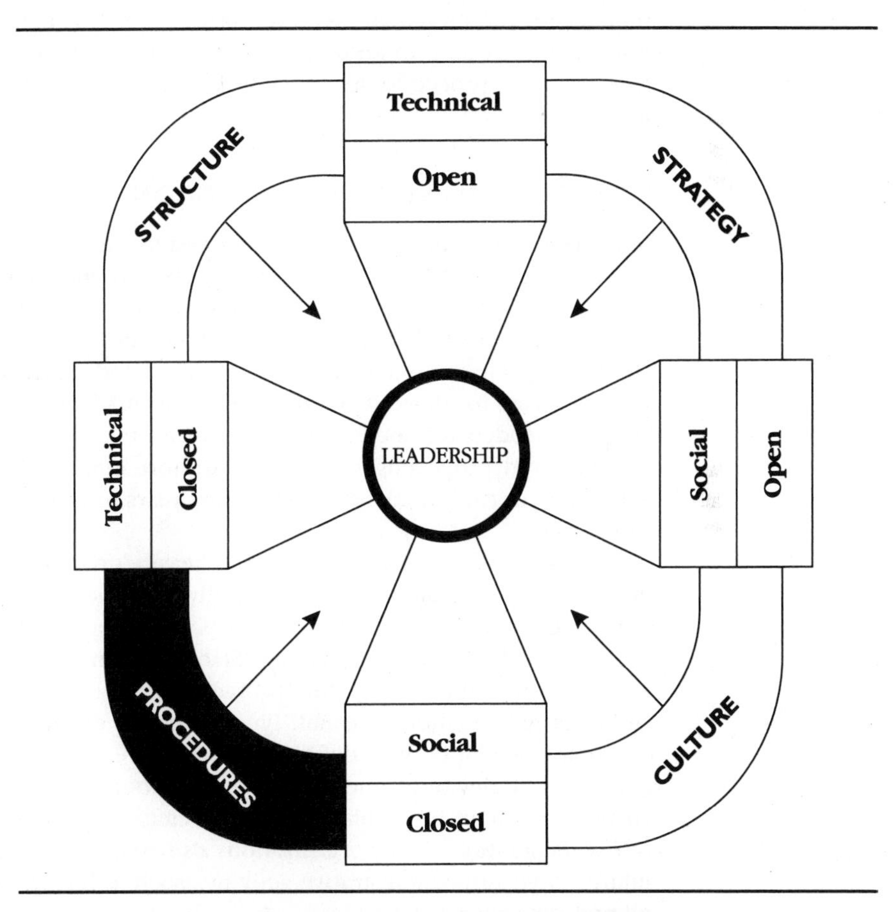

FIGURE 10-2 ***The Management Systems Model***

delayed because of the difficulty in transporting mail through the city streets, what with traffic jams.

When a system cannot effectively regulate itself, it also has less capacity to achieve its purposes. Systems with control problems suffer inconsistency in the quality of products and services. Customers of McDonald's restaurants or Disney's amusement parks expect a certain level of service. This expectation arose—and continues to be met—because the inputs and transformative processes in the system are designed to produce standardized products and services, regardless of the location. McDonald's restaurants effectively integrate routine technology and precise, well-documented procedures to achieve this standardization. One former employee recalls, "They call[ed] us the Green Machine 'cause the crew had green uniforms then. And that's what it is, a machine. You don't have to know how to cook, you don't have to know how to think. There's a procedure for everything and you just follow the procedures" (in Garson, 1988).

PROCEDURES MUST FIT THE SITUATION

Organizations produce expected positive results, such as high-quality products and valuable services. Unfortunately, they also produce unexpected results—both negative effects such as pollution, uninspired employees, and frustrated customers and positive results such as high morale or useful norms and values. These unexpected outputs must be recognized and analyzed before programs can be initiated to adjust the inputs and transformation processes that produce these unintended results. For example, many business firms have begun to redesign word processing stations to be more comfortable and healthy for users. Some organizations have taken many years to recognize the importance of this issue.

To regulate themselves, systems rely heavily on cybernetic negative feedback controls. This approach is especially valuable when the system is dynamic because in dynamic situations it is much more difficult to predict exactly what will occur in the future. Stable systems may be able to reduce uncertainty by anticipating future problems and then preparing to respond to the forecasted situation or event. Procedures are created on the assumption that situations will be well defined and repetitive and that a programmed response can be devised to help bring the situation under control. Procedures can be very useful for stabilizing specific elements of organizations, such as technical operations. Many organizations also establish standard operating procedures to govern administrative activities such as hiring, billing, purchasing, and performance evaluation.

GLOBAL SYSTEMS SPOTLIGHT

UPS: The Tightest Ship in the Shipping Business

United Parcel Service (UPS) formed as a messenger service in 1907 and has grown to become the world's largest package delivery company. UPS's numbers are impressive; over 65,000 delivery drivers, over 11.5 million parcels delivered per day worldwide, and over $13 billion in revenues. Today UPS delivers air freight around the world to Japan, Eastern Europe, the United Kingdom, Germany, Canada, and Puerto Rico. In October 1990 UPS formed a joint venture with the Yamoto Transport Company of Japan. UPS Yamoto forecasts 25 percent annual growth through 1995, and sales volume of ¥50 billion. UPS has also negotiated a joint venture with Sovtransavto, the largest transport company in the USSR.

"UPS maintains rigid control over nearly every aspect of its operations. Each task from picking up or delivering parcels on a route to sorting packages in a central hub is carefully calibrated according to productivity standards. Workers know precisely what is expected, and deviations are tolerated only rarely. The combination of employee ownership and tight operating controls has worked so well that change has been slow coming to UPS. . . . But UPS is now in the throes of major change, forced into the high-tech era by competitors who have been making effective use of gadgetry to woo new business. High-flying Federal Express has cornered 57% of the rapidly expanding air-express market by picking up packages the same day an order is placed, knowing precisely where the package is en route, and then delivering it to its destination the next morning. Customers have come to expect such prompt service, and UPS, a distant second to Federal Express with about 15% of the air-express business, is now moving ahead to add to its air fleet and to match technically advanced rivals with such devices as electronic scanners in sorting centers and on-board computers in its delivery vans" (*Fortune*, p. 56).

By the end of 1991 UPS had moved in to compete in the global arena by investing over $5 billion in its fleet of 125 large cargo-carrying jets such as Boeing 747s and 757s and DC-8s. In 1991 it also announced that it would outfit all 65,000 drivers with hand-held computers.The 11-by-14-inch delivery information acquisition device (DIAD) has a scanner, keypad, and screen and will capture customer signatures. The hand-held computers will help UPS track packages. The combination of tight operational controls, employee incentives, and excellent service have made UPS the dominant competitor in moving into full global competition in the mid-1990s. "After winning a superb reputation as conservative team players, the owner-managers at United Parcel Service are turning aggressive. Watch out, Federal Express!" (*Fortune*, p. 56).

Source: Based on "Big Changes at Big Brown." *Fortune* (January 18, 1988), p. 56; "UPS and Yamoto Forge International Cargo Inroads," *Tokyo Business Today* (May 1991), p. 35; "Cargo: Solid Footing," *Air Transport World* (September 1991), pp. 66–72; and "Hand-Held Computers Will Help UPS Track Packages," *Network World* (May 6, 1991), pp. 15–16.

THE TECHNICAL CORE

In some organizations the technical system is dispersed throughout the system. For example, in a hospital the transformation of inputs takes place in surgical operating rooms, the physical therapy area, the psychological treatment center, and the office of nutrition. In other organizations, especially manufacturers, the most critical transformations occur at a single location, often referred to as the **technical core.** The technical core of a brewery is its brewing vat, and of a restaurant the kitchen, and at a telephone company, the switchboard (whether human-operated or computerized).

▶ SECTION 2
TECHNOLOGY

~ THE STUDY OF TECHNOLOGY

Technology is generally considered as the means used to transform inputs into outputs. The narrowest definitions of technology focus on the mechanical and workflow dimensions. Broader views, such as Charles Perrow's (1970), propose that technology is whatever is used to make transformations, including knowledge. Most research on technology has focused on differentiating among types of technologies. In general, differentiations are made on the basis of the complexity, workflow variability, and interdependence of the tasks involved in making the transformations (Fry, 1978).

The contingency theorists understood the practical ramifications of changes in demand for products, both in terms of demand levels and innovations. The research in this area centered on the relationship between technology and two major factors: structure and uncertainty. The early work by Woodward (1965) set the foundation for future work by attempting to distinguish among different classes or types of technology: unit and small batch, large batch, mass production, and continuous processes. Woodward suggested that effective performance in the organizations she studied could be attributed to alignments between various structural features of the organizations and the type of technology. Although the validity of these conclusions has been frequently questioned, the effect of the work was to mobilize interest in understanding technology.

The Aston studies (Hickson, Pugh, and Pheysey, 1969), conducted by a group of scholars from the University of Aston, England, sought to determine the effects of uncertainty generated by workflow. They wanted to understand how various degrees of linkage or interdependence among tasks generated uncertainty. The Aston group's focus was on determining the extent to which operating variability (the production of standard versus nonstandard outputs) was related to autonomy, centralization, and span of control (Fry, 1978). Their findings diminished the role of technology as a factor in organizational design, but raised new interest in the relationship between organization size and structure.

Interest in the role of work design in creating interdependence and ultimately uncertainty was reinforced by the seminal writing of James D. Thompson. The concern for establishing classifications of various types of technology was continued in Thompson's classic work, *Organizations in Action* (1967). Thompson generalized the concept of interdependence to a framework that could be applied to specific operations or used to characterize the overall functioning of entire organizations. Thompson differentiated among three basic forms of technology: long-linked, mediating, and intensive. He proposed that each form of technology would create its own unique type of interdependence.

The search for a link among technological uncertainty, structures, and procedures found its broadest application in the work of Charles Perrow. Perrow (1970) viewed technology in nonmechanical terms, from the perspective of knowledge used to transform raw materials into a newer form. Perrow defined technology in terms of the human contributions of thought and action, whereas his predecessors emphasized the mechanized aspects of work. Since the focus of technology had now shifted toward the human perspective, the factors being studied reflected that bias. Perrow maintained the interest in uncertainty, but focused on the interaction between variability in workflow and the human actions taken to accomplish a task. **Workflow variability** was defined in terms of the frequency that exceptions to the norm are encountered by workers in the pursuit of completing their task. The more variability in the flow of work, the less predictable work becomes.

When unexpected tasks arise, the likelihood increases that such tasks may not have a readily available solution. Under these circumstances, a search for a solution must commence. Perrow (1970) proposed that the type of technology is partly determined by the nature of the search process for solutions. The search for a solution to problems created by nonstandard or exceptional circumstances is likely to be relatively complex. The extent to which the search process is clear and structured is known as **problem analyzability**. The interaction of these two factors—task variability and the nature of the search process for solutions—can be visualized by constructing a typology of technologies based on the amount of uncertainty they generate (see Figure 10-3).

Fry (1978) found that Perrow's technological model was "significantly related to dimensions of organization structure, especially at the workflow level" (p. 194). The interest in exploring the relations among technology, uncertainty, and structure have framed the way managers viewed the task of organizational design over a quarter century. Interest in managing the effects of uncertainty has shifted more recently toward creating new forms of technology, such as flexible manufacturing cells, robotization, and just-in-time inventory programs. The emphasis has moved toward revising the way that technology is used as a strategic weapon. As a result, there is a greater emphasis on creating structures that allow rapid changes in manufacturing strategies and promoting high-speed practices of quick entry into and exit from markets. Consequently, greater importance is placed on developing entirely new prod-

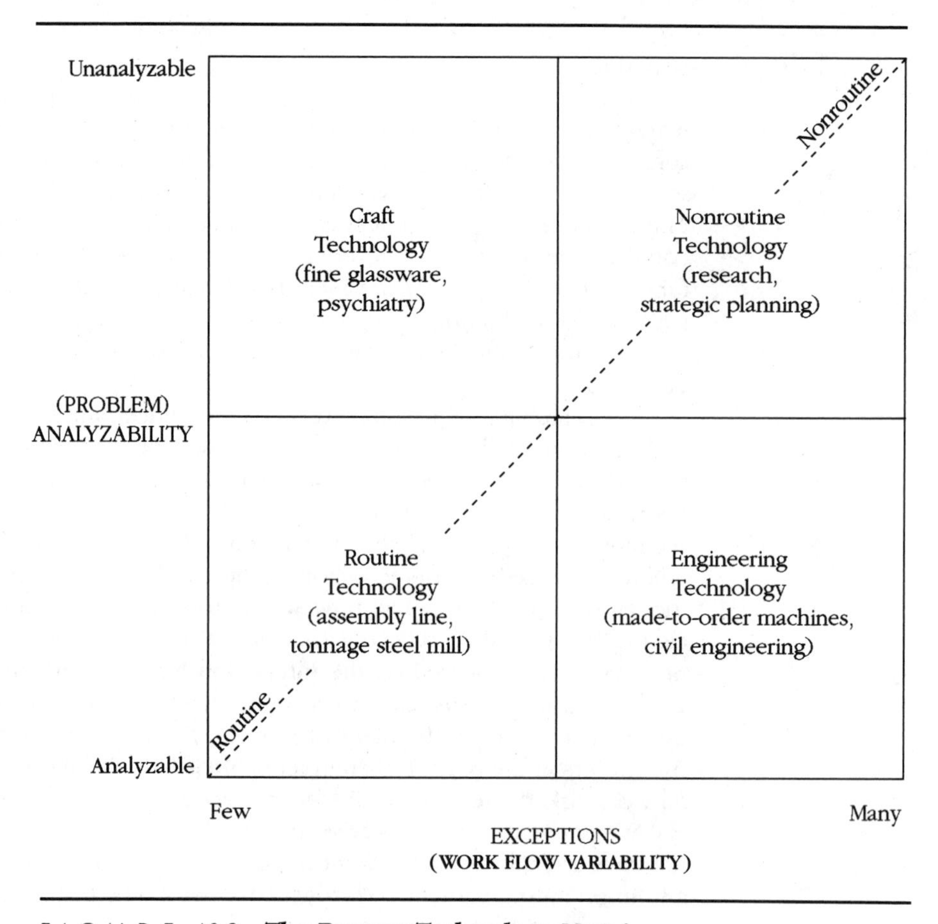

FIGURE 10-3 ***The Perrow Technology Matrix***

Source: C. Perrow, *Organizational Analysis* (Monterey, Calif.: Brooks-Cole, 1970), p. 78.

ucts and markets to stimulate continuous growth for companies that take advantage of the multiplicity of opportunities present in the global economy.

AN INTEGRATIVE PERSPECTIVE ON TECHNOLOGY

All of the perspectives previously discussed are incorporated into the model of technology developed by Randolph (1981). Randolph's model integrates the three major dimensions of technology together (see Figure 10-4). The first two dimensions—problem analyzability and task predictability—can be readily understood. Task interdependency relates to the extent of the coupling between the various subsystems in an organization.

Thompson (1967) studied technology in relation to how varying types of interdependence influenced organizations. He identified three basic types of

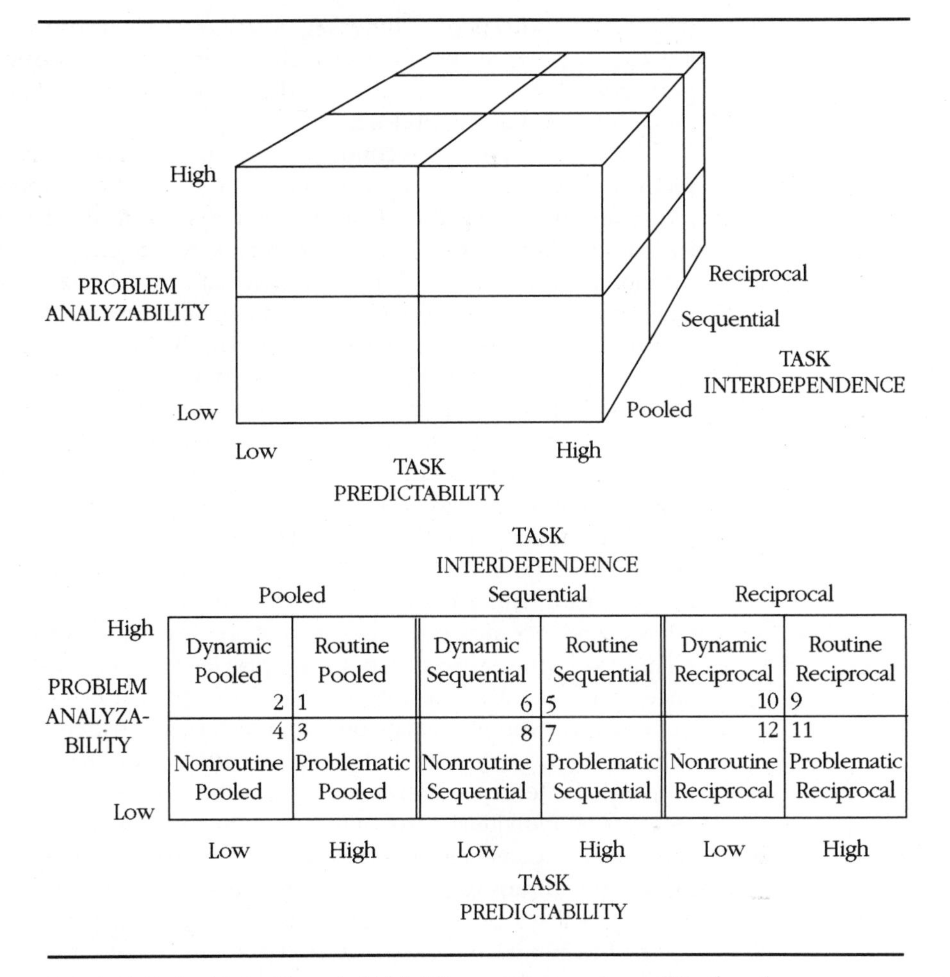

FIGURE 10-4 ***Randolph's Three-Dimensional Technology Typology***

Source: A. Randolph, "Matching Technology and the Design of Organization Units," *California Management Review* (1981), *23,* no. 4, p. 41.

technology that could be differentiated on the basis of the type of interdependence present. The three types of technology are:

1. Long-linked
2. Mediating
3. Intensive

Long-linked technologies structure tasks in a sequential manner where each step in the process is dependent on the one prior to it. Typically such technologies are assembly line-like, yet in some office environments this may not be obvious. For example, processing an insurance claim follows such a pattern of sequential tasks. More often this type of technology would be like that seen in a soda pop bottling facility or a traditional auto assembly line. The interdependence in this type of technology is clearly **sequential interdependence** and is created by standardized operating procedures and the work design that has been established.

The second type of technology that Thompson identified is **mediating technology.** In this type of technology the inputs and outputs of the organization are joined together. Usually this takes the form of bringing together various clients who wish to be interdependent. The classic types of mediating technologies can be found in the activity of a telephone utility, a postal service, an employment service, and even a dating service. The type of relationship that exists here is known as **pooled interdependence.** Here the clients are brought together, buyers and sellers, users and providers, and the organization provides a brokerage-like function.

Finally, the remaining type of technology that Thompson identified is known as **intensive technology.** Organizations that are characterized by intensive technology have a vast repertory of resources and abilities but the specific combination of them that is actually used depends on the needs of the client. This type of technology can be found in a medical center, a construction company, and a fire department. In this type of technology the type of interdependence that exists is known as **reciprocal interdependence.** It is reciprocal in the sense that both the provider and client continuously influence each other. For example, the needs of a patient in a hospital will dictate the treatment the physicians offer and in turn the patient's response will depend on the efficacy of the treatment that has been provided.

Pooled interdependence exists when the functioning of the whole system depends on individual parts. When coupling is sequential, the outputs of one subsystem become the inputs of another. When coupling is reciprocal, the inputs and outputs of each subsystem also serve as the inputs and outputs for the other subsystems, forming a series of feedback loops.

In this model technology can create uncertainty, thus destabilizing the system, in three basic forms:

1. When the patterns of causality within a system are unknown, the type of task interdependence will be unclear.
2. When the dynamic patterns of workflow are unknown, it will be difficult to identify task variability.
3. When there is a gap between knowledge of a problem and the information needed to solve the problem, the solution cannot be devised.

TECHNOLOGY AND UNCERTAINTY

Each technology type creates its own unique blend of uncertainty and risk of system failure in each subsystem. The environment created by technology type is suited to the use of certain managerial tools. In general, highly certain environments are routine and recurrent and thus are compatible with the use of procedures. On the other hand, procedures are not always effective control devices in uncertain environments. In those situations, cybernetic controls, loose structures, and heuristics are more appropriate. **Heuristics,** you will recall, are rules of thumb that will lead to exploration and discovery of new ideas to reduce the level of uncertainty in a situation. For example, work design problems may be addressed by a heuristic that says, "continually optimize the relationship between human factors and technical factors."

March and Simon (1958) have related various conditions of uncertainty to types of responses. Under routine conditions, performance programs can be designed and activated; they simply resolve that specific situation. Procedures are the basis of such performance programs. On the other hand, highly uncertain situations require problem-solving responses, which emphasize searching for alternative solutions.

A word processing center in a financial services company is an example of a routine type of technology. There will be relatively few unexpected variations in the task, and the exceptions that occur usually have well-defined solutions. As a consequence, managers will be able to design procedures that simply and efficiently reduce the margin of error in the unit.

Procedures can be very effective if matched to the level of uncertainty and carefully implemented. For example, in a dynamic environment, such as the air freight industry, inventory levels can be effectively managed by using economic order quantity (EOQ) models and just-in-time inventory plans. According to Jack Grayson of the American Productivity Center, "The true philosophy behind JIT is continuous improvement of everything, and the elimination of all waste of time, people, effort, and materials" (in Sugarman, 1989, p. 16). These systems are largely controlled by procedures. The minicase discusses the approach taken by a freight company, Skyway Freight Systems.

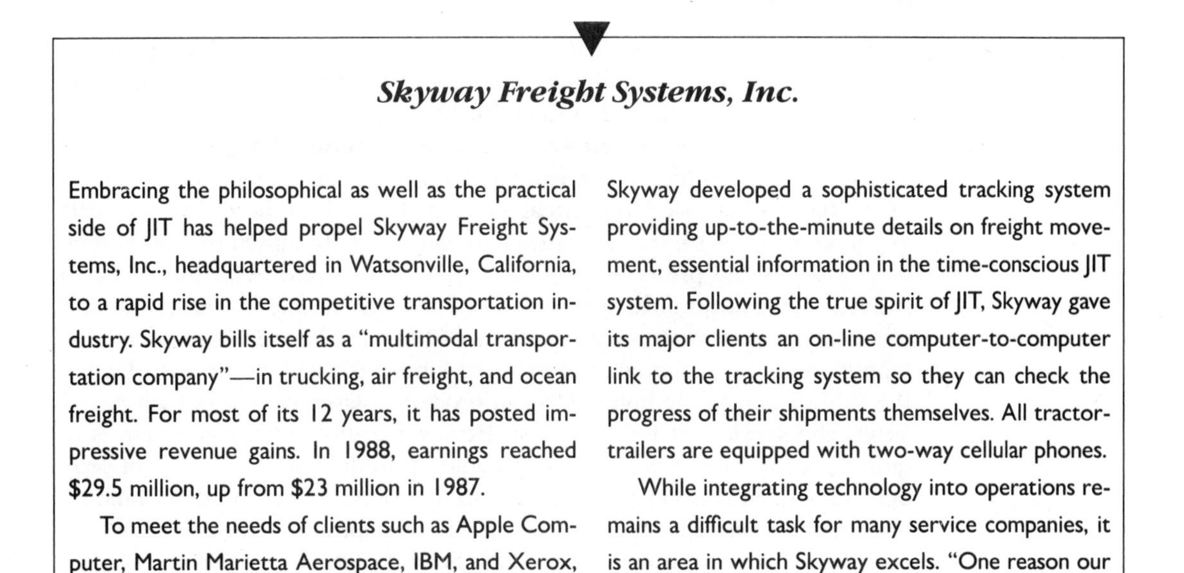

Skyway Freight Systems, Inc.

Embracing the philosophical as well as the practical side of JIT has helped propel Skyway Freight Systems, Inc., headquartered in Watsonville, California, to a rapid rise in the competitive transportation industry. Skyway bills itself as a "multimodal transportation company"—in trucking, air freight, and ocean freight. For most of its 12 years, it has posted impressive revenue gains. In 1988, earnings reached $29.5 million, up from $23 million in 1987.

To meet the needs of clients such as Apple Computer, Martin Marietta Aerospace, IBM, and Xerox, Skyway developed a sophisticated tracking system providing up-to-the-minute details on freight movement, essential information in the time-conscious JIT system. Following the true spirit of JIT, Skyway gave its major clients an on-line computer-to-computer link to the tracking system so they can check the progress of their shipments themselves. All tractor-trailers are equipped with two-way cellular phones.

While integrating technology into operations remains a difficult task for many service companies, it is an area in which Skyway excels. "One reason our revenue-per-employee is so high is the productivity gains from 11 years of fine-tuning our computer systems," says Jim Watson, Skyway president and co-founder along with chief executive officer Robert Baker. "We constantly ask how we can add more revenue but not necessarily equate it with an equal percentage rise in employees," Watson says.

Source: A. Sugarmann, "Skyway Freight Embraces JIT Philosophies to Propel Forward," *Management Review* (1989), p. 16.

Some systems have relatively lower levels of stability, primarily because of the effects of human factors on decision making. There are many highly publicized examples of technologies that have become destabilized and out of control—the Exxon *Valdez,* the Chernobyl nuclear power plant, and the Union Carbide plant at Bhopal are but three. Perrow (1984) details a number of these situations, which he calls "normal accidents." Normally, lack of system balance and self-regulation do not have such dramatic consequences, yet these cases clearly demonstrate that all systems have normal variations in their ability to regulate themselves. The effective implementation of procedures plays an essential role in maintaining the balance of such systems.

SECTION 3
PROCEDURES AS A MANAGERIAL TOOL

REGULATING THE OPENNESS OF THE TECHNICAL SYSTEM

Technical systems may be characterized by the extent to which they interact with their environment, including that part of the environment that is within the organization. Technical systems that permit relatively little interaction with the environment are **closed technical systems.** When there is a relatively high level of interdependency between the subsystem and its environment, there is an **open technical system.** From the time of the Industrial Revolution to the 1980s companies have generally attempted to seal off technical systems and preserve their basically closed nature. With the advent of **flexible programmable systems** (FPS) and just-in-time inventory systems, this goal has become less feasible. Starr (1989) describes FPS as systems that "apply computer technology to drive all kinds of machinery in the office, in the factory, and in the home. FPS can produce a large variety of high-quality goods and services at low unit cost. However, a substantial investment is required" (p. 167). FPS are systems capable of producing many types of products; they must continually interact with their environment to determine the requirements and specifications for the newest product. Some examples of firms that have successfully applied FPS (Schonberger, 1986) follow:

- Vought Corporation has invested $10 million in a flexible manufacturing system that is capable of producing "531 different parts, usually in batches of one." In this system parts are not made in advance for inventory; they are made one at a time when needed.
- General Electric Corporation built a $16 million flexible manufacturing system in Erie, Pennsylvania, for manufacturing railroad locomotives. The system has cut lead time for production from sixteen days to sixteen hours.

In closed technical systems the transformation process is relatively well isolated from environmental disturbances and under full control of managers because they decide on its configuration, segmentation, scheduling, costs, and so on. Such a system is cyclical: the same inputs are transformed with the same set of technological operations into the same outputs. Even if the transformative process is somewhat flexible and allows for online changes of the parameters of the output (for example, an auto assembly line that allows variations in type of engine and gearbox), the logic of routine operations does not change. The flexibility just consists of diversity built into the technology through additional routines. Potential disruptions of operations relating to inputs, outputs, and the transformative processes are also mostly predictable, well structured, and cyclical. All these features enable the system to maintain its balance primarily through application of routine programs consisting usually of a series of logical steps. For example, someone who posts checks for an investment firm might follow the set of procedures described in Figure 10-5.

The level of routinization and stability of operational procedures depends primarily on the stability of the transformative processes (technology) and the extent to which the system is open to environmental influences. The more stable the transformative process, the more appropriate are standardized procedures. The explanation of such a relationship takes into account three elements. First, procedures ensure the continuity of materials, finances, information, human inputs, and acquisition of resources and the regularity of

---

Step 1. Make sure that the check is signed.

Step 2. Make sure that the check is dated properly (neither postdated nor stale dated).

Step 3. Make sure that the written and the numeric amounts on the check correspond.

Step 4. Make sure that the check is not drawn on a foreign bank.

Step 5. Make sure that the check is made payable to the company or is duly endorsed.

Step 6. If the check meets all these criteria, post it.

Step 7. If the check does not meet one or more of the above criteria, put it into a separate file marked "Rejected Items."

Step 8. At the end of each day, send the file "Rejected Items" to Unit Z (second floor, room 202), which handles such checks.

FIGURE 10-5 ***Procedure for Posting a Check***

operations. Second, standardization minimizes the costs of operations. Instead of developing customized solutions and interventions, procedures are simple and inexpensive because they require workers to expend less time, attention, and sophisticated skills. Finally, stability of transformative processes allows for the accumulation of knowledge, through experience, about potentially disruptive events. As these events occur over time, they are classified into separate categories (for example, machine breakdown, customer complaints, budget deficits, energy shortages) and new procedures are engineered to deal with them. Each time such an event occurs, the appropriate standardized procedure is triggered to respond. The costs of intervention are very low because the procedures state clearly every detail of a complete program of action in order to solve the problem.

PROCEDURES AS A STABILIZING TOOL

Operational procedures not only aim at maintaining the internal technical equilibrium of the organizational system, but also contribute to a sense of social stability. For example, when there is a disciplinary problem with an employee most companies have a number of procedures that must be followed, including providing a verbal warning to the employee, a written warning, a suspension, and ultimately termination or transfer. Routine, stable procedures make technical and economic sense primarily if they address events within the managerial control of the organization. Control through procedure is particularly used in situations involving routine technical systems with high risks. For example, nuclear power technology is relatively stable, but the consequences of a system failure could be severe. Thus nuclear power plants must develop strictly enforced procedures to reduce the odds of ever experiencing a system failure. These steps consist of redundant controls, precise procedures, and extensive training programs for plant operators.

PROCEDURES AND THE EFFECTS OF THE ENVIRONMENT

Procedures can also be used to control the boundaries of technical systems in order to increase managerial control over events that may disrupt equilibrium in that system. For example, inclement weather may cause air-traffic controllers to implement conservative procedures for landing or to reroute air traffic to another airport that is not experiencing the same problem. Managers play a critical role in determining the extent to which they will allow environmental factors to influence the technical system. They may selectively regulate and allocate the flow of factors—materials, energy, people, information, and money—imported from the environment. In stable systems these actions can all be implemented and controlled through the use of procedures. External forces such as sales demand or regulatory issues can also be dealt with through the application of procedures to particular situations.

Environmental changes can create threats or opportunities that may call for a routinized response. The manager's role is to determine the extent to which the system will respond to the change by altering or creating procedures. Managers judge when to open the system to external forces and when to reduce their impact. They can simply perform this role by indicating which procedures should be implemented at any given time.

When a self-regulated balance is achieved with both the external environment and the other systems in the organization, the necessary flows of information, money, and products or services will take place to preserve the balance of the larger system. For example, payment by consumers to producers allows producers to pay accounts payable to suppliers. Resources are then available for the suppliers to make transformations, and so on. Porter (1980) views the productive system as being linked to both suppliers and customers. He points out that customers, producers, and suppliers all exist in a chain of interdependence. Producers are suppliers for customers and customers for suppliers. For this system to remain in balance, several practical criteria must be met:

1. Timeliness and predictability. The inputs, resources, and outputs must be available at the planned time. If the time is not prearranged through agreement, it must be predictable.
2. Quantity. The inputs, resources, and outputs must be available in both sufficient and precise quantities to fulfill potential user needs.
3. Specifications and standards. The inputs, resources, and outputs must meet all necessary specifications such as size and shape to fulfill potential user needs.
4. Function. The inputs, resources, and outputs must meet all performance criteria in terms of durability, strength, flexibility, and so on, to fulfill potential user needs.

USING PROCEDURES IN DYNAMIC ENVIRONMENTS

In many environments constant disruptions of internal equilibrium are a norm rather than an exception. This turbulence may be sufficiently unique to reduce the effectiveness of procedures as tools for regulation and control. In agriculture and recreation, for instance, weather may throw the industry into turmoil. The power and petroleum industries are heavily influenced by the political and economic stability of the oil-producing nations. The airline industry is experiencing the tumult associated with long-term changes in its structure. As the case shows, the turbulence caused by many mergers and acquisitions was sufficiently complex to make many procedures inoperative or ineffective.

Northwest Airlines

On October 1, 1986, Northwest Airlines acquired its competitor Republic Airlines, Inc. The merger resulted in numerous customer service problems and employee difficulties. During the period when 331 aircraft were redeployed and new schedules for over 1,600 daily flights were created, traffic exceeded forecasts and resulted in overwhelmed computer systems. The effect was to create general chaos in reservations and seating plans. The problems in coordinating operating procedures caused losses of baggage and common flight delays. As these operational problems continued, airline travelers began to turn to competitors for more reliable service.

Although environmental changes may make procedures difficult to implement, procedures can nevertheless be used to reduce the uncertainty associated with such change. Most systems are subject to unanticipated disruptions at one time or another. Such disruptions provide justification for the use of three specialized types of procedures to counterbalance the destabilizing effects: buffering, compensating, and regulating procedures.

BUFFERING PROCEDURES

Buffering procedures shield the organization from potential disturbances. Organizations may try to avoid shocks resulting from delays or improper deliveries by stockpiling inventories of essential inputs. Hospitals, for example, maintain redundant supplies of oxygen, energy (backup power generators), and drugs and instruments vital to operations and continuous care of sick. Airlines keep stocks of spare parts and human resources to ensure that planes can continue to fly.

In the same vein some organizations protect their transformative process itself. Preventive maintenance procedures (for example, specifying the exact timing for the overhaul of aircraft engines) protect organizations from machine shutdowns and breakdowns. Quality control procedures ensure the minimization of costs due to the need to handle defective parts. Training of personnel might also be standardized (in terms of timing, frequency, and content) to ensure constant upgrading of employee skills otherwise left to chance.

Finally, organizations shield their output side. The most typical example is an inventory of finished products, enabling organizations both to smooth over fluctuations of demand and ensure continuity of deliveries in the event of disturbances in transformative processes.

Buffering procedures stabilize organizations and help maintain internal equilibrium. They prevent destabilization even in case of external disturbances. However, they are costly. Any stockpiling of materials, inventories, or preventive programs add to the general costs of operations. Therefore, techniques such as OR and systems analysis are used to analyze the optimal point between the cost of these procedures and the cost of their absence.

COMPENSATING PROCEDURES

Compensating procedures are oriented toward eliminating disturbances or their negative effects on the system's balance. Compensating procedures are triggered primarily after the occurrence of a disturbance at the input side, during transformation, or at the output side.

Examples of a disturbance at the input side are the delivery of materials that do not meet quality standards, or an excess of patients waiting for treatment in the hospital's emergency room. Organizations should be prepared for such contingencies and react. Otherwise the disturbance might have dire effects, building up disruptions within the organization and in its environment and creating a snowball effect. For instance, low-quality materials introduced into the transformative process will result in higher costs, changes in operations, and low quality of output. A long line of patients badly needing immediate assistance might result in unnecessary deaths or health complications and legal suits against hospital. Compensating procedures could include rejection of inputs or preventive changes in technology, getting more physicians to the emergency room or transporting patients to another hospital.

Disturbances might also occur during the transformative process. The introduction of new technology might demand rapid training and upgrading of skills of the work force. A surgeon might discover that the case at hand is more complicated than previous diagnosis suggested. Lectures might spur new interests in students that go beyond subjects the teacher wanted to cover. These and similar problems arise often during the course of repeated transformations. As the knowledge about their frequency and content is accumulated, standard answers are developed and implemented.

Compensating procedures can also be applied on the output side. Organizations compensate for lower quality of products or surplus of goods in the marketplace with lower prices or special warranties. The reverse is true as well—shortages of brand goods of exceptional quality will result in increased prices.

Compensating procedures should result in the stability of an organization, just as buffering procedures do. But while the final result might be the same, the means are different. Buffering procedures shield an organization from disturbances or their effects; compensating procedures are primarily used when a disturbance actually occurs. Then that disturbance is classified and a proper response delivered.

REGULATING PROCEDURES

Regulating procedures help to maintain system balance in spite of disturbances and problems. They are the most complex, multifunction procedures, and often are built of a series of buffering, compensating, and additional routines. Production planning systems and human resource management are typical examples of such procedures. For example, a human resource management system performs at least three functions. First, it encourages people with particular skills and needs to join the organization (buffering). Second, it ensures employees' compliance with organizational rules to promote conformity and motivation (compensating). Third, if regulates disturbances like occasional lack of conformity and motivation by selectively applying tangible and intangible benefits and penalties (regulating). Production planning ensures a timely and balanced flow of operations assigned to particular workstations. Regulation is easy as long as production goes exactly according to the plan. However, some problems will occur with predictable frequency because the productive capabilities of people differ, operations have limits on their flexibility, operations may break down, special orders take precedence over planned production, and so on. Therefore, planning procedures become more complex than the simple assignment of people and jobs to workstations according to the technological demands. They must regulate desirable flows and address the necessity to maintain balance in the event of different contingencies.

FITTING THE PROCEDURE TO THE SITUATION

The three types of procedures and typical practical examples are shown in Figure 10-6. The schematic also indicates the most typical relations between procedures and the social and technical balance of the organization.

Operational procedures are repetitively applied to typical problems and disruptions of internal balances. However, the standardization of procedures and their typical application at the most routinized and stable level of operations do not necessarily mean that managers are limited by their rigidity. Managers might be quite flexible for at least three reasons.

First, it is true that operational procedures might be triggered in many instances automatically and used as algorithms (for example, automatic reordering in a grocery store when inventories fall below predetermined levels). Nevertheless, managers always retain their discretionary power. They might block the implementation of a procedure or decide to solve a problem in a

	INTERNAL SOCIAL EQUILIBRIUM	INTERNAL TECHNICAL EQUILIBRIUM
Buffering procedures	Hiring procedures	Inventory control
COMPENSATING PROCEDURES	Training	Maintenance and repairs of the machines
Regulating procedures	Evaluation and reward procedures	Planning and production control

FIGURE 10-6 ***Selective Examples of Operational Procedures***

Source: A. K. Kozminski and K. Obloj, *An Introduction to Organizational Equilibrium* (Warsaw: Polish Scientific Publications, 1989), p. 250 (in Polish).

different way. When this happens, the procedure-following
is replaced by problem solving. So while procedures are
minimizing cost and attention, managers can always overru
dent to do so.

Second, different organizations faced with similar dist
will devise different ways to deal with them. Two basic choices are
in-house procedures or to acquire them externally through vendors, such as consultants. The discussion of procedures has tacitly assumed that they are an intrinsic part of the organizational performance and are followed by employees. However, many organizations rely on contracts with external suppliers to perform repetitive or routine operations. For example, most publishing houses outsource printing and binding. This step eliminates the most standard managerial operations while retaining full control over nonstandardized aspects of business—product development, production, and sales. Outsourcing printing is particularly useful for textbook publishers because they do not require printing in a steady flow. They tend to release most books in February or March for selection by professors for the following fall semester. A textbook publisher that also did printing would likely have times when the presses would remain idle or inventory would remain in storage. While outsourcing is generally common, though, some publishing companies have found it economical to print books in-house. They may publish books for nontext markets, such as tax and legal reference books, with relatively stable demand that can even out the fluctuations raised by the textbook cycle.

Third, while the problems and disturbances that two different organizations face might be identical, the procedures designed to buffer, compensate, or regulate them might be very different. There is no direct correspondence between the form and content of a disturbance and the design of an organizational response.

DIFFERENTIATING AMONG PROCEDURES, RULES, AND ALGORITHMS

Both rules and procedures are mandates or commands for action, but they differ. A rule is a predetermined course of action that governs behavior in a specific situation. For example, "Safety glasses must be worn in this area" is a rule. Procedures, on the other hand, require that a number of steps be followed in sequential order. The sequence of events followed to admit a patient to a hospital is a procedure. Procedures are based on the conventional wisdom within the organization regarding the best overall way to accomplish a task. They may take into account economics, technological constraints, and human abilities to carry out such directives. In this regard, procedures attempt to optimize the performance of a number of criteria.

The intent of rules is to prevent personnel from exercising discretion in favor of a predetermined judgment made to cover recurring routine situations. Rules, when properly applied, have the benefit of saving time and money and, in some situations, removing the potential for personal conflict (Thibault and

Kelley, 1959). Of course, rules can also stifle individual initiative, creativity, and a sense of control if they are perceived as being excessive, senseless, or outdated. Bureaucracies have become infamous for their red tape and instilling a sense of helplessness among employees. Peters (1988) observes, "Sad to say, rule books are only referred to in order to slow action, defend turf, and assign blame. Have you ever heard of anyone going to a rule book to figure out how to speed things up?" (p. 455). Rules and procedures can be effective tools for integrating behavior and controlling technical operations efficiently, but the economies must be balanced against their human impact (see Figure 10-7).

While rules are the building blocks from which procedures are constructed, procedures are the foundation for algorithms. **Algorithms** are discrete sets of instructions that, when followed precisely, guarantee attainment of a desired outcome. Landa (1974) has defined an algorithm as "a precise, generally comprehensible prescription for carrying out a defined sequence of elementary operations in order to solve any problem belonging to a certain class" (p. 11).

Algorithms, unlike procedures, must be stated so precisely that there is only one possible interpretation of their intent. Algorithms differ from proce-

---

Rule

A. Yes, accept
B. No, deny

Procedure

1. Turn on VCR
2. Turn on television
3. Turn TV to channel 3
4. Pick up remote control
5. Select desired channel on VCR
6. Acquire favorite snack
7. Sit back and enjoy

Algorithm

1. Greet customer
2. Acknowledge customer's request to cash check
3. Determine size of check:
   a. $20 or less
   b. $100 or less
   c. more than $100
4. Request appropriate form of identification

Check Size/ID Matrix

Check Size	*Required ID*
a. $20 or less	Valid driver's license
b. $100 or less	Driver's license and credit card
c. $101 or more	License, and bank ID credit card

5. If person offers proper ID, cash check
   If customer fails to provide proper ID, call bank manager
6. Thank customer for their business

FIGURE 10-7 ***Comparison of Rules, Procedures, and Algorithms*** dures in that their exclusive focus is to dictate a single means to satisfy the technical requirements of the task. In practice, algorithms are used to program actions for recurring, routine activities. For example, a bank teller should follow a particular formula for cashing a customer's check. Figure 10-8 explains the characteristics of algorithms.

In contrast to rules and procedures, algorithms include situational contingencies. They are able to handle a broader range of situations than procedures. Yet procedures are attuned to a wider variety of needs.

EVALUATING PROCEDURES AND OTHER OPERATIONAL CONTROLS

Once procedures and other operational controls have been established, they are often taken for granted. The power of their influence on organizations—and specifically the technical system—is generally minimized or altogether ignored. This is partly true because procedures normally operate silently and unobtrusively. Yet, their subtlety in operation masks the fact that they are potentially very powerful influences on behavior in organizations.

Over time people can become accustomed to following given procedures, yet they forget that collectively operational controls limit their freedom of choice substantially. On the positive side, this restriction on choice promotes coordination, stability, and predictability. Conversely, such controls may dampen innovation, reduce organizational responsiveness to change, and over the long term contribute to the formation of a culture with restrictive world views.

A classic case of this pattern could be seen at IBM during the mid-1970s. When presented with the possibility of manufacturing personal computers (PCs), IBM's executive board scoffed at the suggestion that a real computer maker would consider making what appeared to them as toys. The finely tuned organization at IBM was effective in turning out mainframes, but restricted the ability of its members to see future opportunities. The result was a delay in entering the PC market that afforded smaller competitors the time to get established.

There is clearly a need for organizations to judiciously use the five sys-

1. You need only read those parts of the algorithm relevant to the specific problem you are faced with solving.
2. You need only read each part once.
3. You do not have to remember prior decisions made in the algorithm.
4. The language of each part is simple.
5. You are confident that you have the right answer at each step and at the conclusion.

FIGURE 10-8 ***Characteristics of Algorithms***

Source: I. Horabin and B. Lewis, *Algorithms* (Englewood Cliffs, N.J.: Educational Technology Publications, 1978), p. xi.

temic tools. An imbalance in their application will have a ripple effect on the other tools, as well as on performance. The potential value of each tool, at any given time, must be evaluated in the context of its relationship to how the other tools are currently being used. Procedures must not be implemented and forgotten while attention is shifted to the other tools. There must be a systemic effort to continually align them with the rest of the system.

Although control-oriented tools are more closely related to the technical system, they can clearly influence the social system. This is particularly true in regard to culture. Restrictive controls can be removed when a strong culture exists. This is because the culture—through its norms, peer pressure, and effect on shaping beliefs—can significantly control people's behavior. The next chapter explores how culture can be used as a systemic managerial tool.

~ SUMMARY

Procedures are another systemic managerial tool that can be used in conjunction with the elements of the management systems model to manage organizations. Procedures are most effective when employed for the purposes of regulating, buffering, stabilizing, and closing the technical subsystem of organizations. Procedures are very powerful in determining behavior within human systems because they are often taken for granted and followed automatically. One of the most important dimensions of the technical system is its technical core, the site of the transformations that contribute value to customers, and ultimately, the organization. The remainder of the technical system is composed of tasks and processes that are coupled to create the framework of the system.

~ KEY TERMS AND CONCEPTS

procedures
input-transformation-output system
technical system
technical core
technology
workflow variability
long-linked technologies
sequential interdependence
mediating technology
pooled interdependence
intensive technology
reciprocal interdependence
heuristics
problem analyzability
closed technical systems
open technical systems
flexible programmable systems (FPS)
buffering procedures
compensating procedures
regulating procedures
algorithm

~ QUESTIONS FOR DISCUSSION

1. How does the use of procedures for regulating an organization's technical equilibrium differ from the way structure is commonly used?

2. Is it possible to rely heavily on procedures and still maintain high levels of creativity in an organization?
3. Many people are familiar with the many procedures used by such government agencies as tax or motor vehicle departments. Their use has become synonymous with the derisive terms *red tape* and *bureaucracy.* How can a manager of a public agency determine the point at which the use of procedures has become excessive?
4. How would the training for people who work with nonroutine forms of technology differ from those whose work is routine?
5. Acme Aerospace has just opened a new division that specializes in high-tech optical devices for aerospace applications. You have been assigned the job of managing the development of a new camera-telescope for use on a research satellite that will survey the outer reaches of the solar system. The mirror for the telescope must meet specifications to the 1/1,000th of a millimeter. The product must have an innovative design, but still be durable. The total production run is one unit. How will you ensure the quality and integrity of the product? What blend of the following methods will you use: cybernetic controls, inquiring system, procedures, algorithms, and TQM?

REFERENCES

Cavaleri, S. (1988). "An Empirical Assessment of the Relationship Between Technology and Training in Organizations," unpublished doctoral thesis. Troy, N.Y.: Rensselaer Polytechnic Institute.

Fry, L. (1978). "A Rationalization and Test of the Technological Imperative Through a Typology of Technology-Structure Relationships," unpublished doctoral dissertation, Ohio State University.

Garson, B. (1988). *The Electronic Sweatshop.* New York: Simon and Schuster.

Hickson, D., Pugh, D., and Pheysey, D. (1969). "Operational Technology and Organizational Structure: An Empirical Reappraisal," *Administrative Science Quarterly, 14,* no. 3, pp. 378–97.

Horabin, I., and Lewis, B. (1978). *Algorithms.* Englewood Cliffs, N.J.: Educational Technology Publications.

Koestler, E. (1945). *The Yogi and the Commissar.* London: Jonathan Cape.

Koontz, H., O'Donnell, C., and Weinrich, H. (1980). *Management,* 7th ed. New York: McGraw-Hill.

Kozminski, A., and Obloj, K. (1989). *An Introduction to Organizational Equilibrium.* Warsaw: Polish Scientific Publications.

Landa, L. (1974). *Algorithmization in Learning and Instruction.* Englewood Cliffs, N.J.: Educational Technology Publications.

March, J., and Simon, H. (1958). *Organizations.* New York: John Wiley & Sons.

Perrow, C. (1970). *Organizational Analysis: A Sociological View.* Monterey, Calif.: Brooks/Cole.

Perrow, C. (1984). *Normal Accidents.* New York: Basic Books.

Peters, T. (1988). *Thriving on Chaos.* New York: Harper and Row.

Porter, M. (1980). *Competitive Strategy.* New York: Free Press.

Randolph, A. (1981). "Matching Technology and the Design of Organization Units," *California Management Review, 23,* no. 4, pp. 39–48.

Schonberger, R. J. (1986). *World Class Manufacturing: The Lessons of Simplicity Applied.* New York: Free Press.

Starr, M. (1989). *Managing Production and Operations.* Englewood Cliffs, N.J.: Prentice-Hall.

Sugarman, A. (1989). "Skyway Freight Embraces JIT Philosophies to Propel Forward," *Management Review,* p. 16.

Thibault, J., and Kelley, H. (1959). *The Social Psychology of Groups.* New York: John Wiley & Sons.

Thompson, J. D. (1967). *Organizations in Action.* New York: McGraw-Hill.

Woodward, J. (1965). *Industrial Organization: Theory and Practice.* London: Oxford University Press.

11 Culture as a Systemic Tool

Culture is to the organization what personality is to the individual—a hidden yet unifying theme that provides meaning, direction, and mobilization.

Ralph Kilmann (1984)

1. To understand how the various elements of an organization interact to bring a culture to life.
2. To recognize the two archetypal cultures that are likely to form in different types of organizations.
3. To learn how managers can play a role in supporting the development of culture within an organization.
4. To describe how culture acts to regulate the social system of organizations.
5. To learn to use culture to understand the mental models that will predominate in a given system.
6. To explain the role played by culture as a systemic managerial tool in the management systems model.

SECTION I
UNDERSTANDING CULTURE

INTRODUCTION

Culture is a systemic tool that is closely aligned with the social system of organizations (Figure 11-1). It is one of the more flexible systemic tools in that it can be used in a variety of ways depending on the needs of the system. Yet, culture is also difficult to define, unwieldy to manage, and, at times, unpredictable in its effect. When the force of an organization's culture can be harnessed, however, it can be an extremely valuable resource.

Organizational culture normally develops and gains richness over extended periods of time. An organization's culture develops within the larger context of the **societal culture,** the civilization. Prior to the expansion of global businesses, organizational cultures and societal cultures shared the same roots. Today, in a world of global corporations, this pattern is less likely to be found. One of the most interesting challenges for European managers in the 1990s will be the integration of the different societal cultures found in the European Community. The global systems spotlight takes a look at this issue.

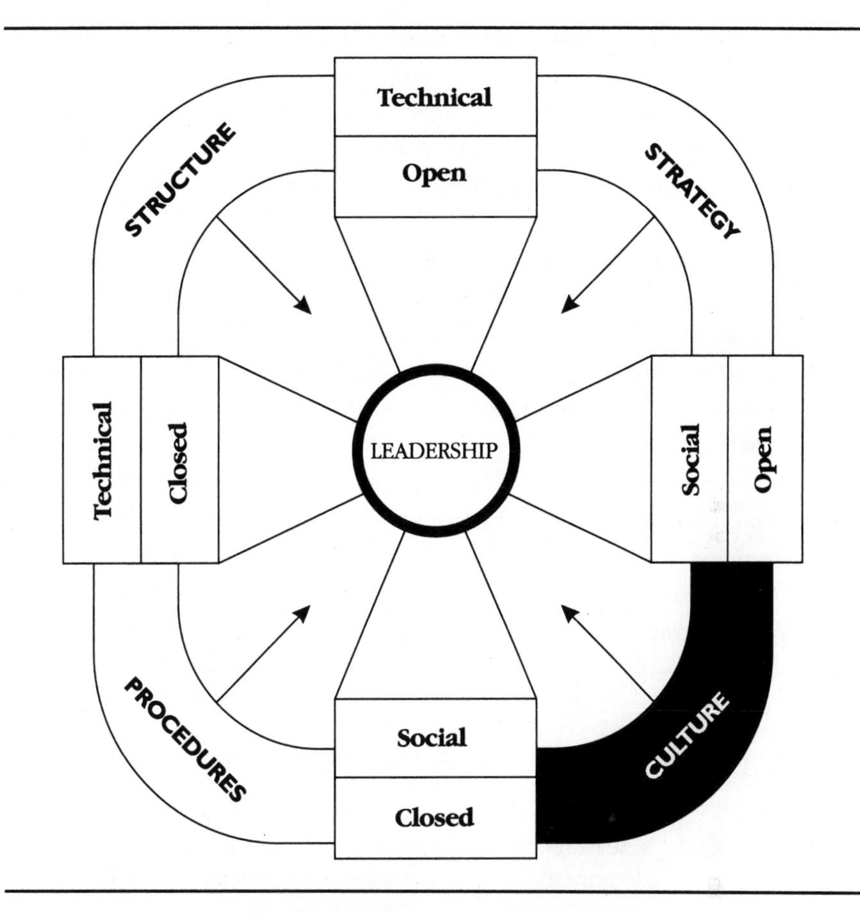

FIGURE II-1 *The Management Systems Model*

GLOBAL SYSTEMS SPOTLIGHT

Organizational versus Societal Culture in Europe

The integration of Europe's market and the assimilation of Eastern Europe into it will be an arena for the competing cultures of management. In general, there are four main management cultures: German, French, British, and Eastern European. These four cultures present striking contrasts in management styles, performance targets, corporate strategies, and the way workers and shareholders are treated.

German-style corporate culture influences Denmark, the Netherlands, and, to a lesser degree, Belgium. Managers in these relatively prosperous economies concentrate more on revenues than on profits. The companies found in these economies are generally less profitable than in other European countries because of lower inflation, closer links with banks, and, most importantly, a tendency to relatively more generosity to employees. In France and Ireland the corporate cultures value revenues and profitability in a similar way, but still place considerable emphasis on employee rewards relative to returns to investors. The British, Italian, and Spanish corporate cultures are in many ways modeled after the standard corporate culture found in the United States. This culture values profits more highly than revenue. Investors are very influential through an active stock exchange, and they are favored relatively more than employees.

Finally, the Eastern European corporate culture is naturally very different because it evolved (in several different countries) under communist rule. The managers of these firms became oriented more toward production and resource utilization than revenues and profits. Employee compensation is of the utmost importance, as it was the main way to keep social order in the industry. The investors' interests—be the investors the state or private—were and still are largely dismissed. The managerial culture of Eastern firms could be aptly labeled a *technocracy.*

These striking cultural differences have to be resolved if European economic unity is to be anything more than a fond dream. The gulf between the different ways of corporate life is so important that it might be the major source of cultural conflict and managerial challenges in a fully integrated European economy.

CULTURE AS A MANAGERIAL CONCEPT

The topic of organizational culture has a long history and many roots. The research conducted many years ago by the human relations school of theorists demonstrated that human values, norms, and rules of conduct play a critical role in organizational performance. However, even before this approach to management became popular in the late 1950s, a number of prominent scholars appreciated the important role of norms and values in management. Barnard (1938) stressed the significance of group cultural values and norms. He argued that a leader—say, the chief executive officer of an organization—must establish and communicate a system of values among organizational members. His ideas were later expanded upon by sociologist Philip Selznick (1948), who described how norms and values help to define the identity of organizations. The role of the leader, in Selznick's opinion, is to cultivate and defend those values. Jaques (1952) probed the topic of organizational culture in great depth, giving the following description of its essence:

> The culture of the factory is its customary and traditional way of thinking and of doing things, which is shared to a greater or lesser degree by all its members, and which new members must learn, and at least partially accept, in order to be accepted into service in the firm. Culture in this sense covers a wide range of behavior: the methods of production; job skills and technical knowledge; attitudes towards discipline and punishment; the customs and habits of managerial behavior; the objectives of the concern; its way of doing business; the methods of payment; the values placed on different types of

work; beliefs in democratic living and joint consultation; and the less conscious conventions and taboos. Culture is part of second nature to those who have been with the firm for some time. (in Denison, 1990, p. 28)

Prior to the 1980s, organizational culture was generally viewed as an impractical, theoretical pursuit of scholars rather than as an important dimension necessary for managing an organization. In that decade, however, culture emerged in the mainstream of managerial thought and practice. Its significance was heralded by the widely popular book *In Search of Excellence* by Peters and Waterman (1982). This book popularized the concept of the "excellence culture" of high performing organizations such as the Disney Corporation, IBM, and McDonald's. For one of the first times in history, writers claimed that organizational culture was a possible cause of top performance.

Organizational culture erupted with so much power in the 1980s for three main reasons. First was a growing interest in, and even wonderment at, the economic success of Japan and Japanese firms. Japan, a country destroyed during World War II and without any significant natural resources, became an economic superpower during the late 1970s. Many scholars have searched widely to explain this success. One of the most widely accepted explanations was offered by William Ouchi. Ouchi compared American and Japanese businesses to isolate their differences. He found that the most distinctive feature of Japanese corporations is an organizational culture based upon involvement and participation of all members (Ouchi, 1981). While American firms had skillfully developed such hard systems tools as strategy and structure, the Japanese firms developed soft systems tools—work skills, managerial style, communications networks, and, most importantly, a social system based on shared values.

The second reason for the growth in attention to culture was the performance of U.S. businesses. Simultaneous to the rapid rise of the Japanese businesses, American firms, particularly manufacturers, suffered loss of market share. Increased bankruptcies in the United States and Western Europe created a fervent interest in understanding the nature of organizational success. The fundamental question of what makes a firm a winner took on prime importance for practitioners and researchers alike. The search for new recipes for success yielded the development of a number of core concepts relating to organizational culture. Many of the practices that had somehow been ignored as being trivial, such as rituals, myths, and the creation of heroes, now were viewed in a new light. Shared norms and values and a consistent vision expressed in stories, myths, rituals, and heroes were considered keys to success and excellence. Peters and Waterman (1982) stressed this perspective:

As we worked on research of our excellent companies, we were struck by the dominant use of story, slogan, and legend as people tried to explain the characteristic of their own institutions . . . in an organizational sense, these stories, myths, and legends appear to be important, because they convey the organization's shared values, or culture. Without exception, the dominance and coherence of culture proved to be an essential quality of the excellent companies. (p. 75)

The final reason for the emergence of the concept of organizational culture in the 1980s complements the second reason. For many years positivism dominated the study of organizations. Positivists assume that the necessary facts that characterize a situation may be identified and measured through the use of logic. Rigidly designed research approaches relying on the language of statistics permeated the study of management. Even in relatively soft fields such as motivation and leadership, research variables were quantified, measured, and correlated with the help of hundreds of statistical techniques. As a result, qualitative variables, which were difficult to observe and measure, were not taken seriously by many researchers. However, this bias eventually created a backlash; increased dissatisfaction with such approaches led to a significant shift in emphasis that occurred in the late 1970s and early 1980s. A qualitative methodology, which advocated the use of soft methodological tools like unstructured interviews, case studies, and comparative research, became popular. Naturally enough, these tools proved to be useful in describing and explaining qualitative, ill-structured aspects of organizational performance—such as norms and values, the traits of organizational culture.

The emergence of the organizational culture school in the 1980s was virtually inevitable. The various shifts that led to the emergence of culture as a core concept complemented each other. Scholars and managers had the same basic goal: to identify a recipe for success. In tandem with the increased acceptance of qualitative scientific tools, organizational culture entered the mainstream of managerial thought. The trend was set with three books that examined the challenges that Japanese corporations posed for American companies: *Theory Z* (Ouchi, 1981), *The Art of Japanese Management* (Pascale and Athos, 1981), and *In Search of Excellence* (Peters and Waterman, 1982). However, in spite of the enormous number of books and articles published in this domain during the 1980s, it is difficult to generalize about the findings. Relatively comprehensive studies of organizational culture have only been undertaken recently, and the concept of culture is still rather vague. Even more importantly, its effects on organizational performance are not fully known.

The concept of organizational culture has been defined in a variety of ways by different scholars. Gagliardi (1986) proposes that culture is a "coherent system of assumptions and basic values which distinguish one group from another and orient its choices" (p. 119). Sathe (1985) defines culture as "the set of important assumptions (often unstated) that members of a community share in common" (p. 10). William B. Rener, vice-chairman of the Aluminum Company of America, offers a particularly interesting definition: "Every textbook offers a definition, but I like a simple one: culture is the shared values and behavior that knit a community together. *It's the rules of the game;* the unseen meaning between the lines in the rule-book that assures unity" (in Kilmann, 1984, p. 92). In these, and other definitions, we can find the three most common elements of **organizational culture:**

- Basic, **tacit assumptions**—beliefs that organizational members hold about themselves and others, existing relations and their meanings, and the nature of organization and its environment.

- Organizational members' **values and norms,** which are the visible core of culture and which regulate mission, goals, patterns of communication, and evaluation.
- **Artifacts,** which are the most visible and transparent elements of a specific culture, such as language and jargon, stories and myths, ceremonies, rituals, and managerial practices.

These elements are organized hierarchically: tacit assumptions are the basis for creation of values and norms, which manifest themselves in organizational jargon, stories, ceremonies, rituals, and behaviors (see Figure 11-2). This hierarchy is important from the systems point of view because it illuminates problems both with using culture as a managerial tool and with changing culture. In order to create, use, or change an organization's culture, a manager must operate on all three levels and align them. Organization change efforts that ignore the comprehensive hierarchical aspects of culture are less likely to achieve the desired change.

Prior to examining the various ways that culture develops and how it might be used as a systemic tool of management, it is useful to take a brief look at the organizations that use it as a dominant tool. Two of the main types of such organizations are ideological organizations and professional organizations.

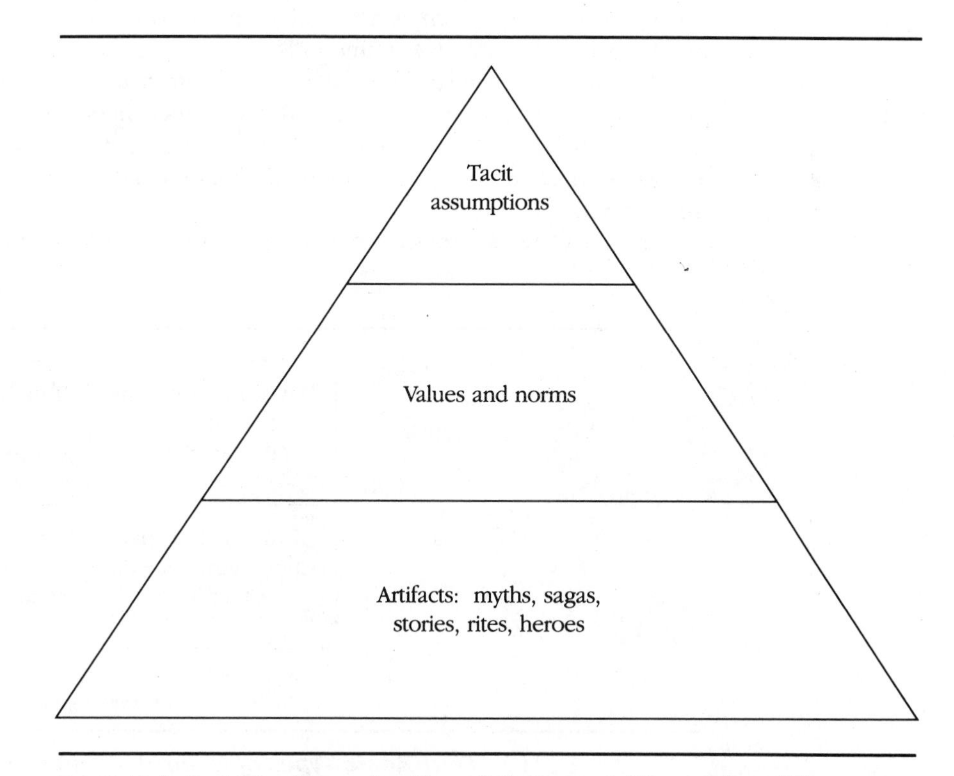

FIGURE 11-2 ***Organizational Culture: A Hierarchy of Elements***

Deal and Kennedy (1982) have also identified four types of cultures. Their typology is based on two main factors: the risk inherent in the company's business, and the speed of feedback regarding results of their decisions. Deal and Kennedy suggest that each type of culture has its own unique strengths and weaknesses, which are suited to some tasks better than others. (See Figure 11-3.)

IDEOLOGICAL ORGANIZATIONS

Webster's New World Dictionary defines *ideology* as "the body of ideas on which a particular political, economic, or social system is based." By **ideological organizations,** we mean organizations driven by an inspiring mission or vision. They state their mission in terms of ideas and values that are continuously and clearly communicated to their members. Such organizations stress common values and norms, because those commonalities are the glue that keeps them organized. Such organizations are cohesive, in that members share a common, overarching goal. However, cohesiveness should not be mistaken for conformity.

The classic examples of ideological organizations include religious organizations such as the Roman Catholic Church (especially its different orders, the Dominicans, Franciscans, Jesuits, and so on), radical political parties, voluntary associations (such as Alcoholics Anonymous), and socioeconomic institutions like the kibbutzes found in Israel.

The main mechanisms used to achieve coordination in such organizations are the standardization of values and the indoctrination of their members. These mechanisms ensure that a relatively homogeneous and pervasive culture is created. In such settings the culture becomes a major frame of reference for all members.

From a systems perspective, culture is often used as a substitute control

Risks		Feedback Rate: Slow	Feedback Rate: Fast
	High	Bet Your Company Culture (Exxon, NASA)	Tough-Guy Macho Culture (Police depts.)
	Low	Process Culture (Chubb Insurance Co., CPA firms)	Work Hard/Play Hard Culture (Frito-Lay, Mary Kay Cosmetics)

FIGURE 11-3 ***Four Basic Types of Organization Culture***

Source: Based on T. Deal and A. Kennedy, *Corporate Cultures: The Rites and Rituals of Corporate Life* (Reading, Mass.: Addison-Wesley, 1982).

mechanism to replace such typical systemic tools as structure and procedures. In ideological organizations the structure generally is decentralized, simple, and flat. Members are grouped in small subsystems, do not hold clear-cut positions, and engage in various tasks on an ad hoc basis. Ideological organizations place less interest in formal planning and control mechanisms; instead, they favor strengthening core values and norms. A continual process of adjusting the culture often substitutes for traditional coordination mechanisms and structures. Although the effects of culture on performance are a matter for debate, the attraction to using culture as a managerial tool is strong. The reason is very pragmatic: control through an invisible, common ideology is cost effective and powerful compared to clumsy traditional mechanisms of coordination and control through formal structure and procedures.

This type of pattern is clearly illustrated by the Roman Catholic Church. It has only five layers of structure, with the Pope at the top and priests at the bottom—fewer layers than a typical mid-sized business. The most important subsystem of this organization is the local parish, usually a small, autonomous group of priests responsible for services in a village, city, or district. Since priests' roles are interchangeable, they can substitute for each other in the event of almost every imaginable contingency. Their responsibilities are situationally defined and dynamic. The rule of thumb that guides their decisions is very simple and deceptively general: "Love God, and do what you like." There is no need for extensive controls or coordination mechanisms because freedom of choice is limited by values, norms, rules—most of them more implicit than explicit. Deal and Kennedy (1982) stressed that the Roman Catholic Church has maintained its sway for centuries not by strategic planning or hierarchy "but through one of the strongest and most durable cultures ever created" (pp. 194–95). In many respects, this prototype of an ideological organization represents a potentially viable model for the successful business firm of the future. Some business organizations (for example, many Japanese firms) already display characteristics of ideological systems or are trying to introduce them with varying degrees of success.

PROFESSIONAL ORGANIZATIONS

The second type of organizational culture is the **professional organization:** universities, hospitals, consulting, and craft-production manufacturing firms are examples. The most important feature of such organizations is the way they select and train their human resources. Professional organizations hire their members from a select population that has been socialized and trained according to precise rules stressing the importance of unique skills and professional ethics. The professional code of ethics of physicians is embodied in the Hippocratic Oath. Similarly, in some European countries scholars getting their Ph.D.s take an old, traditional Latin oath that stresses the importance of pursuing the scientific truth.

Professional organizations maintain their culture by ensuring that consistency and continuity exist within the values of their leaders. This applies equally to all such organizations, whether physicians in a hospital or the faculty of a university. These organizations seek to reinforce the consistency of culture through constant training and indoctrination of personnel. Mintzberg (1983) has noted:

> There typically follows a long period of on-the-job training, such as internship in medicine and apprenticeship in accounting. Here the formal knowledge is applied and the practice of the skills perfected, under the close supervision of members of the profession. On-the-job training also completes the process of indoctrination, which had begun during the formal teaching. Once the process is completed, the professional association typically examines the trainee to determine whether he has the requisite knowledge, skills, and norms to enter the profession. (p. 191)

Positive feedback operates throughout the entire process because the dominant coalition of professionals maintains its position thanks to the commonality of skills, values, and norms. The stronger they are, the stronger and more influential will be the coalition and thus the stronger its dominance of the system.

SECTION 2
USING ORGANIZATIONAL CULTURE

DEVELOPING THE CULTURE IN AN ORGANIZATION

An anecdote about an American firm trying to build its culture indicates how some executives view the culture-building process. The story goes that an American executive attended a seminar that promoted the relationship between culture and corporate performance. Upon returning to the company, the enthusiastic executive called in one of his vice presidents and assigned him the task of building the company's culture. He then instructed the VP to report back in a month as to whether the goal had been accomplished. In reality, of course, culture building is a long-term process that is relatively difficult to manage because it revolves around concepts that are soft and over which managers have only limited control.

To describe the evolution of organizational culture, we will use the framework of four major stages developed by Pasquale Gagliardi (1986) (see Figure 11-4).

CHOOSING VALUES

In the first stage of development, when an organization is created, its founder makes value commitments by choosing the type of business and major strategic objectives. These choices are based primarily on the founder's beliefs and assumptions regarding how to succeed. This process generally takes into ac-

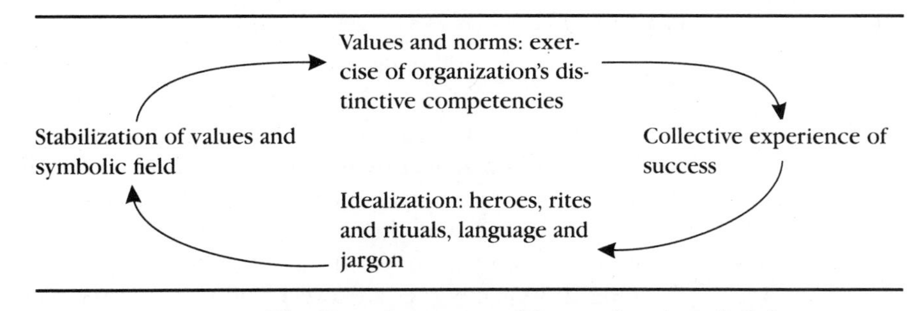

FIGURE 11-4 ***The Development of Organizational Culture***

Source: Adapted from P. Gagliardi, "The Creation and Change of Organizational Cultures: A Conceptual Framework," *Organization Studies* (1986), 7, no. 2, p. 125.

count the type of environment the organization will be performing in, competitor strengths and weaknesses, and customer preferences, among other factors. "These character defining choices are often not made verbally, they might not even be made consciously. . . . The institutional leader is primarily an expert in the promotion and protection of values" (Selznick, in Peters and Waterman, 1982, p. 281). During this stage of development, organizational members might not know about these choices and may not share the leader's ideas. Nevertheless, founders usually have the power and ability to shape the actions of employees and orient the organization in the desired direction.

Throughout the course of history famous founders have left their cultural imprint on their organizations. For example, Thomas Watson Jr. shaped IBM with his persistence on respect for the individual, technological innovation, and customer service. Ray Kroc's obsession with quality, service, cleanliness, and value in McDonald's was the guiding light for all actions by employees. William Hewlett and Dave Packard of Hewlett-Packard maintained their insistence on the primacy of innovation and flexibility in shaping their high-tech company. Donald Burr's legendary efforts to turn PeopleExpress Airlines into a major player in the airline industry were driven by his belief in the needs to unleash the power of the individual and to offer value to customers through full commitment of its members.

REINFORCEMENT BY SUCCESS

In the second stage of cultural evolution in organizations, the guiding norms and values are strengthened by the achievement of success. The validity of the founder's direction is now confirmed by experience and is likely to be shared by most organizational members. Those that cannot adapt to or accept the particular organizational pattern usually exit the firm. The psychological costs of participation for the remaining members are relatively low because—due to success—the organization's survival is not in question. Indeed, until there are shifts in the competitive environment the firm is likely to continue along the established lines, which led to its success.

IDEALIZATION IN ARTIFACTS

In the third stage of cultural development, the norms and values undergo idealization and further reinforcement by the emergence of artifacts such as jargon, heroes, and rituals. Organizational jargon, catchy phrases, and buzz words are created almost by themselves. Examples are Tandem Corporation's "Tandemize it—means make it work"; and Polaroid's "The bottom line is in heaven." The jargon and phrases help to create company uniqueness and communicate the company direction and key values to all its members.

Organizational heroes are a very important component of a culture, since they personify the bedrock beliefs of the company and its members. They symbolize values, norms, and strengths of the company in a form that is clear and easy for all to understand. In the United States, such legendary figures as Henry Ford of Ford Motor Company, Pierre du Pont of du Pont, Mary Kay Ash of Mary Kay Cosmetics, Ray Kroc of McDonald's, Lee Iacocca of Chrysler Corporation, and many others have become part of the industrial folklore and heroes of their companies. In Europe, Giovanni Angelli of Fiat, Carlo de Benedetti of Olivetti, Jan Carlzon of Scandinavian Airlines System, Percy Barnevik of Asea Brown Boveri, and Pehr Gyllenhammer of AB-Volvo, just to mention a few, achieved similar global recognition and hero status. In Japan, Yohei Mimura, chairman of Mitsubishi, and Akiro Morita of Sony Corporation are typical symbolic figures—heroes of their companies and to some degree of their nation. Every successful organization also has lower-profile heroes who are generally more human and approachable than their executive counterparts. Veteran salespersons, clerks, or manufacturing workers who were responsible for extraordinary efforts are likely to fill these roles.

Heroes provide role models of virtuous behavior and inspire others by setting high standards of performance. They translate the organization's mission, values, and norms from a concept into a reality for both the outside stakeholders and employees. Myths and legends are built around heroes in the process of creating a rich organizational history. Such a legacy provides a powerful indicator of what is and what is not important in organizational life and performance in each company. They help to support the organization's claims that its culture is unique and in some way superior to that of competitors. In these simple, yet powerful, ways heroes shape and reinforce the organization's culture by illustrating to others that the sometimes lofty ideals are attainable.

Finally, organizations develop rituals to periodically reinforce the culture. The role and importance of rituals is well argued by Deal and Kennedy (1982): "Without expressive events, any culture will die. In the absence of ceremony or ritual, important values have no impact. Ceremonies are to the culture what the movie is to the script, the concert is to the score, or the dance is to values that are difficult to express in any other way" (p. 63). The richer the shared history, the more elaborate the rituals. Nothing is left to chance in the creation and staging of such ceremonies and settings. The design of offices, the format for meetings, awards presentations, new product introductions, and training programs are elevated to special status. They communicate that employees are important, managers care, and the organization is special; they also imply that everybody should try harder. Consider, for example, the high drama of a Mary Kay Cosmetics sales seminar.

Mary Kay Cosmetics

At these sales meetings, classes are conducted and the hundreds of Mary Kay salespeople learn something. But what they wait for are the different awards nights—more than thirteen hours of performances that educate, motivate, inspire, and entertain. One night the spotlight is on the team leaders and future directors who have been dubbed cultural heroes. They parade across the stage in bright red jackets, passing the microphone to tell stories about how they personally achieved success a la Mary Kay herself. To those in the audience the message comes through loud and clear: these people made it and so can I. In another ceremony, awards for the best sales are given—pink Buicks and Cadillacs. One year the cars simply "floated" down onto the stage from a "cloud"—a weighty touch of hoopla that produced overwhelming response from the crowd. But the biggest night is a five-hour spectacular when the company crowns the Director and Consultant Queens (supersaleswomen) in each category. They're given diamonds and minks and surrounded by a court of women who have also achieved terrific sales. At the end of this extravaganza, everyone understands that the challenge of the company is in sales.

Source: T. Deal and A. Kennedy, *Corporate Culture: The Rights and Rituals of Corporate Life* (Reading, Mass.: Addison-Wesley, 1982), p. 74.

While the culture at Mary Kay may not be equally effective at every organization, it clearly meets the needs of the people who work at Mary Kay. Cultures, like personalities, are unique; what is attractive to one person is inappropriate for another. This is an important dimension of culture. People who do not fit in usually leave, and the culture becomes self-reinforcing. Whether this benefits the organization is unclear. An organization will not enjoy long-term benefits from a culture primarily driven by conformity and win-lose forms of competition that drive out people who are innovative collaborators.

STABILIZATION OF VALUES

Organizational artifacts such as jargon, symbols, heroes, rituals, and myths about these artifacts evolve over time, creating an organizational history with a clear message that identifies the causes of success and desirable patterns of behavior. In the fourth stage of culture development, these values and norms are increasingly taken for granted. Over time such memories gradually are solidified in members' thinking patterns and become embedded in their beliefs and assumptions about the way things work. If the organization is successful, a full circle is created because the idealization of the collective experiences and artifacts that have evolved reinforces the original values and norms that characterized the early organization. The ideals that were once the corporate visions and missions become unquestioned assumptions that guide behaviors and actions of members.

MAINTAINING ORGANIZATIONAL CULTURE

Once the culture is established, it needs constant reinforcement to perpetuate itself, especially in regard to new members of the organization. Organizations choose different approaches to this problem, but recent research (Ott, 1989; Pascale, 1985; Sathe, 1985) shows four activities common to companies that maintain strong and cohesive cultures.

HIRING PROCEDURES

Inculcating the culture in employees starts before new employees are even hired. Organizations preselect their members by explicitly stating their preferences and expectations and by pursuing applicants who already possess unique, desirable skills and a set of values consistent with the organization's. As previously noted, this process is typical of ideological and professional organizations; more recently, however, it has been adopted by a wider variety of firms. They use extensive, sophisticated selection techniques and enact the hiring process as a relatively complex ritual. These practices help to assimilate candidates with similar values and to screen out applicants who are not compatible with the organization's expectations and who may resist the second step: socialization.

THE SOCIALIZATION PROCESS

Socialization refers to the process by which new members learn and internalize the assumptions, values, norms, and rules of conduct that comprise the organization's culture. Many American organizations rely on two of the simplest possible techniques. The first technique applies to members, who are given detailed manuals of the company's policies and procedures and simultaneously pressured with overloaded work levels. Under stress and pressure, new members quickly learn what is important and what kind of behavior is expected. Japanese corporations, by contrast, rely on a technique of moving managers around to achieve socialization. New members under the guidance of a manager performing the role of mentor rotate frequently among various departments and assignments within the organization. This strategy exposes new members to different task demands and inculcates common norms and values at the same time. Rarely, if ever, do Japanese companies provide their employees with detailed manuals of policies or procedures. One manager of a Japanese steel mill explained this approach: "Our values and norms are so important to us that we never put them in writing. New employees must feel them, learn them, immerse in them and understand them. If they cannot learn to read our culture by themselves they must leave the firm."

The second technique of socialization popular in U.S. corporations occurs in later stages, when members are constantly trained and indoctrinated; they are "processed" (Ott, 1989). People whose work styles or beliefs deviate from the norm and cannot accept the culture or match well with the system are (subtly or otherwise) forced to leave or exit voluntarily.

REINFORCEMENT OF THE EXISTING CULTURE

In the third step of maintaining culture, the culture is reinforced by repeated exposure of the organization's members to core ideals and information, controls, and rewards. Such structures and procedures are critical in shaping behaviors and providing systemic justifications for these activities. In general, these structures and procedures operate on the basis of cybernetic feedback principles. Expected and desired behaviors are rewarded (positive feedback) while undesirable and deviant behaviors are punished, either directly or indirectly (negative feedback). Direct punishment might take the form of supervisor's rebukes, a lowered bonus, or a delayed promotion. Indirect punishment might take the form of demands for detailed justification of decisions (a lot of paperwork), peers' comments, or subtle changes of a person's placement in the organizational information network, among other possibilities.

It is worth mentioning that if an organization's reward, control, and information procedures are not consistent with its declared values and norms, the culture begins to break down. Otherwise two different and even contradictory sets of values and norms are created: real values, which regulate day-to-day behaviors and actions (values-in-use), and symbolic, officially declared, espoused values (Argyris and Schon, 1978). If these procedures are consistent with cultural values and norms, however, new members will be able to identify with the organizational culture. They will move from mere acknowledgment and acceptance of the status quo to greater commitment and identification. Identification with an organizational culture lowers the stress and psychological costs of participating in the system.

ADOPTING CULTURAL ARTIFACTS

The fourth and final step in the maintenance of culture involves members internalizing organizational language, myths, sagas, stories, legends, ceremonies, rites, rituals, and heroes—all the artifacts that have already been mentioned. These symbolic signals provide members simultaneously with points of reference, memories, and directives. The artifacts reinforce the values and norms, providing coherent symbols that are both easy to remember and rich in meaning.

The processes of creating and maintaining organizational culture are illustrated in Figure 11-5. The former model, which illuminated the creation of organizational culture, is expanded here by the addition of the self-reinforcing cycle that maintains that culture. These two circles amplify each other and provide the inertia for a natural dynamic of cultural development. This cyclical dynamic can be reversed to lessen the impact of their elements. This reversal process relies on the power of negative feedback and signals of cultural decline.

PRACTICAL APPLICATIONS OF CULTURE

At the beginning of the chapter we stressed that the topic of organizational culture gained recognition thanks to the influence of Japanese management practices. Research shows that in larger Japanese corporations, culture plays an important role in maintaining both internal and external equilibria. Therefore, it is valuable to consider briefly the main elements of Japanese organizational culture: lifetime employment, a zigzag patterning of promotions based on ability and seniority, and consensus-seeking decision making.

The norm of lifetime employment reflects the long tradition of merchants' houses, which were family-owned trade and production companies that hired employees as young apprentices and expected them to stay for life. Postwar developments, especially the limited labor market of skilled professionals, the limited mobility of the labor force, and the development of trade unions relating to the specific firms rather than industries, supported the adoption of lifetime employment (Clark, 1979).

Larger firms usually hire their members right after graduation and then offer them the stability of long-term employment. This practice is supported by internal reward and promotion processes. Employees are paid a flat wage with a bonus given once or twice yearly. The size of the bonus reflects the

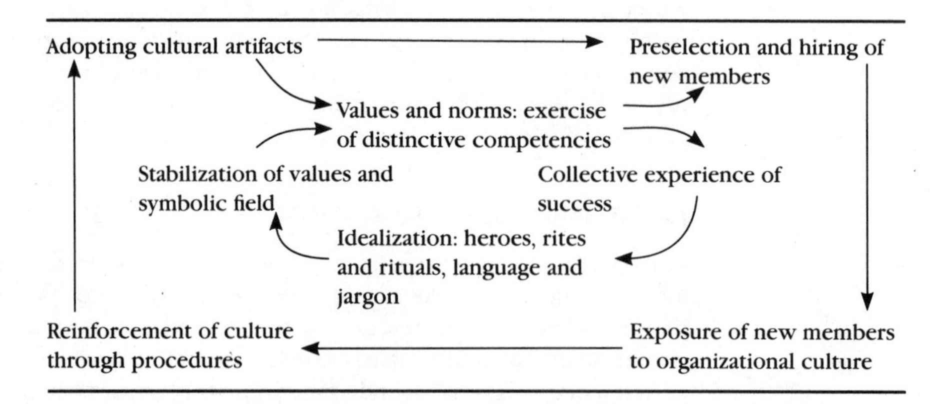

FIGURE 11-5 ***The Feedback Loops Creating and Maintaining Organizational Culture***

profitability of the corporation as a whole. This reward system creates strong motivation for employees to take actions consistent with the interests of the enterprise.

Promotions are determined by age and ability. The ability-seniority principle is very important because it rationalizes the lifetime employment concept. The longer an employee stays with the firm, the greater the probability that he or she will enjoy subsequent promotions. Each lifetime employee must move every few years to a different department, which excludes rigid specialization. Employees are given responsible tasks to perform, sometimes even official titles, but most of their posts during the first ten years are temporary (Nonaka, 1990). Along with age come automatic promotions, but only to the level of department head. Any further vertical movement in the hierarchy is determined by abilities rather than seniority. Figure 11-6 presents four typical career paths within Fuiji Xerox. This unusual promotion practice ensures that employees are exposed to many different aspects of the firm's operations. Furthermore, it limits the likelihood that younger employees will try to get on the fast, vertical track of promotions. The constant flow of employees and changing job assignments make the structure of such organizations very flexible and flat.

Lifetime employment and the ability-seniority promotion system along a zigzagging career path ensure the continuity and commonality of values and norms throughout the organization. These strategies also generate an internal environment where the harmony of interests and expectations is strongly supported by decision-making processes (Ouchi, 1981).In general, Japanese decision making is quite slow because the Japanese strive for consensus. However, this approach is reputed to engender high levels of commitment from

NAME	EDUCATION AND MAJOR	CAREER PATH
Yoshida Hiroshi	University, education	Technical service staff—personnel—product planning—program management
Fujita Ken'ichiro	University, commerce	Marketing staff—product planning—product management
Suzuki Masao	University, mechanical engineering	Design—research—design
Kitajima Mitsutoshi	University, electrical engineering	Technical service staff—quality guarantee—production

FIGURE 11-6 ***Sample Career Paths at Fuiji Xerox***

Source: Ikujiro Nonaka, "Redundant, Overlapping Organization: A Japanese Approach to Innovation Process," *California Management Review (1990), 32,* no. 3, p. 36.

participants. The prime Japanese decision-making process is known as ***ringi.*** It has three main characteristics that distinguish it from other forms: it is bottom-up, group oriented, and consensus seeking.

The decision-making process usually starts with the formulation of a problem encountered in operations. Informal discussions, consultations, and persuasion take place. Written descriptions of the problem and proposed solutions begin to circulate in the organization. Each manager usually feels free to comment on and add to the proposal. At the same time, managers become more informed about the problem and possible solutions. As a rule, divergent interests must be taken into account and compromises sought. By the time the decision is made, virtually all members of the organization who may be affected by the decision are able to voice their opinions and influence the final solution.

The whole *ringi* process is slow, but the homogeneity of managerial background, training, perspectives, values, and norms compensates to increase its speed and efficiency. Consequently, the culture is reinforced by the *ringi* process because that process amplifies the importance of all the factors that made it possible: cooperation, group orientation, common values, core norms, informality, face-to-face relations, and lifetime employment (Pascale and Athos, 1985; Clark, 1979; Ouchi, 1981).

This brief overview suggests that the culture found in Japanese corporations is a potentially powerful tool for systemic management. Culture is instrumental in regulating the internal and environmental social interfaces of the organization. (Of course, the success of Japanese corporations must be viewed in the broader context of Japanese society.) The social and technical dimensions of equilibrium are closely interrelated, and principles of lifetime employment, flexible career paths, and *ringi* decision making allow managers to balance almost every important element of performance throughout the system.

Conceptually, we can distinguish the social, technical, internal, or external roots of these principles. In practice, all these distinctions become blurred. Lifetime employment, while serving to stabilize closed social systems, also has an important impact on relations with the environment in the form of hiring policies, public relations, and so on. Flexible career paths, which ensure homogeneity of cultural norms across the organization, also have direct and indirect impacts on the operation of the technical core of the organization. *Ringi* decision-making procedures ensure a good fit between internal and external dimensions of equilibrium. Therefore, while the assumptions, values, norms, and artifacts of Japanese culture constantly reinforce internal harmony, they are also skillfully used to achieve a delicate balance of social-technical relations within the organization. Quality circles are a well-known example of a sociotechnical mechanism strongly supported by Japanese organizational culture. The norms that prize high quality and continuous improvement mesh with the quality circles to fully balance the system both internally and with respect to the environment.

The cultures of various individual Japanese firms are strongly reinforced by similar cultural principles that prevail throughout the huge Japanese conglomerates known as ***keiretsu.*** *Keiretsu* (literally meaning "societies of business") are cohesive groups of enterprises supported by one major bank that trade with each other. Their relationships are further tightened by maintaining close buyer-supplier relations and interlocking ownerships. Among the most well known of these *keiretsu* are Sumitomo, Mitsubishi, Mitsui, Dai Ichi Kangyo, Fuyo, and Sanwa. Collectively these groups earned 18 percent of the total net profits of all Japanese business during the late 1980s. They had almost 17 percent of Japan's total sales, held over 14 percent of its total paid-up capital, and employed almost 5 percent of its workforce (Ferguson, 1990).

These examples provide evidence to support the implications of the management systems model—that culture is an effective tool for implementing strategy. Whenever a new strategy is developed—whether it relates to the introduction of a new product or service, or penetration into new markets—an organization's culture can either support or hinder the change. Such was the case at a prestigious life insurance company. The company decided in the mid-1980s to diversify into new forms of insurance. However, over the years it had built such a great reputation for life insurance that its agents functioned more as order takers than as salespeople. Upon entering the new market, the agents were so ill-equipped to compete head-to-head with more aggressive competitors that the company failed in its efforts to successfully expand its operations.

CULTURE AND PERFORMANCE

Despite the seeming attractiveness of culture as a managerial tool, the relationship between culture and organizational performance is still sufficiently unclear, warranting caution. Yet mounting evidence attests to the value of culture as a systemic managerial tool. Denison (1990) studied the relationship between cultural and behavioral characteristics and financial performance in thirty-four organizations. The results "provide compelling evidence that it is quite possible to use cultural and behavioral measures to predict the performance and effectiveness of an organization over time" (p. 83). Denison (1990) views organizational effectiveness as being the product of the interaction of four cultural factors: adaptability, mission, involvement, and consistency (Figure 11-7). Additionally, there are some conceptual models and qualitative accounts of the relationship between culture and performance. For example, Sathe (1985) proposes that culture acts directly to mediate the fit between required and actual organizational behavior patterns to influence the level of organizational effectiveness.

From the perspective of the MSM, the belief that culture and performance are related is intuitively attractive. A number of arguments can be presented to defend this position. On a fundamental level, culture can be seen as operating much in the same fashion as procedures, particularly in terms of the way that procedures provide a context for decisions and limit freedom of choice. Culture also defines the situation for choice makers, promoting certain alter-

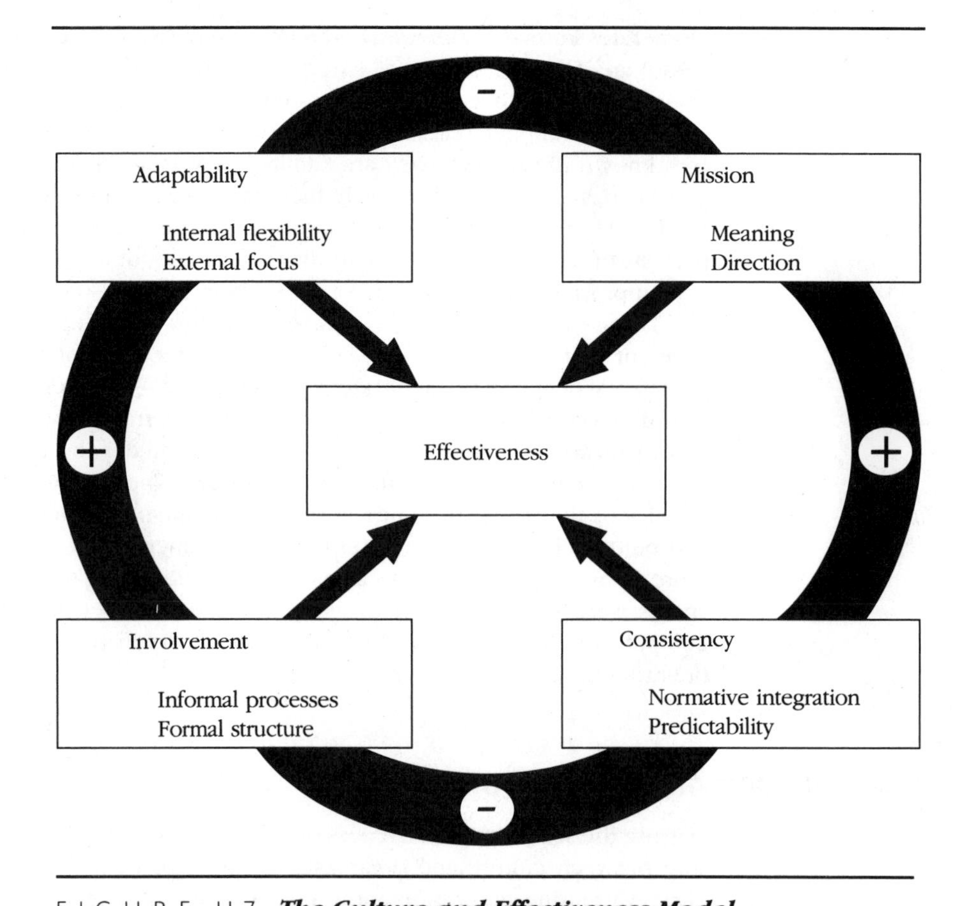

FIGURE 11-7 ***The Culture and Effectiveness Model***

Source: R. D. Denison, *Corporate Culture and Organizational Effectiveness* (New York: John Wiley & Sons, 1990).

natives over others. This occurs through limits created by organizational values and norms and through the creation of a self-selected pool of people who remain in the organization and enter positions of power.

CULTURE AS A SYSTEMIC MANAGERIAL TOOL

Our analysis of the formation, maintenance, and exploitation of organizational culture leads to four conclusions about the role of culture in sustaining a system's equilibrium.

First, organizational culture generates forces that can powerfully influence organizational equilibrium, especially in the social system. This power stems from the fact that culture dictates to organizational members what is important and expected. It also establishes limits on behavior in terms of what is desirable and what is acceptable. Through influential core values, role models, rites, and practices, culture captures members' imagination and channels their behaviors toward the organization's goals. It also has a direct impact on technical operations. This is because the cultures found in the best companies often include such values as constant innovativeness (3M Company or Sony) or market share leadership (Matsushita or General Electric).

Second, organizations build cohesive cultures primarily to regulate social systems. Internalized assumptions, norms, and values are powerful instruments for ensuring involvement, integration, and harmony among organizational members. In every organization it is possible to recognize the unique, characteristic patterns of rules, training, thinking, language, and performance standards that exist within the various subsystems (Neuhauser, 1988). Strong organizational cultures can bring those varied interests and values into a cohesive whole and ensure unity of direction, enhance information processing, and regulate decision-making processes.

Third, strong and cohesive organizational cultures produce systemic effects throughout entire organizations by balancing the relations of various subsystems. This dimension of culture is most apparent in the culture of Japanese firms in features like lifetime employment, zigzagging career paths, and *ringi* decision making. Similar conclusions have been reached by Peters and Waterman (1982) in their study of the best-run American companies and Denison (1990) in his study of relations between organizational culture and effectiveness.

Fourth, organizations differ in terms of the influence of their culture and how that influence is exploited by managers. It is natural when an organization is started and operates in largely informal, non-standardized ways to rely on culture as a soft tool of management. As the organization grows and matures, managers typically develop harder tools like procedures to achieve systemic equilibrium. Some organizations, however, follow both tracks: retaining and developing soft and hard tools. Japanese firms are probably the most illustrative case.

The strength of culture as a systemic management tool lies in its capacity to effectively stabilize the entire organization. Anchored deeply in the social system, culture has multiple and diverse balancing effects on the overall operation of a firm. It acts mainly by regulating organizational equilibrium. It allows for the preservation of an organization's identity in spite of incremental changes that might be introduced. As previously noted, culture is primarily oriented toward achieving equilibrium in the social dimension. But through the behavioral conditioning of organizational members, culture also indirectly influences the internal and external technical equilibria of a system.

This strength of culture might become a major weakness if an organization experiences unexpected change. Culture is difficult to create, and it is resistant to change. Internalized assumptions, values, and norms are much more difficult to change than technology, structure, or strategy—even in the midst of organizational decline and crisis. As Gagliardi (1986) aptly observed:

> The experience of failure does not in itself lead the organization to explore routes which are different from those sanctioned by the group's basic values and point of view, just as the failure to catch fish in the Mediterranean would

> never in itself have induced the sailor of olden times to go beyond the Pillars of Hercules in search of fish. For these reasons, many organizations will die rather than change, and in this sense we may say that *organizations do not learn from negative experiences.* (p. 129)

The inertia of organizational culture is therefore a double-edged sword. Under favorable conditions, it helps to maintain a productive equilibrium. When the criteria of equilibrium must be transformed, however, the culture may become the major barrier to innovative change. This issue will be discussed in greater detail in the next chapter, on leadership as a systemic tool of management.

SUMMARY

Organizational culture is a complex pattern of tacit assumptions, values and norms, and artifacts developed within the organization. The creation of culture is a dynamic learning process. Organizations and managers differ widely in the extent of their ability to exploit this tool. When properly used, culture becomes a powerful tool enabling managers to maintain balance, primarily in the social dimension of organization. Its effectiveness is greatly enhanced if it fits into an overall approach aimed at the systemic application of tools such as strategy, structure, procedures, and leadership.

Unfortunately, relatively little is known about the specifics of how to manage culture. Even less is known about the relationship between culture and organizational performance. Yet myriad accounts of the importance of culture have emerged from high-performing organizations. Part of this difficulty arises from the fact that culture has yet to be precisely defined. It is part of the soft side of organizations, along with leadership, that is more challenging to understand. Its similarities with leadership will become more evident in the next chapter.

KEY TERMS AND CONCEPTS

societal culture
organizational culture
tacit assumptions
values and norms
artifacts
ideological organization
professional organization
ringi
keiretsu

QUESTIONS FOR DISCUSSION

1. A new manager of a major division of Acme Corporation went to a seminar on culture and was so impressed by its potential value to the firm that she decided to attempt to transform the current culture. Where should she start and how long will it take?

2. Japanese firms are noted for their rich organizational cultures. To what extent should companies in Europe and North America attempt to duplicate their model of corporate culture?
3. The United States is a cultural melting pot. To what extent does this inhibit attempts to create cultures that equal Japanese firms in richness? To what extent is this a barrier to culture formation in global organizations?
4. Identify at least three common rituals or practices in an organization with which you are familiar. What do these rituals symbolize about the organization? What do they say about its values?
5. You have been promoted to a senior executive position at Acme Industries. The company is only five years old, but people are complaining that it does not have a culture. The CEO has turned to you for some action, asking that you establish a meaningful ritual that will bring the 125 employees together around a common theme. Do you take action? If so, what type? Do you respond that you will not do it, or ignore the request and hope that the CEO will forget?

REFERENCES

Argyris, C., and Schon, D. (1978). *Organizational Learning.* Reading, Mass.: Addison-Wesley.

Barnard, C. (1938). *The Functions of the Executive.* Cambridge, Mass.: Harvard University Press.

Clark, R. (1979). *The Japanese Company.* New Haven, Conn.: Yale University Press.

Deal, T. E., and Kennedy, A. (1982). *Corporate Culture: The Rites and Rituals of Corporate Life.* Reading, Mass.: Addison-Wesley.

Denison, R. D. (1990). *Corporate Culture and Organizational Effectiveness.* New York: John Wiley & Sons.

Ferguson, C. H. (1990). "Computers and the Coming of the U.S. *Keiretsu,*" *Harvard Business Review,* July–August, pp. 55–70.

Gagliardi, P. (1986). "The Creation and Change of Organizational Cultures: A Conceptual Framework," *Organization Studies, 7,* no. 2, pp. 117–34.

Jaques, E. (1952). *The Changing Culture of a Factory.* Hinsdale, Ill.: Dryden Press.

Kilmann, R. (1984). *Beyond the Quick Fix.* San Francisco: Jossey-Bass.

Mintzberg, H. (1983). *Power in and Around Organizations.* Englewood Cliffs, N.J.: Prentice-Hall.

Neuhauser, P. C. (1988). *Tribal Warfare in Organizations.* New York: Ballinger Publishing.

Nonaka, I. (1990). "Redundant, Overlapping Organization: A Japanese Approach to Innovation Process," *California Management Review, 32,* no. 3, pp. 27–35.

Ott, S. (1989). *The Organizational Culture Perspective.* Chicago: Dorsey Press.

Ouchi, W. G. (1981). *Theory Z.* Reading, Mass.: Addison-Wesley.

Pascale, R., and Athos, A. (1985). *The Art of Japanese Management.* New York: Simon and Schuster.

Peters, T. J., and Waterman, R. H. (1982). *In Search of Excellence.* New York: Harper and Row.

Sathe, V. (1985). *Culture and Related Corporate Realities.* Homewood, Ill.: Richard D. Irwin.

Selznick, P. (1948). "Foundations of the Theory of Organization," *American Sociological Review,* February, *13,* pp. 25–35.

12 Leadership as a Systemic Tool

Leaders do not show the same concern for process as managers. They are more concerned with the environment or culture within the organization. Managers consider changing process as innovative, but leaders view changes in the culture of the organization as innovative.

Richard Cyert (1991)

1. To understand the differences between transactional and transformational styles of leadership.
2. To comprehend how the challenge facing the systemic leader differs from those facing the traditional leader.
3. To recognize the potential value of the various leadership perspectives.
4. To explain how transformational leadership is especially relevant to the process of managing from a systemic view.
5. To realize that leadership is a managerial tool for changing organizations in the present and the future.
6. To realize how leadership differs from other tools in the management systems model.

SECTION I
TRANSACTIONAL LEADERSHIP

INTRODUCTION

There has always been an easy understanding among scholars and practitioners that there is a difference between good management and leadership. Leadership seemed to be a mystical property of charismatic individuals. What set Napoleon Bonaparte, Peter the Great, George Washington, Mao Zedong, Charles de Gaulle, Mohandis Gandhi, Martin Luther King, John F. Kennedy, and Lech Walesa apart from the crowd of typical politicians and social activists? Why have Thomas J. Watson Jr. (IBM), Akiro Morita (Sony), Jack Welch (GE), and Jan Carlzon (SAS) gained worldwide recognition as leaders of their firms, while thousands of other CEOs are considered just good managers? These questions have excited great interest among both scholars and the general public. They have also triggered a wave of recent research intended to unravel the mysteries of the sources of effective leadership.

The major perspectives that emerged from research into leadership focused on the role of power, personality traits, interpersonal behavior, and situational factors. Together these research approaches are part of a broader paradigm known as *transactional leadership.* The first part of this chapter will explore this view and discuss its strengths and weaknesses. Then the discussion will examine a more modern approach, known as *transformational leadership,* which offers great potential as a systemic tool for managing the equilibrium of an organization by creating a transcending vision of the future environment and developing a compelling sense of mission that engenders the support of members of the organization. Finally, this discussion will lead to an integrated approach proposing that leaders establish a dual role, first designed to maintain organizational equilibrium and second aimed at periodical shifts that disturb equilibrium to direct the organization to a more sophisticated level.

Leadership also plays a central role in the management systems model. Leadership is the key force that integrates the various systemic tools of strategy, structure, procedures, and culture (see Figure 12-1). Since these tools are complementary—substitutes or adjuncts to each other—leaders must determine which combination will yield the desired results at a given moment in time. Organizations can achieve short-term success by judicious use of the other systemic tools. However, over the longer term these tools must be effectively integrated through leadership. Consequently, leaders may find themselves challenged to determine when to fully exercise their leadership and when to yield in favor of alternative tools. Asserting one's leadership and choosing the right time to attempt to lead is also critical to a leader's success. An example of the importance of judgment in leadership is found in Poland's leader Lech Walesa:

GLOBAL SYSTEMS SPOTLIGHT

In the summer of 1980, unemployed electrician Lech Walesa jumped over the fence of the Gdansk shipyard in Poland in order to join the general strike, which had spread through Poland as a challenge to the ruling Communist Party. The direct result of the famous strike was a series of negotiations and compromises that established the first independent workers trade union in this communist country: the Solidarity union. During 1980 and 1981 Walesa skillfully led Solidarity from one clash with the Communist Party to another, careful to avoid giving the communists any reason for reprisal.

In a brief time, Walesa began to display many of the traits characteristic of leaders: he was dominant, self-confident, and skillful in dealing with difficult political situations. His power as the leader of Solidarity was not threatened during this time, and he expertly used his charisma and symbolic rewards to gain the loyalty of a large group of close supporters. However, due to the Soviet Union's insistence, the Polish Communist Party leadership imposed martial law in December 1981 and dissolved Solidarity. The activists were imprisoned, the union's assets were taken, and any attempt to challenge communism was severely and quickly repressed. From 1982 to 1989, Walesa remained a symbolic, heroic figure who did not relent to the communist regime and rejected any offers made by the communists. His image as an emerging great leader was enhanced when he was awarded a Nobel Peace Prize in 1984. He served as a beacon of hope for the future for the majority of Polish society. He boldly articulated a brave, new vision of a free Poland that was almost impossible to imagine after decades of communist domination.

By 1988 Solidarity had gone underground, but was gradually gaining strength while the influence of the ruling Communist Party began to diminish. At this point Walesa led a successful general strike, which resulted in an offer to negotiate a historic compromise between the communists and the Polish people. Walesa served as the spearhead in the effort to build a new vision of Poland's future. His steadfast belief in a free Poland helped create the momentum for change through a number of new initiatives that slowly began to build the power necessary to break the communist hold over Poland. The ultimate result of the process led by Walesa was the first democratic government in Eastern Europe, which was established late in 1989. Walesa's leadership climaxed with the presidential elections in December 1990—won by Walesa.

As the leader of the opposition Walesa clearly demonstrated both transactional and transformational dimensions of leadership. The task that he faces as president of contemporary Poland is very different from his earlier challenges. He is not leading people against a common enemy anymore. Furthermore, he does not have direct influence over the course of events in Poland, as a democratically chosen Parliament actually holds the reins of power. Only time will tell if Lech Walesa will be able to maintain his position of leadership in the situation he worked so hard to achieve.

TRADITIONAL CONCEPTS OF TRANSACTIONAL LEADERSHIP

Until recently most efforts by scholars to understand leadership effectiveness centered exclusively on the leader. The influence of the emerging social sciences, especially psychology, came to affect the research on leadership between 1900 and the late 1940s. Research shifted toward examining leader actions taken in relation to followers. During this period, leader attitudes and behaviors were seen as the stimulus for initiating change in organizations. An area of common ground among people who have studied leadership is that leaders are able to either cause or facilitate changes in the behavior of followers through the exertion of personal influence. Despite that agreement, there is considerable debate about the precise means by which leaders exert such influence.

Leadership effectiveness has traditionally been framed in terms of the abil-

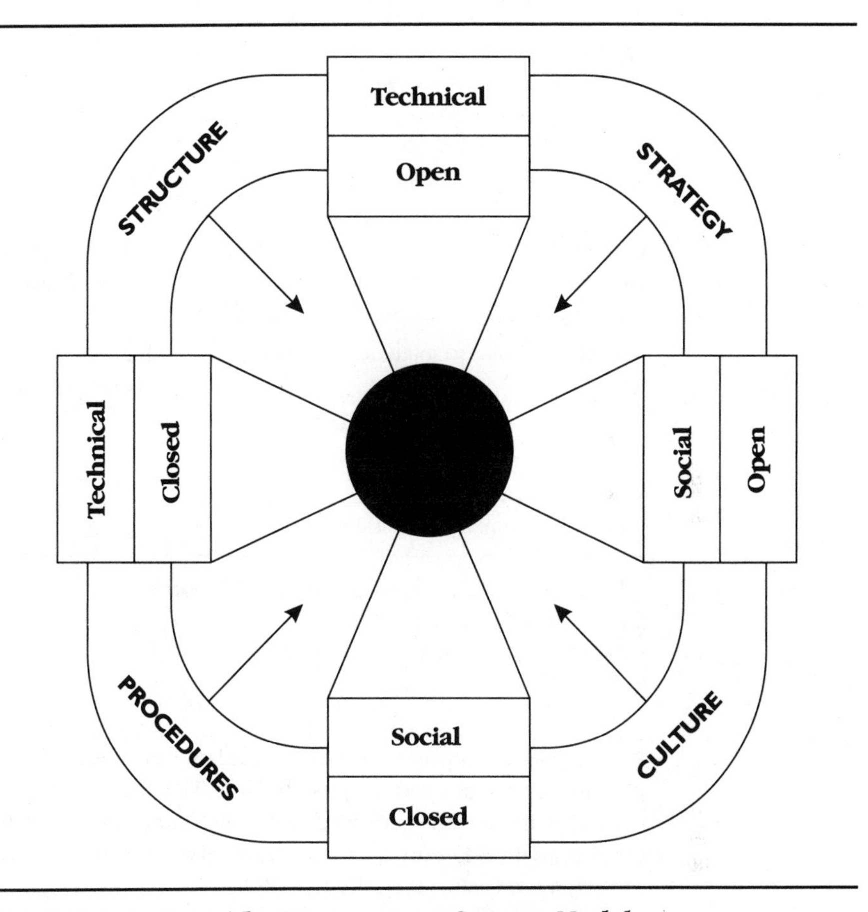

FIGURE 12-1 ***The Management Systems Model***

ity of leaders to achieve goals. Goals seem to be attractive criteria to use for evaluating leaders since they are relatively simple to understand and measurable. From the perspective of the management systems model, however, goal attainment is not a good indicator of organizational health. We propose that effective leaders are those who can balance equilibria of key systems while simultaneously transforming them. Before we can explore this perspective, however, we must examine more fully the traditional view of leadership.

THE TRAIT PERSPECTIVE

The trait perspective is the oldest leadership approach. Underlying it are several powerful assumptions that have significantly contributed to the mysterious and almost magical perception of leadership (Bennis and Nanus, 1985):

1. Leaders are born, not made.
2. Leadership is a rare skill based on personality traits.
3. Leaders are charismatic.

TRAITS	SKILLS
Adaptable to situations	Clever (intelligent)
Alert to social environment	Conceptually skilled
Ambitious and achievement-oriented	Creative
Assertive	Diplomatic, tactful
Cooperative	Fluent in speaking
Decisive	Knowledgeable about group task
Dependable	Organized
Dominant (desire to influence others)	Persuasive
Energetic	Socially skilled
Persistent	
Self-confident	
Tolerant of stress	
Willing to assume responsibility	

FIGURE 12-2 ***Traits and Skills Found to Be Characteristic of Successful Leaders***

Source: Gary A. Yukl, *Leadership in Organizations* (Englewood Cliffs, N.J.: Prentice-Hall, 1981), p. 70.

With these assumptions in mind, scholars have focused on the question: what personality traits make a good leader? The major problem with such an approach is its limited generalizability. Reseach into different individual leaders came up with different, and quite extensive, sets of characteristics. An example of such a list is shown in Figure 12-2.

POWER AND LEADERSHIP

One of the earliest approaches to leadership was developed by French and Raven (1959). They identified five major sources of power that leaders can potentially draw upon to enhance their leadership position vis a vis followers:

1. *Legitimate power* is created by the position occupied by a manager in an organizational hierarchy. It refers to a manager's right (authority) to define tasks that subordinates must perform because they have an obligation to comply with legitimate orders.
2. *Reward power* depends on the leader's control over valued rewards such as money, promotion, and recognition. Employees comply with legitimate orders in the belief that their performance will be rewarded. People who control valued outcomes for others have reward power.
3. *Coercive power* reflects the flip side of the rewards. While compliance and good performance is rewarded, lack of discipline and poor performance can be punished. The potential threat of using this tool induces fear in followers. To a degree, this will increase the leader's ability to control sub-

ordinates' behavior. It is not an effective tool in business organizations when carried to its extreme.

4. *Referent power* is based on a leader's reputation and charisma, personal characteristics that appeal to followers. These forces may engender the admiration and loyalty of followers. Referent power increases as a leader's behavior becomes a model for the behavior of others.
5. *Expert power* is based on the perception of followers that a leader possesses superior experience, knowledge, or skills. If subordinates believe that they can learn or otherwise gain from the expertise of a leader, the probability of their compliance and admiration is increased.

Both rewards and punishments can also be formalized and regulated by systemic forces, such as organizational structure and procedures. They can also be imbedded informally in the culture and regulated through norms and traditions. Ceremonies can be institutionalized forms of reward (Peters and Waterman, 1982). The major insight gained from this line of research resides in understanding the different sources of influence and how they can be combined to get results.

The power perspective also has weaknesses. The sources of referent and expert power are partially a function of a manager's prior experience—by already being a manager one gains access to the institutional support that can increase a leader's effectiveness. This is true of legitimate power as well, for it is usually gained through outstanding prior performance. Therefore, a circular pattern of self-fulfilling cause and effect emerges: a leader may draw on expert, referent, and legitimate power, but in order to acquire those powers, he or she must demonstrate leadership.

THE BEHAVIORAL PERSPECTIVE

The behavioral perspective on leadership concentrates on building profiles of which leader behaviors are the most effective. It proposes that leadership styles are positively related to organizational performance. The concept of *leadership style* is based on the seminal thinking of Lewin and Lippitt (1938), the first to suggest that behaviors could be classified based on how leaders involved themselves with work-related and people-related issues. For the first time leadership was visualized as existing on a continuum. In this case the continuum is in terms of either work-people orientation or autocratic-democratic dimensions (see Figure 12-3).

A **leadership style** is a general pattern of behavior that favors either tasks

Work-oriented	⟷	People-oriented
Autocratic	⟷	Democratic

FIGURE 12-3 ***Continuum of Leadership Types***

or people in decision-making activities. Numerous researchers have attempted to determine whether a linkage existed between leadership styles and organizational performance (Bass and Valenzi, 1974; Likert, 1967; Vroom and Yetton, 1973; Yukl, 1981). Among the most publicized of these studies were those conducted at Ohio State University and the University of Michigan.

The Ohio State University team, identified a continuum with extremes that they labeled as consideration and initiating-structure styles. A considerate style of leadership is concerned with creating and maintaining a friendly and rewarding working climate. Considerate leaders care about the comfort, well-being, respect, and dignity of subordinates. A leadership style that emphasizes initiating-structure is oriented toward clarity in the way tasks are defined, and standards of performance and procedures are established. The most effective

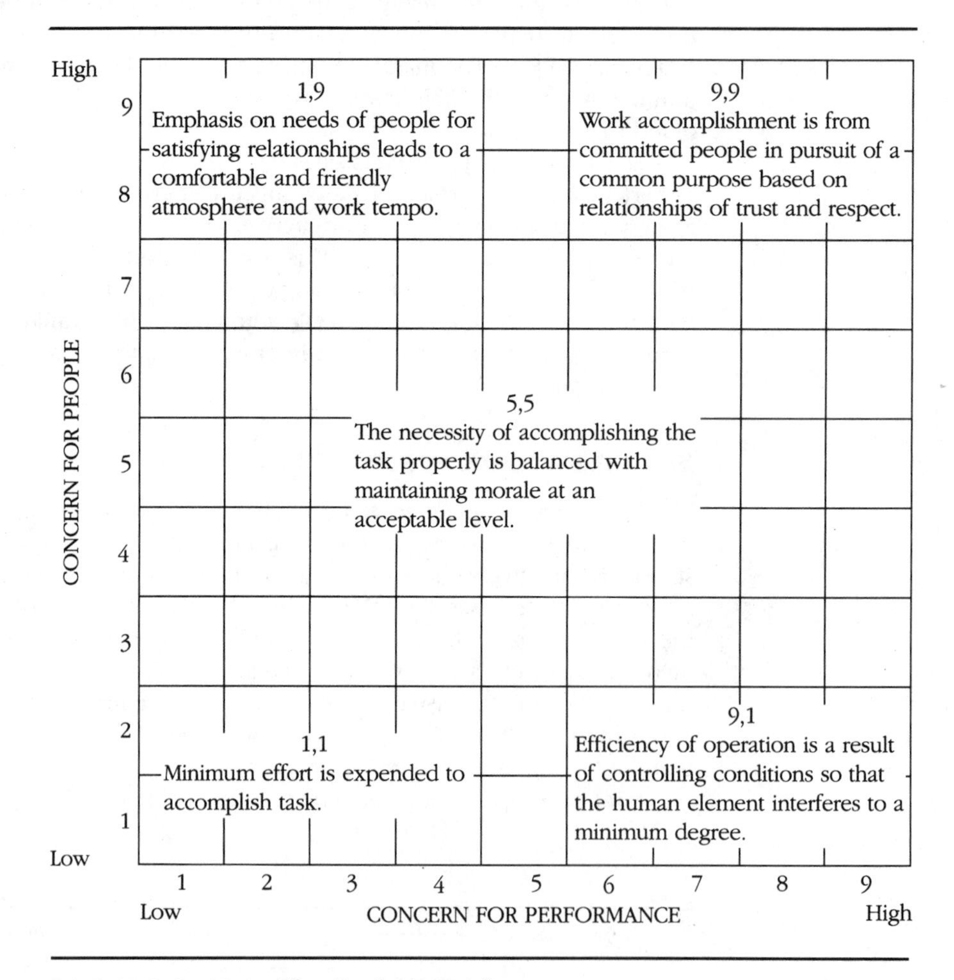

FIGURE 12-4 ***The Cockpit Grid***

Source: R. Blake, J. S. Mouton, and A. McCanse, *Change by Design* (Reading, Mass.: Addison-Wesley, 1989), p. 9.

style was thought to be one combining the high-performance orientation of strong concern for both structure and people (Schriesheim and Bird, 1979).

This concept was embellished by Blake and Mouton (1964) in their well-known model, the Managerial Grid, and later (1989) in its revised form as the **Cockpit Grid** (Figure 12-4). The Cockpit Grid grew out of a series of seminars and research the authors conducted with cockpit crews at United Airlines (UAL). The grid provides a visual mechanism whereby leaders can better understand the interaction between concern for people and concern for performance, and how this interaction influences the choice of leadership style. This conceptual tool also allowed leaders to compare their leadership style against the grid and its ideal, the (9,9) style. The (9,9) style exists when leaders fully integrate concern for people with concern for performance. The grid approach is not only an approach to leadership. It has also been promoted by its designers as a managerial tool for organization development and penetrating organizational culture (Blake, Mouton, and McCanse 1989).

Research conducted at the University of Michigan under Rensis Likert (1967) followed a similar theme in the search to link group effectiveness with leadership style. The Michigan team identified two distinct leadership styles, job-centered behavior and employee-centered behavior. The team's initial finding was that groups with employee-centered leaders were more productive and enjoyed a better quality of work life.

The behavioral perspective on leadership is attractive in its simple persuasiveness. More importantly, it makes a meaningful attempt to balance the oft-opposing forces of the social and technical aspects of management. Its prime weaknesses have been noted by advocates of the contingency approach to leadership: it does not account for the diversity of effective leader behaviors or of largely situational circumstances such as the type of task, technology, or environment. The behavioral theory of leadership continues the faulty premise of the trait approach, which is that there is one best way to lead.

THE CONTINGENCY PERSPECTIVE

The contingency approach to leadership dominated organizational and management theories from the late 1960s to the early 1980s. This perspective on leadership—known as **situational leadership**—begins with one major assumption: *it all depends.* There is no single universally effective style; no leadership traits, sources of power, or behaviors always lead to success. The implication is that either leaders must vary their tactics to fit the situation or they must be matched to assignments based on the congruence between their style and the job requirements. The contingency approach proposes that to ascertain the most effective style for a given situation it is necessary to assess the key elements that shape the forces in operation at a given time. This general approach is reflected in the Fiedler (1967) and Vroom-Yetton (1973) models.

FIEDLER: LEADER MATCH

Fiedler contributed to the contingency approach to leadership by constructing a model that establishes a link between key situational elements and leadership styles. The Fiedler model attempts to match the personal style of the leader with three situational variables: leader-member relations, task structure, and position power of the leader. Fiedler's approach is based on the following beliefs:

1. All situations vary in the extent to which they support the leader's ability to influence.
2. The leadership styles that are effective in favorable situations are not effective in unfavorable circumstances.
3. Leadership effectiveness is promoted when the leadership style is congruent with the variable elements of the situation.
4. Leadership style is an enduring human characteristic that is not easily changed. Therefore, it is most useful to match the leader and the situation.

Figure 12-5 summarizes the major points of Fiedler's model. It shows that task-oriented managers are more likely to be effective leaders in very favorable situations and in the least favorable situations.

The Fiedler model is a useful tool for learning to conceptualize leadership issues from a contingency perspective. There is considerable doubt, however, about its validity and its use as a tool for practitioners.

VROOM AND YETTON: SITUATIONAL LEADERSHIP DECISIONS

The Vroom and Yetton contingency model (1973) is also based on the idea that effective leadership depends on situational conditions. In this case, the

Situational Characteristics	1	2	3	4	5	6	7	8
Leader-member relations	Good	Good	Good	Good	Poor	Poor	Poor	Poor
Task structure	High	High	Low	Low	High	High	Low	Low
Position power	Strong	Weak	Strong	Weak	Strong	Weak	Strong	Weak
Effective leadership style	Task oriented		Relationship oriented				Task oriented	

Very favorable ⟷ Very unfavorable

FIGURE 12-5 ***Fiedler Contingency Model of Leadership***

Source: F. Fiedler and M. Chemers, *Leadership and Effective Managememt* (Glenview, Ill.: Scott, Foresman, 1974), p. 80.

variables are the nature of the problem, the amount of information available to support a high-quality decision, and the level of participant input and of participants' acceptance of organizational goals. Based on these three factors Vroom and Yetton define five leadership styles along a continuum ranging from authoritarian to democratic leadership. Authoritarian leaders solve problems in a rational fashion by using the information available at the time of the decision. Democratic leaders share problems with subordinates and act as facilitators of the problem-solving process. The team of subordinates is asked to generate several alternative solutions, evaluate them, and select the most appropriate. The remaining styles on the continuum are hybrids of these two pure approaches.

The Vroom-Yetton model is a rational decision-making approach to leadership. As is the case with any rationalized view of human behavior, it is limited because it ignores so many other contingencies. Yet the Vroom-Yetton model stands out for its integrity of design and its general acceptance as an effective model of the cognitive processes involved in leadership and decision making.

SYSTEMS-ORIENTED CONTINGENCY PERSPECTIVES

Presently there is an absence of any truly systemic models of leadership behavior available. However, within the contingency perspective on leadership are several systems-oriented theories that warrant further discussion. In general, these models pay greater attention to the role of the environment in determining leader behavior, in contrast with traditional contingency approaches. Fiedler and Vroom and Yetton, for instance, define the environmental contingencies in terms of the **task environment.** This definition is narrowly focused, concentrating only on jobs and small groups within organizations. The systems-oriented approaches of Osborn and Hunt (1975), Bass, Valenzi, and Farrow (1977), and Kerr and Jermier (1978) attempt to extend the consideration of the environment beyond the boundaries of the organization. Kerr and Jermier have also attempted to design a model that explicitly accounts for the systemic interaction among leadership determinants.

These systems-oriented contingency approaches attempt to associate the concept of the environment with leadership effectiveness. Although none of these models is widely cited in current management literature, they represent a line of thought that has expanded the concept of leadership to be more systemic and open.

Osborn and Hunt (1975) sought to explain the dynamic behavior of leaders. Labeled the *adaptive-reactive theory,* their model perceives that leaders adapt to the system and react to pressures from subordinates. The system is comprised of two basic elements, micro and macro factors. Micro factors primarily relate to tasks and the people who perform them. Macro factors include a broader set of forces including environment, size of the organization, technology, and structure. Overall, the results of attempts to verify this theory have provided general support for it.

Osborn, Hunt, and Bussom (1977) attempted to integrate the concept of leadership style with Ashby's Law of Requisite Variety. They reasoned that organizations must have the capability to regulate their own internal workings to accommodate pressure for change created by a dynamic environment. To recognize situations where control of such internal structures and mechanisms was needed, they believed, leaders would need to have a complex appreciation of the environment. The results of their research indicated that effective leader behavior was related to the degree of match with environmental circumstances.

In a similar fashion, Bass, Valenzi, and Farrow (1977) extended the definition of the leadership environment to include economic, legal, political, and social factors. They conceived of leadership dynamics in terms of a core open system based on the relationship between the leader and subordinates. This open system is vulnerable to the effects of power and information emanating from the environment. Their research focused specifically on whether leadership styles correlated with the way managers perceived the environment. They found that managers who placed greater importance on the role of economic environmental factors were more directive while those who emphasized the role of legal, political, and social environmental forces were more consultative.

An interesting and provocative systems-oriented approach to leadership was developed by Kerr and Jermier (1978). Their view is that organizational effectiveness largely depends on a host of organizational characteristics, including leadership. More importantly, they say that various factors within organizations can be substitutes for leadership, a view compatible with the management systems model described in these pages. Much as control can be achieved through procedures or via the skillful use of culture, leadership must be viewed as one element within an overall system. The factors that Kerr and Jermier have identified as **substitutes for leadership** are the degree of formalization and specialization, task characteristics, and subordinate features (see Figure 12-6).

The primary function of the two typical leadership styles (instrumental, or task oriented, and supportive, or people oriented) can be neutralized or substituted with other instruments of management. *Neutralizers* counteract the effectiveness of leadership; *substitutes* have the potential to render leadership impossible or unnecessary. Thus the same results can be achieved, irrespective of the leader's styles and behaviors, if situational characteristics (subordinates, tasks, and organization) meet the requirements shown in Figure 12-6. The Kerr and Jermier model has not been extensively tested to determine its accuracy.

The major strength of the systems-oriented contingency approaches is the recognition that leadership is not a simple phenomenon and that effective leadership has no single cause. In the quest to match performance, leadership style, and a variety of other factors, research has increasingly leaned toward uncovering complex and dynamic relations that offer the most comprehensive explanation of leadership.

SUBSTITUTE OR NEUTRALIZER	SUPPORTIVE LEADERSHIP	INSTRUMENTAL LEADERSHIP
Subordinate Characteristics		
1. Experience, ability, training		Substitute
2. Professional orientation	Substitute	Substitute
3. Indifference toward rewards offered by organization	Neutralizer	Neutralizer
Task Characteristics		
1. Structured, routine tasks		Substitute
2. Feedback provided by task		Substitute
3. Intrinsically satisfying tasks	Substitute	
Organizational Characteristics		
1. Cohesive work group	Substitute	Substitute
2. Leader lacks control over rewards	Neutralizer	Neutralizer
3. High formalization (explicit goals, plans, responsibilities)		Substitute
4. Inflexibility (rigid rules and procedures)		Neutralizer
5. Leader located apart from subordinates, limited communication possible	Neutralizer	

FIGURE 12-6 ***Substitutes for and Neutralizers of Leadership***

Source: Adapted from S. Kerr and J. M. Jermier, "Substitutes for Leadership: Their Meaning and Measurement," *Organizational Behavior and Human Performance* (1978), *22,* no. 3, pp. 375–403.

APPRECIATIVE MANAGEMENT

The majority of leadership models emphasize the cognitive dimensions of leading. Their thrust is to prescribe how leaders may redirect their actions vis a vis followers to effect change. The redirection of leadership has often been discussed as a contingent process independent of the leader's own values and beliefs. Shrivasta, Fry, and Cooperider (1990) have called for a shift of mind among executives to achieve a more "appreciative" posture. **Appreciative management** seeks to integrate the mind and soul of the leader with core values and beliefs of the leader, the organization, and the society in general. Appreciative management is based on a systemic, holistic paradigm that requires that leaders reshape their own mental models by considering the fundamental assumptions and world views that underlie those mental models. These scholars argue, "We must recognize . . . that the whole pattern, the whole underlying premise of Western science, has masked questions of human value and has lead inexorably to the kinds of problems and dilemmas we now face. It is impossible to create a well-working society on a knowledge base that is fundamentally inadequate, seriously incomplete, and mistaken in its basic assumptions" (Shrivasta, Cooperider, and Associates, 1990, p. 12). In essence, appreciative management is a way of knowing the world, a mental framework, that can enable leaders to align themselves with others in a mutually beneficial way.

EVALUATING TRANSACTIONAL LEADERSHIP PERSPECTIVES

Each school of thought on leadership has its own limitations. The contingency perspective is limited because its methods of measurement are generally rather subjective and general. While this approach does provide a powerful indication that in relatively stable environments, leadership styles should fit the situation, in relatively dynamic environments, such as those widely experienced in the global economy, these prescriptions are not viable. The idea that effective leaders must adapt their styles to the requirements of situation assumes several unrealistic notions such as the beliefs that leaders are capable of performing with chameleonlike flexibility in dynamic environments or that organizations have unlimited ability to freely move leaders around to provide good fit between leader style and task environment. Furthermore, the approach ignores the potential role played by neutralizers and substitutes as a leadership tool.

In summary, nearly half a century of research on leadership offers some important insights into the nature of effective leadership. First, a leader should have versatile sources of power. Authority, based on position in a hierarchy, is not enough—a leader should also know how to combine the use of referent, expert, reward, and coercive powers. Second, leaders should have special personality traits making them capable of energizing followers. Third, leaders should skillfully balance their task and people orientations, integrating these two major elements of behavior. They should structure tasks to balance technical and social forces and support people by taking into account their expectations and preferences. Fourth, effective leaders must adjust their style to the contingencies and demands of the situation. They must take into account the structure of tasks, problems at hand, types of personnel, and such characteristics of the organization as levels of specialization, standardization, and formalization as well as the environment.

The leadership perspectives that have been discussed to this point are founded on a general world view known as **transactional leadership** (Burns, 1978). Transactional leaders rely on the skillful use of power, rewards, and contingencies to maintain or marginally improve performance and to ensure compliance with organizational policies. To be effective the transactional leader attempts to fulfill the expectations of followers while simultaneously meeting the demands of situation. Bass (1985) comments:

> Transactional leaders serve to recognize and clarify the role and task requirements for the subordinates reaching the desired outcomes. This gives the subordinates sufficient confidence to exert the necessary effort. Transactional leaders also recognize what the subordinates need and want and clarify how these needs and wants will be satisfied if the necessary effort is expended by

> the subordinates. Such effort to perform or motivation to work implies a sense of direction in the subordinate as well as some degree of energization. (pp. 12–13)

Transactional leaders are instrumental in their approach. They focus on structure, rewards, punishments, and procedures to maintain and improve conditions that create desired behaviors (Nadler and Tushman, 1990). In some respects transactional leaders are mundane. That is, they engage in a mechanical process of structuring, manipulating goals, matching structure and procedures, and ensuring compliance with directives to get things done efficiently.

The polar contrast to this type of leader is the transformational leader, who does not emphasize strict compliance from followers or require it (Kuhnert and Lewis, 1987). Building on recent developments in the study of organizational culture and strategy, the transformational leader adopts a different mental model. This leader is oriented toward envisioning a future that will be compelling to others while working to empower and energize organization members to work freely to pursue goals in their own way. Simply, **transformational leadership** pulls others continuously toward an ideal future. In many respects, the evolution of leadership from transactional to transformational leadership represents a transfer from hard systems assumptions toward soft systems thinking. In the next section we will examine transformational leadership in greater detail.

SECTION 2
TRANSFORMATIONAL LEADERSHIP

CHARACTERISTICS OF TRANSFORMATIONAL LEADERS

Transformational leadership restores the role of the leader to a dominant position in two respects. First, the transformational leader's frame of reference is not a particular situation or style. Rather it is a congruent set of creatively followed **core values.** These guiding standards include such factors as integrity, respect, trust, and fairness (Kuhnert and Lewis, 1987). Transformational leadership normally deals with broad issues that are of major importance to the organization. For routine, daily circumstances, such leaders may use an approach that centers to a greater extent on building the capacities of individuals, known as the microskills approach to leadership (Cavaleri and DeCormier, 1991).

Second, transformational leaders recognize the need for **second-order change,** or innovation. Indeed, the ultimate test of the extent to which a leader practices transformational leadership is the value placed on managing such change. The pursuit of such change allows the system to cope with rapid environmental changes and respond to customers in a more timely fashion than can competitors. In short, transformational leaders transcend existing definitions of what the organization actually is and work to create a new way of being for the organization. This new way of being will include an altered equilibrium at a more sophisticated level.

Several critical elements make transformational leadership possible. There are three closely related major components of transformational leadership (Kotter, 1990; Nadler and Tushman, 1990; Tushman, Newman, and Romanelli, 1986) (see Figure 12-7):

1. **Transcending,** the ability to grasp the nature, scope, and importance of forces leading to second-order change.
2. **Envisioning,** developing a compelling vision of the organizational future and translating it into a mission anchored in core values and standards.
3. **Energizing,** inspiring members of the organization to institutionalize the innovative change needed to implement the vision.

TRANSCENDING

Transformational leaders transcend traditional analysis of environmental potentials and other possible futures. They have an extraordinary ability to look ahead and spot new trends and patterns, which can be translated into opportunities. Such leaders exploit opportunities and even transform threats into points of organizational leverage. Transformational leaders are capable of defying tradition and skillfully riding the waves of discontinuity, much like a champion surfer. They thrive on dynamic conditions within their industries. Many ominous situations that threaten average performers—such as shifts in product life cycles, deregulation, and technological breakthroughs—are converted into opportunities by the transformational leader.

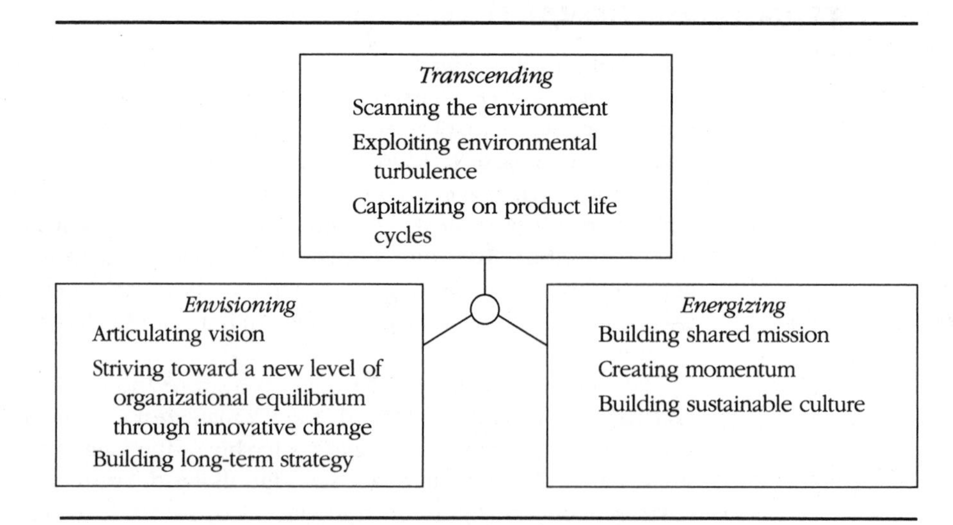

FIGURE 12-7 ***The Critical Factors of Transformational Leadership***

ENVISIONING

Transformational leaders supply their organizations with a bold, exciting, almost unattainable mission. This sense of purpose is built on a set of core values and standards. The leader's imaginative analysis of future opportunities must be followed by an articulate process of envisioning. Transformational leaders of Komatsu (Ryoichi Kawai), GE (Jack Welch), Apple (John Sculley), Asea Bown Boveri (Percy Barnevik), SAS (Jan Carlzon), and Zanussi (Carlo Verri) have the courage to think the unthinkable, dream the unattainable, and develop a new, bold vision of the future. The mission statements developed by transformational leaders have three important features that help to reconcile the major contradictions that will emerge between stability and change, certainty and risk, crisis and success:

1. A mission translates a leader's vision of the future into a clear sense of direction. It almost naturally coordinates a scope and timetable of necessary activities. A mission is a benchmark, a simple beacon for the organization's strategies (for example, Bristol-Meyers Squibb's mission to be "number one by 2001," IBM's mission that "IBM means service," or toy producer Lego's motto, "Only the best is good enough"). The mission introduces certainty of direction in the turbulent, risky, and often confusing environment.
2. A mission challenges the existing organization by demanding that a new level of equilibrium be established. A new fit must be created among environment, strategy, structure, procedures, and culture, based on an alternative set of values and standards. However, a transformational mission does not demand immediate change of all existing managerial tools. Rather it builds on the particular strengths of some management tools and aligns others to synthesize a new logic of performance.
3. A mission provides an organization with a consistent long-term strategy. A value-based strategy lessens the importance of short-term difficulties and reframes them in the context of long-term success. Without this explicit frame of reference, short-term considerations, pressing contingencies, environmental fluctuations, and so on would take precedence over the strategic challenge to build a new level of organizational equilibrium.

Quinn (1988) suggests that master managers are able to harmonize the paradoxical forces that operate in complex situations. As Quinn observes,

> There is a marked difference between the novice and the master. The novice does not understand the transformational cycle, but is limited by reliance on deductive rules, fear of failure, and hesitancy to engage risk and uncertainty. In contrast, the master moves confidently through the cycle, each time achieving a greater understanding of the unknown, a fuller appreciation of excellence and how to create it. Thus the master gains a complex, holistic, and paradoxical understanding of the cycle. (pp. 25–26)

Gardner (1989) also suggests that leaders can be differentiated from managers by their ability to intuitively appreciate the nonrational aspects of situations and their ability to think in terms of longer time frames.

A truly visionary, transformational mission stabilizes dynamic organizations. It introduces a sense of certainty in the turmoil resulting from transition to new positions in products, markets, technologies, and so on. It simultaneously works to create a culture where the status quo is not accepted and innovative processes emerge. One example of such an approach can be found at the computer firm, NeXT, Inc., led by former Apple Computer founder and CEO Steve Jobs.

Steve Jobs and NeXT, Inc.

Steve Jobs savored the moment. In the crisp light of an overhead projector, the founder of NeXT, Inc., placed a transparency of the company's bank balance. All eyes sought a single entry: payment from International Business Machines Corp. for a licensing agreement. It was worth millions.

A cheer rose from the 100-plus NeXT employees crammed into the fledgling computer firm's Palo Alto R&D building that fall day in 1988. As they toasted each other with champagne-filled plastic cups, any doubts they had about the company's future disappeared.

The meeting, however, was more than a confidence booster for NeXT. It was also a vivid example of Jobs's unique management style. To company insiders it is known as the "open-system," industry slang for a computer design that is compatible with hardware made by others. It's a metaphor: The fact is that within NeXT there are no secrets. Equally important, nothing learned inside the company is leaked to the outside world. To this day, no one outside the company, other than top IBM executives, knows how much the IBM payment was worth.

And that's the way Steve Jobs wants it to be. No longer the youngster who built Apple Computer Co. into a billion-dollar technology leader, he's older now and has learned many lessons, some of them the hard way. He realized that if NeXT was to succeed, as much thought had to go into the company's organization as its product. The result, as might be expected when Jobs is involved, is a little unusual.

The traditional management-by-objectives theory that sapped Apple's creativity has been tossed out. Instead, Jobs is running NeXT by coveting values, such as innovation, excellence, and productivity over results. If everyone pursues his values, positive results will inevitably follow. And he is making sure the entire company is guided by similar principles as he brings in new talent, so NeXT won't trip at its first spurt of growth.

The cornerstone of this management system is NeXT's unrelenting emphasis on internal communication. "Steve recognized something unique was being created at NeXT. He isn't simply devoted to developing a product," contends Phillip Wilson, NeXT's vice-president for human resources. "He is really interested in creating the first company for the 21st century."

It may sound like a New Age management primer, but a close look inside NeXT shows Jobs's approach is yielding surprising results. At company-wide monthly meetings and annual retreats, Jobs listens to suggestions and complaints and, in turn, goads employees into matching his high standards. What he's doing is also shrewd. By letting the process permeate every level of the organization, from hiring to strategic planning, Jobs is not merely building a company, he's molding a team. He's winning a degree of participation and commitment other chief executives would envy. Jobs is betting that these qualities are NeXT's ticket to becoming a billion-dollar company.

Source: M. Rothman. "Who's Running NeXT?," *California Business,* April 1990.

Steve Jobs's leadership of NeXT clearly is based on the belief that using soft systemic managerial tools such as values and culture building will produce hard results over the long run.

ENERGIZING

Transformational leaders are successful in energizing and inspiring organizational members to share and follow the alternative mission. Aligning mission, strategies, structures, and procedures is not sufficient to transform an organization. A transformational mission gives members of an organization a dream, which can boost morale, energize them, and serve as a cohesive force during a protracted period of transition to future levels of equilibrium. It offers a picture of a desirable future, raises expectations of success, and provides a possible escape from difficult times. Transformational leaders enroll, enable, and encourage people. They unleash human energy, initiative, and creativity while at the same time striving to build the level of commitment to the mission. The process of energizing and building commitment has three main components (Itami and Roehl, 1987):

> First, while the mission might emanate from the top, in order for successful transformation to be realized, it must be shared. The great companies are particularly successful in building common visions through strong values and norms expressed in mission statements, credos and even songs (e.g., Matsushita employees sing every day about "sending our goods to the people of the world, endlessly and continuously, like water gushing from a fountain"; IBM also has what is known as the IBM "fight song").
>
> Second, leaders must create momentum by giving people freedom of action, offering opportunities to take fuller responsibility and enjoy the taste of success. These beliefs normally translate into more self-directed work activity and a change from traditional managerial roles. It also means flatter organizational structures, more teams-based decisions, and greater sharing of information.
>
> Third, leaders must create a thick organizational culture in order to sustain momentum and peoples' commitment. Senge (1990) notices that visions and energy often die prematurely because people become discouraged by the difficulty in implementing a mission, overwhelming demands of current reality, and the lack of clear connection between mission and actions. In order to

sustain a forceful momentum, leaders must operate at the limits of organizational consensus and abilities by eliminating unsuccessful practices, promoting new values, and establishing new challenges.

General Electric tries to maintain momentum and support organizational culture *development* by special training sessions called *Work-Outs,* where cross-functional teams cooperate to address actual business problems, future threats, opportunities, and directions. At the heart of the Work-Out program is CEO Jack Welch's idea of empowerment.

GE's Work-Out Program

Work-Out has a practical and an intellectual goal. The practical objective is to get rid of thousands of bad habits accumulated since the creation of General Electric. . . . We want to flush them out, to start a brand new house with empty closets, to begin the whole game again. The second thing we want to achieve, the intellectual part, begins by putting the leaders of each business in front of 100 or so of their people, eight to ten times a year, to let them hear what their people think about the company, what they like and don't like about their work, about how they're evaluated, about how they spend their time. Work-Out will expose the leaders to the vibrations of their business—opinions, feelings, emotions, resentments, not abstract theories of organization and management. . . . The ultimate objective of Work-Out is so clear. We want 300,000 people with different career objectives, different family aspirations, different financial goals, to share directly this company's vision, the information, the decision-making process and the rewards. We want to build a more stimulating environment, a more creative environment, a freer work atmosphere, with incentives tied directly to what people do.

Source: N. Tichy and R. Charan, "Speed, Simplicity, and Self-Confidence: An Interview with Jack Welch," *Harvard Business Review* (September–October 1989), pp. 116, 118.

Transformational leadership differs significantly from transactional leadership in terms of its objectives and means. The instrumental and control orientation of the transactional leader is replaced by processes designed to enable and energize organization members. Leaders effect such transformation of the organization by introducing second-order changes and breaking traditional frames of reference. Transformational leaders must be credible, with the power to influence others in order to instill a challenging vision and translate it into organizational values and norms that members share. The transformational leader is a seemingly paradoxical combination of considerate coach, inspiring teacher, stimulating liberator, and restless achiever.

THE CHALLENGE FOR LEADERS

Kouzes and Posner (1988) have developed a practical explanation for leadership effectiveness that has become widely accepted by practitioners. They propose that effective leaders are able to transform challenging opportunities into their favor and ultimately toward success. They have outlined a number of steps that leaders can take to transform challenges into opportunities:

1. Look at every job as if it were an adventure.
2. Treat every new assignment as if it were a turnaround.
3. Ask other people to join in solving problems.
4. Break free of routine that is unproductive and try something new.
5. Institutionalize processes for collecting innovative ideas.
6. Put idea gathering on the personal agenda.
7. Renew work teams through career development and new hires. (pp. 68–69)

SECTION 3
LEADERSHIP AS A SYNERGETIC MANAGERIAL TOOL

A THIRD VIEW OF LEADERSHIP

We began our analysis of leadership with a review of the major research perspectives: trait, behavioral, and contingency perspectives. These approaches converge in the paradigm of the transactional leader who operates in a relatively stable environment. This prototypical leader is largely concerned with aligning people with tasks, procedures, and structure. We then discussed the transformative paradigm of leadership, which restored the role of leaders to a more direct action-oriented position. The transformational leader provides a major force in directing and driving organizational changes. Now we address how these two major leadership paradigms complement each other, for a systemic form of synergy can be achieved by their convergence. This interaction can be viewed in terms of its function in the management systems model.

Transactional leadership is based on three conceptual pillars: organizational structure, procedures, and environmental stability. Activities such as assigning tasks, distributing information, enacting rewards and punishments, and invoking procedures are all transactions between leader and followers. Because transactional leadership presupposes a relatively stable business environment, the organization is considered as a relatively closed system. The importance of normally occurring environmental disturbances in this paradigm is diminished in the relatively static concept of the transactional leader. In the face of growing environmental pressures the transactional leader will concentrate on sealing off the organization from these forces and on becoming more tightly controlled and efficient. By contrast, the transformational leader will seek to incorporate the changes into the organization, reframe the purpose of the organization, and transform it to be aligned with the environment. In short, transactional leaders manage organizations as though they are a relatively closed technical system. As a consequence of belief in this world view, transactional leaders seek social compliance and control.

Transformational leadership is based on three very different foundations: strategy, culture, and change. These leaders scan the environment searching for signals, major upheavals, and changes in patterns of relationships among key variables such as products, consumers, and economics. They thrive on seeming discontinuities and chaos (Peters, 1988). The result is that they are capable of creating a vision that transcends the boundaries of normal operations and of developing a long-term mission to direct the future efforts of the organization. They energize people and build a sustainable culture by instituting new sets of norms and values to guide decision making. In short, transformational leaders manage organizations as an open social system with a core of technical operations.

Figure 12-8 vividly illustrates the complementary nature of these two perspectives on leadership.

This perspective argues directly against the current of most recent literature on leadership. In seeking to consistently apply the systems perspective, this view stresses that organizational effectiveness depends on a skillful integration of both transformational and transactional leadership. Transactional leadership centers on increasing organizational stability. This process of striving for stability is anchored in the structure, in procedures, and in incremental adjustments to respond to environmental pressures. Transformational leadership builds on the organization's culture and strategy. This approach is taken

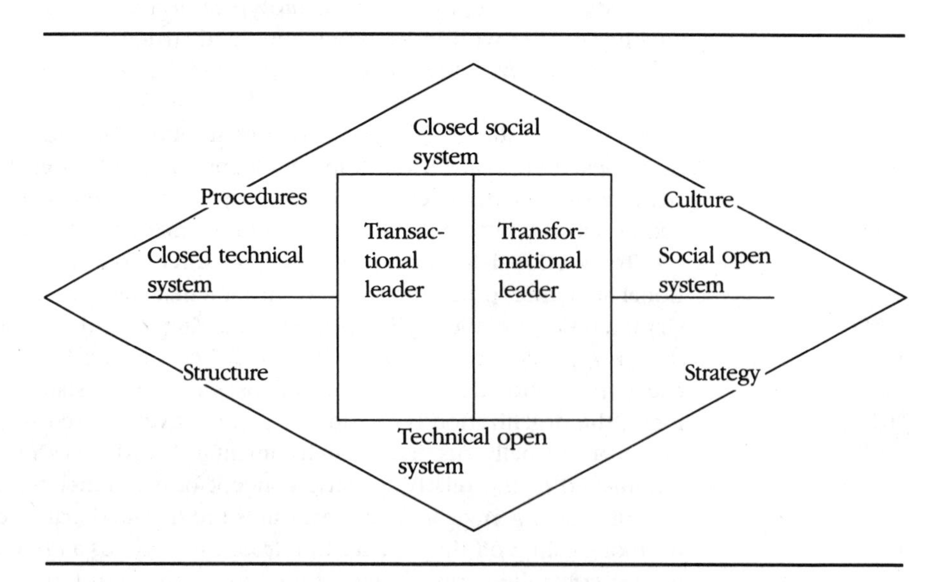

FIGURE 12-8 ***Leadership as a Synergetic Managerial Tool***

to ensure that the organization copes with environmental turbulence and looks beyond the obvious responses to create unique initiatives that react and adjust to change while also seeming to defy the upheaval that often accompanies change. Either leadership perspective, used alone, is insufficient to simultaneously embrace change and provide stability in an environment characterized by global competition, rapid economic and technological changes, and social and political turmoil. The situation at Chrysler in the 1980s clearly demonstrates that transformational leadership alone is insufficient to establish a balanced, vigorous system.

Waves of transactional leadership must be punctuated with islands of transformations, and transformational leadership must be stabilized with attention to the hard systems, transactional instruments of leading such as organization design, unit grouping, and buffering, compensating, and regulating procedures. Only under these circumstances does synergy emerge throughout the system. This important point can best be exemplified with a short overview of the synergetic leadership found in Komatsu, the Japanese earth-moving equipment corporation. The synergistic integration of transactional and transformational leadership is well illustrated in Komatsu's successful transition from being an obscure producer of earth-moving equipment to its current position as a global corporation (Hamel and Prahalad, 1989; "Komatsu Limited," 1988).

GLOBAL SYSTEMS SPOTLIGHT

Ryoichi Kawai and Komatsu Limited

In the late 1960s, Komatsu's sales were around $300 million; sales of the industry leader, Caterpillar, were over $2.4 billion. The company dominated its small domestic Japanese market by offering a narrow range of low-quality products, earning an image as a cheap and nasty producer of heavy equipment. It did not have a strong distribution network and lacked the cost efficiencies that many large-scale manufacturing firms enjoy. To make the situation even worse, the Japanese Ministry of International Trade and Industry (MITI) did not believe that Japan possessed any competitive advantage in this industry. Therefore, in 1963, MITI decided to open the earth-moving equipment industry to foreign capital investment. By 1965 Komatsu faced the imminent threat of intense competition in its domestic market. It seemed that the company had no future, and bankruptcy was its only alternative.

With the survival of the company at stake, Ryoichi Kawai, chairman of Komatsu Limited, developed a bold vision of Komatsu as a future global leader replacing Caterpillar's domination of the industry. Cat became a beacon directing Komatsu's drive toward global competitiveness. The mission-oriented motto that Kawai used to urge his followers onward was simple and powerful: "*Maru-Cat!*"

("Encircle Caterpillar!"). At the beginning Komatsu's mission appeared totally unattainable. What made it conceivable was the crafty leadership demonstrated by Kawai and his executive team in translating the mission. They worked purposefully to translate their visions into long-term strategies of action while building commitment, inspiration, and creative tension among Komatsu employees.

Komatsu has followed this mission for the past thirty years. Simultaneously the mission has been manifested as a series of action plans, clear performance milestones, and a systematic review and evaluation mechanism to track progress toward achieving goals. The mission was broken down to four long-term strategies to surpass Caterpillar: (1) technology development, (2) quality, (3) cost reduction, and (4) diversification of products and markets.

In the 1960s, the company licensed state-of-the-art technologies from International Harvester, Cummings Engine, and Bucyrus-Erie. It quickly proceeded to implement them and establish an R&D laboratory to upgrade them for the future. Total quality management systems were introduced and spread throughout the organization, from procurement to distribution. All personnel, from top management to workers on the assembly line, participated in quality improvement circles. Then project A, aimed at upgrading the quality of small and medium-sized bulldozers, was implemented. Costs were largely ignored—Komatsu's goal was to set the global standard for quality in earth-moving equipment. When this goal of attaining total quality was achieved, a subsequent program to cut costs (without impairing quality) was implemented. Procedures aimed at achieving standardization and unification of parts were implemented. In 1972, Komatsu launched its export program, penetrating quickly into the fast-growing markets of Asia, Latin America, and the Eastern bloc. This program was successfully combined in the late 1970s with a product development program, which resulted in a number of improvements and technical innovations, including the radio-controlled underwater bulldozer and the amphibious bulldozer. In the meantime, Komatsu altered its organizational structure to shift greater attention to international operations and the development of dealer networks. By rapidly responding to external shocks and threats (the oil crisis and the appreciation of the yen against the dollar), Komatsu ruthlessly cut costs. Program V-10, implemented in 1975, was to achieve four main goals: the reduction of costs by 10 percent while maintaining or improving quality; the reduction of the number of parts used by 20 percent; the redesign of products to gain economies in materials and manufacturing; and the rationalization of the manufacturing system.

Unusually bitter strikes and labor-management disputes at Caterpillar's major plant in Peoria, Illinois, in 1979, aggravated by the 1980 layoff of 5,600 employees and financial losses from 1982 to 1984, created an opportunity that Komatsu exploited to increase its market share and expand operations in Europe, Africa, the Middle East, Asia, and Oceania. Its 1979 program, titled "Future and Frontiers," redirected its strategic thrust toward the constant development of new products and new markets. These efforts were supported by a cohesive, thick organization culture. Its successes further energized organizational members and proved that "mission impossible" was actually attainable. Today, Komatsu is the world's second-largest earth-moving equipment company. It has progressively encroached on Caterpillar's market share and has surpassed it in terms of profitability and return on investment.

Source: Based on G. Hamel and C. K. Prahalad, "Strategic Intent," *Harvard Business Review* (May–June 1989), pp. 63–75.

The essential components of this success were both transformational and transactional leadership. Transformational leadership provided skillful analysis of worldwide economic and technological trends in the construction industry, the creation of mission, the development of strategies to exploit these environmental developments, and the cultivation of total commitment from organizational members. Transactional leadership supported these elements with continuous attention to mundane operations, clear programs of action, new procedures, evaluation systems to track progress, rewards for reinforcing desired behaviors, and constant incremental improvements of structure and procedures.

From the perspective of the management systems model, leadership is the central managerial tool, essential to the effective management of organizations. The challenge to leaders is to exercise wisdom in integrating these two styles of leadership into a balanced approach to managing. Furthermore, leaders must learn to temper their desire to lead with the knowledge that other systemic tools in the MSM—such as culture, strategy, structure, and procedures—may singly or in combination prove to be more effective in managing the system than leadership itself. Leaders who are true visionaries must possess the capacity not only to envision a desired future but also to recognize the complexities inherent in using the tools found in the MSM in a kaleidoscopic fashion.

~ SUMMARY

There are many different ways of leading an organization and at least as many theories about which style of leadership works best. When viewed from a systems perspective, leadership is a systemic managerial tool used to coordinate the use of all other tools. To best achieve this purpose, the leader must exercise a combination of transformational and transactional leadership. The effective leader must have more than the capacity for vision and interpersonal skills; he or she must have the ability to envision how to configure the organization in terms of the various managerial tools. The ability to integrate diverse ideas into a common theme, much as a conductor would direct an orchestra, is uncommon. Similarly, leaders must create configurations using the various tools of the management systems model. This integrative dimension is the subject of the next two chapters, on the integrative systems approach to managing through the use of system dynamics and sociotechnical design. The integrative approach is a synthesis of hard, soft, and cybernetic systems thinking. This approach is very rich because, using Ashby's terminology, it has more variety than possessed by any single approach on its own.

~ KEY TERMS AND CONCEPTS

leadership style
Cockpit Grid
situational leadership
task environment

substitutes for leadership
appreciative management
transactional leadership
transformational leadership
core values
second-order change
transcending
envisioning
energizing

QUESTIONS FOR DISCUSSION

1. Are the transactional and transformational styles of leadership mutually exclusive? Explain your answer.
2. Which of the basic systemic approaches provides the basis for transactional leadership? for transformational leadership? How are the two perspectives different in this regard?
3. Are leaders successful because they fit themselves to the situation, because the situation calls for someone with their unique blend of talents, or for some other reason?
4. What capabilities would a systemic leader need to effectively integrate the tools of the management systems model? Can you think of any well-known leaders who come closest to meeting those criteria?
5. At NeXT Inc., Steve Jobs focused on building an organization with certain capabilities. Other leaders, such as Jack Welch at GE, have attempted to focus on both performance outcomes and culture building. Is there a correct balance between these two forces of leading for task performance and building an organization? Explain how it can be achieved.
6. In some regions of the world, organization members typically expect authoritative leadership, or command. In others, employees expect to be drawn to a persuasive vision and be allowed participate in decisions. How can a leader of a global organization with divisions in diverse countries deal with this conflict through their leadership style?

REFERENCES

Bass, B. M. (1985). *Leadership and Performance Beyond Expectations.* New York: Free Press

Bass, B., and Valenzi, E. (1974a). "Contingency Approaches to Leadership," in J. G. Hunt and L. L. Larson (eds.), *Leadership Frontiers.* Carbondale, Ill.: Southern Illinois University Press.

Bass. B., and Valenzi, E. (1974b). "Contingent Aspects of Effective Management Styles," in J. G. Hunt and L. L. Larson (eds.), *Contingency Approaches to Leadership.* Carbondale, Ill.: Southern Illinois University Press.

Bass, B., Valenzi, E., and Farrow, D. (1977). "External Environment Related to Managerial Style," University of Rochester, U.S. Army Research Institute for the Behavioral and Social Sciences Technical Report No. 77–2.

Bennis, W., and Nanus, B. (1985). *Leaders.* New York: Harper & Row.

Blake, R., and Mouton, J. (1964). *The Managerial Grid.* Houston, Tex.: Gulf.

Blake, R., Mouton, J., and McCanse, A. (1989). *Change by Design.* Reading, Mass.: Addison-Wesley.

Burns, J. M. (1978). *Leadership.* New York: Harper and Row.

Cavaleri, S., and DeCormier, R. (1991). "The Microskills System for High-speed Leadership," *Leadership and Organization Development Journal, 12,* no. 4, pp. 9–12.

Cyert, R. (1991). "Academy of Management Newsletter," OMT Division, Arizona State Univesity West, Winter, p. 1.

Fiedler, F. E. (1967). *A Theory of Leadership Effectiveness.* New York: McGraw-Hill.

French, J. R. P., and Raven, B. (1959). "The Bases of Social

Power," in D. Cartwright (ed.), *Studies in Social Power.* Ann Arbor, Michigan: Institute for Social Research.

Gardner, J. (1989). "Mastering the Fine Art of Leadership," *Business Month, 133,* no. 5, pp. 77–78.

Hamel, G., and Prahalad, C. K. (1989). "Strategic Intent," *Harvard Business Review,* May–June, pp. 63–75.

Kerr, S., and Jermier, J. M. (1978). "Substitutes for Leadership: Their Meaning and Measurement," *Organizational Behavior and Human Performance, 22,* no. 3, pp. 375–403.

"Komatsu Limited." (1988). Harvard Business School, case study 9-385-277, Boston, Mass. (revised version).

Kotter, J. P. (1990). "What Leaders Really Do," *Harvard Business Review,* May–June, pp. 103–11.

Kouzes, J., and Posner, (1988). "The Leadership Challenge," *Success,* April, pp. 68–69.

Kuhnert, K. W., and Lewis, P. (1987). "Transactional and Transformational Leadership: A Constructive/Development Analysis," *The Academy of Management Review, 12,* no. 4, pp. 648–75.

Lewin, K., and Lippett, R. (1938). "An Experimental Approach to the Study of Autocracy and Democracy: A Preliminary Note," *Sociometry, 1,* pp. 292–300.

Likert, R. (1967). *The Human Organization: Its Management and Value.* New York: McGraw-Hill.

Nadler, D. A., and Tushman, M. L. (1990). "Beyond the Charismatic Leader: Leadership and Organizational Change," *California Management Review, 32,* no. 2, pp. 77–96.

Osborn, R., and Hunt, J. G. (1975). "An Adaptive-Reactive Theory of Leadership, The Role of Macro Variables in Leadership Research," in J. G. Hunt and L. L. Larson (eds.), *Leadership Frontiers.* Carbondale, Ill.: Southern Illinois Univesity Press.

Osborn, R., Hunt, J., and Bussom, R. (1977). "On Getting Your Own Way in Organizational Design: An Empirical Illustration of Requisite Variety," *Organizational and Administrative Science, 8,* pp. 295–310.

Peters, T. (1988). *Thriving on Chaos.* New York: Alfred A. Knopf.

Peters, T., and Waterman, R. (1982). *In Search of Excellence.* New York: Harper and Row.

Quinn, R. (1988). *Beyond Rational Management.* San Francisco: Jossey-Bass.

Schriesheim, C., and Bird, B. (1979). "Contributions of the Ohio State Studies to the Field of Leadership," *Journal of Management, 5,* pp. 135–45.

Shrivasta, P. (1983). "A Typology of Organizational Learning Systems," *Journal of Management Studies, 20,* no. 1, pp. 7–28.

Shrivasta, S., Cooperider, D. and Associates (1990). *Appreciative Management and Leadership.* San Francisco: Jossey-Bass.

Tushman, M., Newman, W., and Romanelli, E. (1986). "Convergence and Upheaval: Managing the Unsteady Pace of Organizational Evolution," *California Management Review, 29,* no. 1, pp. 29–43.

Vroom, V., and Yetton, P. (1973). *Leadership and Decision Making.* Pittsburgh: Univesity of Pittsburgh Press.

Yukl, G. (1981). *Leadership in Organizations.* Englewood Cliffs, N.J.: Prentice-Hall.

ADDITIONAL READINGS

Bateman, T. S., and Zeithamel, C. (1990). *Management: Function and Strategy.* Homewood, Ill.: Richard D. Irwin.

Hamel, G., and Prahalad, C. K. (1990). "The Core Competence of the Corporation," *Harvard Business Review,* May–June, pp. 79–91.

Hellriegel, D., and Slocum, J. W. (1989). *Management,* 5th ed. Reading, Mass.: Addison-Wesley.

Rothman, M. (1990). "A Peek Inside the Black Box," *California Business,* April, pp. 30–58.

Senge, P. (1990). *The Fifth Discipline: The Art and Practice of the Learning Organization.* New York: Doubleday.

Tichy, N., and Charan, R. (1989). "Speed, Simplicity, Self-Confidence: An Interview with Jack Welch," *Harvard Business Review,* September–October, pp. 112–20.

Tichy, N., and Devanna, M. (1986). "The Transformational Leader," *Training and Development Journal, 40,* no. 7, pp. 27–32.

~ PART ~

IV Integrative Approaches and the Future

A THEME OF THIS BOOK is that organizations often are complex, multidimensional, dynamic systems. Organizations engage in activities that range from the most complex to the relatively routine. As discussed earlier, organizations embody attributes of several well-known objects such as machines, organisms, and games. Some organizations tend to be more like one of these objects than the others, but they all include elements of each. Every metaphor has its own unique ability to capture some part of the essence of organizations. Each represents a unique pattern of how the basic forces within such systems interrelate to create the outcomes that are normally witnessed by managers.

On the simplest level, we might simply visualize organizations as multifaceted entities that actively pursue a set of purposes. Such a view is limited, however. Organizations also have various technical and social dimensions that may find expression in various ways. Some organizations have competing values and beliefs. While they may, over the long run, enrich the system's sense of identity, in the short run they can confuse people about what really matters. Organizations also have needs for growth and innovation, which must somehow be counterbalanced by the need for stability and regulation.

In general, then, organizations must be considered as complex systems that defy comprehensive explanation by any single specific approach. This is partially due to the variety present in all organizations. Ashby's law of requisite variety proposes that only variety can control variety; similarly, we propose that, to be able to explain the workings of a system, a concept must have relatively more variety than the system. Since organizations are systems of technical and social activity, any approach that attempts to explain the operation of organizations must have two fundamental qualities: it must be systemic and it must be capable of addressing the technical and social aspects of organizations.

Earlier, a basic systems typology was proposed in which different systems perspectives were distinguished on the basis of two fundamental dimensions: their relative degree of openness and whether their focus is primarily on the social or technical elements of the system. This typology, in Figure IV-1, identifies where each approach fits.

To gain variety, a systems model can incorporate the essential features of hard, cybernetic, and soft systems thinking. Two perspectives have, in fact, attempted to accomplish this goal with varying degrees of success: system dynamics and sociotechnical systems (STS). Note that they are located at the center of the grid in Figure IV-1. Two chapters in this part examine these alternative systems approaches.

The various perspectives that we will call integrative approaches have several common features. Most importantly, they all contain all of the three basic approaches (hard, soft, and cybernetic thinking). Secondly, they all seek to engage the members of the organization through some form of participation. Third, they all employ hard systems techniques as tools to enhance the situation for people rather than to generate solutions. Finally, they all have an aspect of continuous improvement associated with them. Therefore, we may describe the integrative approach as a multidimensional systems perspective for creating change that uses hard systems tools plus the contributions of other systems perspectives to engage people in a process of continuous organizational improvement. Ultimately, the outcomes of the integrative approaches are the same, in that they focus on the social dimension of the system. They differ, however, in how they attempt to achieve this change. For example, system dynamics is oriented toward helping people learn about how the structure of a system affects its behavior. It relies on active experimentation that leads managers to learn to think systemically. On the other hand, sociotechnical systems focus on matching the organization to its technology, and on improving the quality of work life for its employees (Pava, 1986).

A number of alternative ways of explaining the complexities of systems have emerged in recent years. The common thread among them is that they do not assume equilibrium to be the normal state for a system. Instead, they adopt the perspective that change and transformation are natural processes that should be expected. Although relative stability may be common and convenient in some systems, it is not viewed as characterizing the basic nature

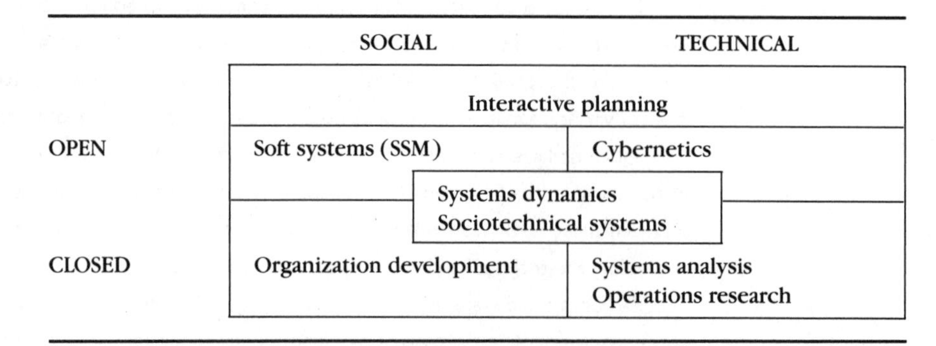

FIGURE IV-1 ***A Typology of Systems Thinking***

of the system. Among the most prominent of these nonequilibrium approaches are chaos theories, self-organizing theories, and the population-ecology and organization ecology perspectives. Chapter 15 will examine the chaos theories in some detail, with reference to organizations.

Integrative and nonequilibrium forms of systems thinking offer new ways of visualizing systems that have fewer restrictions than earlier models. They are capable of explaining a wider variety of systemic phenomena than earlier models. Consequently, they offer greater power for understanding organizations in the complex global systems of the next century.

▲

13 Sociotechnical Systems Design

The product of work is people, as well as goods or services. A society is no better than the quality of the people it produces.

Phillip Herbst (1975)

1. To understand the fundamental principle of sociotechnical systems design—the optimization of human and technical factors.
2. To recognize how this optimization is a means to balance the social and technical systems of an organization.
3. To explain how the sociotechnical systems approach combines hard and soft systems thinking into an integrative framework.
4. To recognize the patterns that have emerged in the development of the sociotechnical systems approach since its origins.
5. To become familiar with the leading sociotechnical systems design experiments conducted around the world.
6. To understand the reasons for adding action-learning and self-design components to the sociotechnical systems approach.

SECTION I
INTRODUCING SOCIOTECHNICAL SYSTEMS DESIGN

AN INTEGRATIVE VIEW

Integrative systems thinking incorporates elements of hard and soft systems thinking, as well as cybernetic thinking. This perspective emphasizes the importance of work design, autonomous work groups, and continuous improvement processes as ways to increase productivity and human satisfaction. However, integrative systems thinking is more than a mere conglomeration of other approaches; it represents a hybrid form of thinking.

Sociotechnical systems thinking is one example of this form of hybrid thinking. The sociotechnical systems (STS) perspective is firmly grounded in many engineering, or hard systems, principles. The design of productive and information systems are based on rationalistic assumptions in the hard systems tradition. Such instrumental approaches emphasize the design of systems to yield uninterrupted work flows and provide high-quality products. This perspective on the design and management of organizations is very machine-like in the way it approaches the technical system. In general, organizations are often regarded metaphorically as a type of machine. Morgan (1986) has observed, "When we talk about organization we usually have in mind a state of orderly relations between clearly defined parts that have some determinate order. Although the image may not be explicit, we are talking about a set of mechanical relations" (p. 22).

The STS perspective departs from exclusive reliance on this mechanical metaphor by creating opportunities for meaningful human involvement in the system. In other words, it also adopts soft systems approaches. This is accomplished by designing elements of the system to be generative, not just instrumental. People are encouraged to participate in work activities that are compatible with both their individual and group needs. For example, at various Volvo plants in Sweden, workers are used as teachers (craftsmen), and learning and education are emphasized (Bernstein, 1988). Such systems are created with the intent of shifting toward being an open system with regard to decision making and employee involvement. This openness results from efforts to empower organization members through continuous learning, autonomous work teams, and the creation of a work environment that offers higher levels of job satisfaction.

Although the sociotechnical approach is generally an integrative approach, it places much greater emphasis on the hard and soft systems approaches than the cybernetic ones. Through the mechanisms of self-regulation, continuous learning and improvement, and self-design of work, various dimensions of cybernetic thinking are being incorporated increasingly by innovative companies into the sociotechnical systems approach.

The effect of implementing STS in organizations is to balance the polar forces of openness-closedness and the social and technical dimensions of the system. The extent to which STS accomplishes such a balance depends on the skill of the people who initiate the change. This chapter will examine the basic concepts and principles that form the foundation for STS and review its benefits and limitations. To get a basic understanding of STS as a philosophy, it is instructive to look at the experiments in improving productivity conducted at the Volvo automobile plant in Kalmar, Sweden, one of the most notable examples of the implementation of STS, where planned efforts were made to redesign work processes and work groups.

GLOBAL SYSTEMS SPOTLIGHT

AB Volvo

The Organization. The AB Volvo plant in Kalmar, Sweden, was one of the first major experiments in STS. A number of conditions that existed in the late 1960s prompted Volvo CEO Pehr Gyllenhammer to initiate a radical redesign of the way automobiles were manufactured. Sweden had one of the best educated populations in the world, and unemployment rates hovered around 1 percent. Although workers were well paid, taxes consumed up to 70 percent of workers' overtime pay. Volvo plants experienced high turnover rates—over 25 percent—in a country where workers had many alternative choices of employment. Wildcat strikes and unacceptable levels of absenteeism in the late 1960s called for a dramatic response.

The Work Environment. Gyllenhammer's goal was to create a factory where people would want to work because quality of work life (QWL) was high. The redesigned Kalmar plant opened in 1974 with the goal of marrying state-of-the-art technology with self-managing work teams integrated by the systems concept. In general, the plant was designed to be a comfortable place for employees to work. Each work area was designed to give the feeling of being in a small workshop. Each area had windows for natural lighting, separate entrances, locker rooms, and comfortable rest areas. The factory was engineered to create independence for work teams by segmenting auto assembly into discrete activities that could be accomplished at different times. The assembly process was segmented by replacing the conventional assembly line with an assembly network.

The Organization Structure. Within the segregated work areas, teams of fifteen to twenty-five people functioned as autonomous self-managing groups responsible for a specific subsystem of auto manufacturing. Group members were cross-trained, and decisions about work assignments were made by collective agreement of the team. First-level supervisors and support technicians serviced several groups, making sure that the necessary resources were available to ensure fewer interruptions in work. Groups determined their work rate, as long as it did not interfere with the workflow of the plant as a whole.

The technical aspects of the plant were designed to take advantage of the current technology. Automobile subassemblies were transported between storage areas and workshops on computerized flatbed vehicles controlled by each team. The position of these trams was adjustable in terms of height and angle of tilt to allow workers easy access to their task. Finally, computerized displays were erected to communicate feedback on current rates of productivity to employees.

Performance. The results of the Volvo experiments in productivity seemed generally positive, although questions remain regarding the cost effectiveness of the system. The high-price, high-quality marketing strategy of Volvo makes such a system feasible. The concept of autonomous work teams seems to be a welcome alternative to the conventional assembly line method for manufacturing automobiles.

Continued Changes. Volvo's experiment in STS

continued with a new variation installed in 1987 in a new plant in Uddevalla, Sweden. The Uddevalla plant manufactures the mid-priced Volvo 740 model. The decision to compete in the mid-priced market required that productivity be raised above the levels achieved in the Kalmar plant. As a consequence, 740s are being mass produced in Uddevalla, but without assembly lines. Autonomous work teams are still employed, but there have been several changes in the way they function. Teams have been reduced in size to seven to ten workers. They are also cross-trained, as were workers at the Kalmar plant, but more repetition is built into work cycles at Uddevalla than at Kalmar. Four cars are assembled per day, and team members work for an average of three hours before they repeat a task. Volvo is making a strong attempt to gain the efficiencies of mass production without the problems of worker boredom and alienation that so often accompany it.

Volvo is working carefully to balance its desire for efficiency with concern for the satisfaction and well-being of its employees. Uddevalla teams work free of supervision and need only go through two layers of management to get to the top. Personnel selection, scheduling, and quality control are all managed by the work teams. The teams are empowered to negotiate weekly production goals with management. Volvo demonstrates its commitment to its employees by investing in sixteen weeks of training for each new manufacturing employee followed by sixteen months of continuing development.

In this plant, which claims to be the world's only automobile factory without assembly lines, output per worker and quality are higher than at any of Volvo's three other plants. The Uddevalla plant turns out 200 cars per week and plans to produce 40,000 cars per year. Volvo's attempt to merge craftsmanship and high-volume production remains one of the most innovative manufacturing strategies in the auto industry.

Volvo's experiments in sociotechnical systems have been adopted in various forms throughout many of the company's other plants, such as those at Koping, Olofstrom, Skovde, Tuve, and Vara. More recent change efforts at Volvo have taken a longer-term perspective, and becoming more oriented toward organization development. New programs are designed to increase workforce involvement and personal growth. The *Full Rolle* ("full speed ahead") and *Dialog* programs have sought to create a common leadership philosophy, improve communications, and emphasize Volvo's core cultural values: quality, care, competence, communication, development, and involvement.

Source: Based on *U.S. News & World Report,* August 21, 1989, p. 42; *Business Week,* August 28, 1989, pp. 92–93; *Ward's Auto World,* January 1990, p. 45; "Job Redesign on the Assembly Line," *Organizational Dynamics,* Spring 1973; R. Griffin, *Task Design: An Integrative Approach* (Glenview, Ill.: Scott, Foresman, 1982); P. Bernstein, "The Learning Curve at Volvo," *Columbia Journal of World Business,* Winter 1988, pp. 87–95.

THE DEVELOPMENT OF THE SOCIOTECHNICAL SYSTEMS PERSPECTIVE

The STS perspective arose largely from work at England's Tavistock Institute in the early 1950s. The Tavistock Institute had been known as a research group that explored various applications of social psychology. A group of Tavistock researchers lead by Kenneth Bamforth and Eric Trist set their first major experiment in an industrial setting in the British coal mining industry.

British coal mines had been known for their autocratic, control-oriented management styles, which resulted in hostile adversarial relations between labor and management. The mines themselves were narrow and dark and required a mining technique that emphasized individuals following rigid, specialized procedures. Called *the long-wall method,* this method had replaced pick-and-shovel work teams when it became possible to create deep shafts and tunnels in the coal fields. Subsequent structural improvements in the mines permitted widening of the tunnels and allowed the assembly of groups of workers. This opening up of the mines permitted men to work interactively rather than as isolated individuals. After observation, Bamforth and Trist realized that when the miners were allowed to have some control over how they approached the task, they were both more productive and more satisfied. The researchers were impressed that a change in the technology and physical environment would trigger changes in human emotions and relationships. The impetus for creating semiautonomous work groups was the product of collective action by the general manager, the union, and the miners. According to Weisbord (1987), the process demonstrated that the miners were capable of making "wise decisions" regarding work arrangements.

From this example the STS perspective emerged to explain that important relations existed between people and technology. According to Trist, people (the social system) and technology (the technical system) were interdependently linked together in a **sociotechnical system.** More importantly, he also recognized that these sociotechnical systems were the product of conscious design rather than the creation of random forces. In many cases the prime criterion for making such design decisions was economic efficiency; in classic early manufacturing systems, such as Henry Ford's assembly line, people and their needs were largely ignored. The STS approach, on the other hand, attempts to create a work system that balances the needs of the organization for efficiency and the psychosocial needs of the workers.

Thus, a sociotechnical system is a planned effort to structure work in a way that jointly considers the human and technical dimensions of the situation. This approach, known as **joint optimization,** is a keystone of the sociotechnical systems perspective. According to Trist (1981), joint optimization "regards man as complementary to the machine and values his unique capabilities for appreciative and valuative judgment. He is a resource to be developed for his own sake" (pp. 42–43). This view represented a major departure in thinking from the prevailing scientific management approaches of the time.

In the conventional, high-speed manufacturing facilities of the early twentieth century, engineers sought to minimize human involvement because people were seen as unreliable and inefficient. Repetitive, simplified tasks, rigid procedures, and tight controls replaced craftsmanship and professionalism in many factories. In their place technology emerged as the source of great gains in efficiency. The belief that technology was the prime factor to consider in management decisions became known as the **technological imperative.** Although the movement toward job simplification and external control of work groups paid dividends in the short run, it also alienated and disempowered people. This paradigm sought to maximize economic efficiency at the expense of people; the STS paradigm attempts to optimize both human and technical factors. Contrast the imbalance shown by the old paradigm with the balance exhibited by the new (Figure 13-1). The core assumptions of these two paradigms are contrasted in Figure 13-2.

The basic assumptions of the scientific management proponents such as Frank and Lillian Gilbreth and the advocates of STS differ in major ways. The Gilbreths' idea was that through specialization and simplification, work could be learned quickly and performed efficiently. While actual practice bore out this theory, in the process individuals became disconnected from the system.

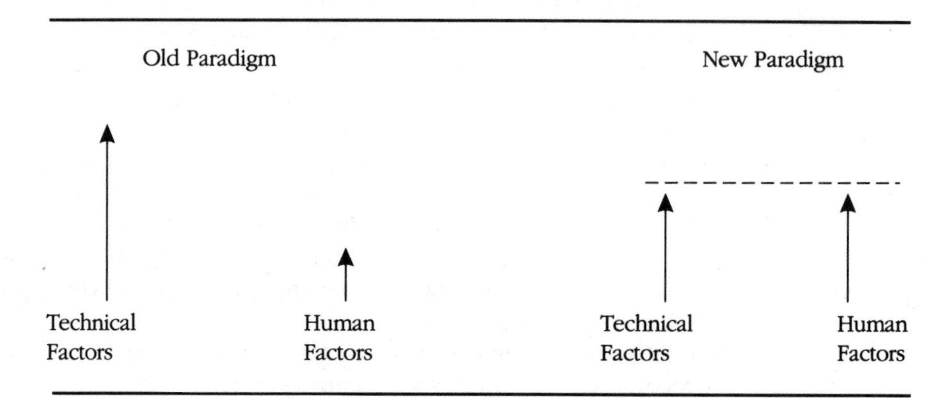

FIGURE 13-1 ***Technical Maximization versus Joint Optimization***

OLD PARADIGM THE TECHNOLOGICAL IMPERATIVE	NEW PARADIGM JOINT OPTIMIZATION
People as extensions of machines	People as complementary to machines
People as expendable spare parts	People as a resource to be developed
Maximum task breakdown	Optimum task grouping
Simple narrow skills	Multiple broad skills
External controls	Internal controls
Tall organization chart	Flat organization chart
Autocratic leadership style	Participative leadership style
Competition, gamesmanship	Collaboration, collegiality
Organization's purposes only	Members' and society's purposes as well as organization's
Alienation	Commitment

FIGURE 13-2 ***Features of the Old and New Paradigms***

Source: E. Trist, *The Evolution of Sociotechnical Systems* (Philadelphia: Center for the Study of Organizational Innovation, University of Pennsylvania, 1981), p. 42.

The individual worker became psychologically disassociated from the end product and the end user, the customer. The minds of workers became disengaged from the creative process of the work itself as the work became progressively mindless. Most importantly, work teams were reduced to aggregates of people doing similar types of work rather than acting as problem-solving teams.

Conversely, the proponents of STS have focused their initiatives on the system level in the organization, rather than on individuals. Trist (1976) has identified a hierarchy of levels of systems that serve as a guide to programs to create sociotechnical systems (see Figure 13-3). We will now examine each of these levels in detail.

SECTION 2
SOCIOTECHNICAL LEVELS

PRIMARY WORK SYSTEMS

AUTONOMOUS WORK GROUPS

Primary work systems are a focused network of activities operating within the boundaries of the larger system, such as a department or a division within an organization. The efforts of the people within a primary work system center on achieving a common goal or purpose. Primary work systems could include a claims unit in an insurance company, a surgical unit in a hospital, or a detective unit in a police department.

One of the critical elements of the primary work unit is the potential for teamwork among its members. This is because STS depends heavily on internally regulated groups known as **autonomous work groups.** Such groups employ the principles of cybernetics (Chapter 6) and the learning organiza-

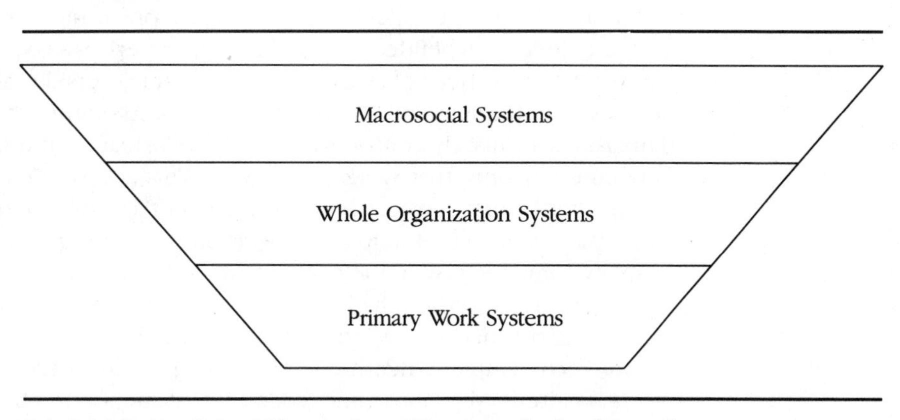

FIGURE 13-3 ***Hierarchy of Sociotechnical Levels***

tion (Chapter 7) to manage themselves. The idea of using cybernetics is to allow group members to monitor variances from desired performance and make adjustments within the group to compensate. For example, in conventional quality control systems workers external to the group regulate quality by identifying deviations from standards and relaying the information back to the group. In autonomous work groups, quality is ensured and controlled within the group by its members. In this way group members continuously learn about the causes of defects and what can be done to prevent or control them. Not only do group members take advantage of a learning environment, but they also are motivated by their sense of control over their environment.

This is the primary cybernetic dimension of the STS perspective. **Self-regulating groups** are able to control their performance only through a continuous flow of information that provides feedback on performance. The cybernetic thinking that underlies self-regulating groups not only provides a strong foundation for quality improvement, it also forms the basis of creating a learning organization. Learning occurs through reiterative cycles of action and experience, reflection, reframing, and new actions. Typically, autonomous work groups have an inherent potential to learn at an accelerated rate due to their design. Autonomous work groups, in order to be effective in achieving self-regulation, must have the capacity for processing information and incorporating new insights into changes in their operations.

A second key element of building the primary work system in the STS is work design. In a sociotechnical system, work design efforts center on individual job engineering as well as designing frameworks for the whole organization. STS interventions with individual jobs generally focus on providing opportunities for people to meet some basic psychological needs while performing the work. While these needs may vary somewhat among individuals, jobs are designed to meet needs that most people share.

CORE CHARACTERISTICS

Hackman and Oldham (1975) proposed one such model that addresses these key points. They argue that the psychological preconditions of high performance are directly dependent on the core characteristics of jobs. Every job is a composite of a variety of attributes. For example, outside sales provides high degrees of autonomy and feedback on results. Assuming that these core job dimensions ultimately contribute to work performance, it is necessary to carefully consider how that work is designed. This approach suggests that work design efforts must consider both work content and job context. For example, when there is a lack of autonomy, decisions pertaining to work methods are made by someone else in the organization. Thus, consideration must be given to the role of the job and the job holder in the organization. Similarly, feedback on work performance may originate from either the task or from the organization. For example, when a computer programmer designs a program that meets its intended purpose, the feedback that the job has been accomplished results from the program running correctly. On the other hand, a customer service representative may depend for feedback on a supervisor's quarterly reports on the number of complaints lodged by consumers. Consequently, jobs must be engineered in a way that accounts for the potential for feedback to emerge from either of these locuses.

The Hackman and Oldham model implicitly argues for the sociotechnical perspective by coupling job characteristics and critical psychological states as cofactors in determining work outcomes (see Figure 13-4). While this perspective is an attractively simple guide for job design, its broader use is limited by its nonsystemic nature. There is a clear need to link jobs with the rest of the system.

The Hackman and Oldham job characteristics model has become a widely cited approach to job design in the psychological literature (Griffin, 1982). Yet

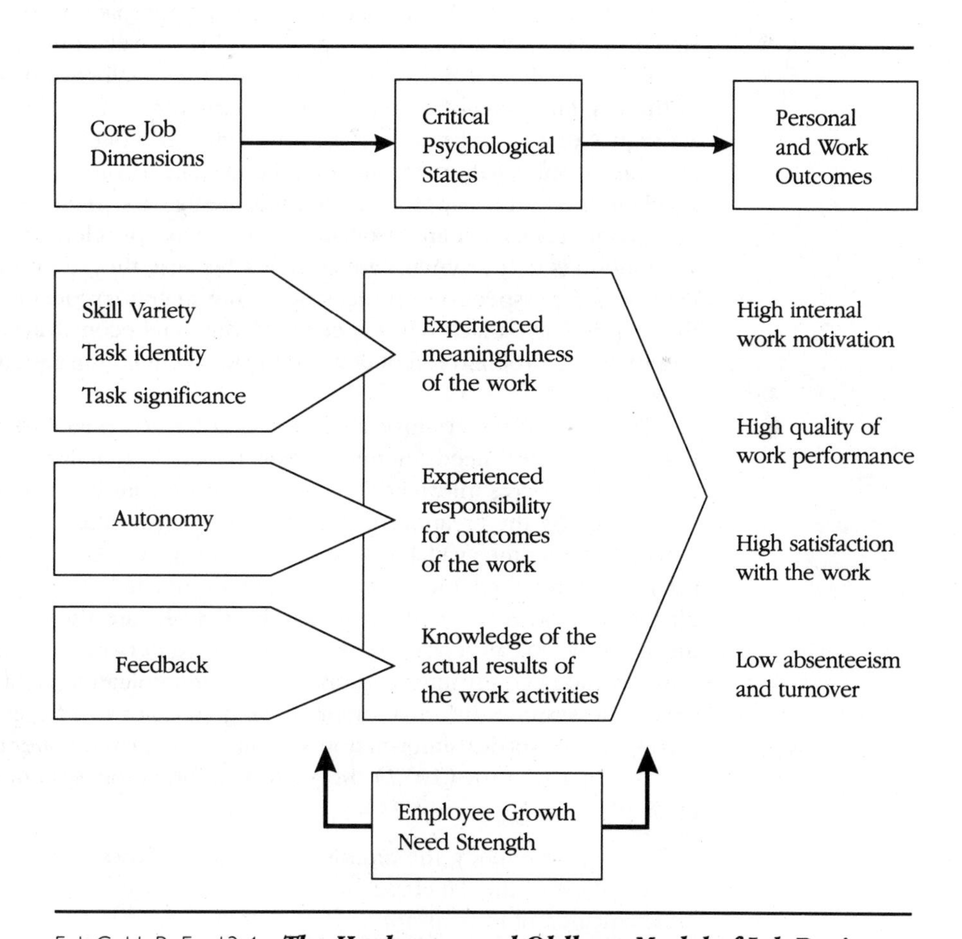

FIGURE 13-4 ***The Hackman and Oldham Model of Job Design***

Source: J. R. Hackman and G. R. Oldham, "Motivation through the Design of Work: Test of a Theory," *Organizational Behavior and Human Performance,* 1976, *16,* pp. 250–79. Reprinted by permission of the author and the publisher.

the authors themselves caution that it would be a mistake to view their model as a "correct or complete picture of the motivational effects of job characteristics" (Hackman and Oldham, 1980, p. 97).

THE INFORMATION PROCESSING LINK

There is little doubt that the STS perspective has contributed to the way work is designed. The creation of autonomous self-regulating work groups brings the opportunity for considerable innovation and collaboration within groups. Yet self-regulating groups must be open to interaction with other groups. The means by which this interaction should evolve are not well defined in classic STS writing. One approach to this issue affords considerable insight into a way of orchestrating group interactions into a collaborative whole.

Galbraith's (1977) information-processing approach to organizational design provides several useful precepts regarding the relationship between tasks, problem complexity, and team interrelationships. Galbraith proposes that one of the key functions of an organization's structure is to transmit information. He argues that organizations often have difficulties because they are not able to transmit information throughout the system to the desired location in a timely fashion. In his approach, the fundamental role of structure is to reduce uncertainty levels that are associated with specific problem areas. The primary assumption is that problems are complex because they are perceived from an uninformed perspective, one lacking in the necessary details to understand the situation, and that when sufficient information becomes available, the complexity of the problem is reduced and the problem is consequently more likely to be resolved.

The goal of uncertainty reduction can be achieved in two basic ways. Either reduce the need for information processing, or increase the system's capacity to process information. Requirements for information processing can be reduced by the creation of slack resources and self-contained tasks. **Slack resources** become available when the performance requirements of an organization are reduced and fewer resources are needed; consequently the need for information is reduced. **Self-contained tasks** are those that concentrate the authority for all related activities in a single person or localized group. In this way the need for inter-organizational communication is dramatically lessened. The system's information-processing capacity can be expanded by investing in the vertical information system or by creating lateral relations. According to Galbraith (1977) the vertical information system has four main elements:

1. Decision frequency, the timing of information flows
2. The scope of the data base
3. The degree of formalization
4. The capacity of the decision mechanism to process information

McDonald's fast-food restaurants have effectively increased information-processing capabilities by installing computerized ordering systems at order counters. Food preparers are able to determine, virtually instantaneously, the demand for each type of food while order takers can find the total cost to the consumer without ever having to enter the price into the computer. The Volvo plant at Kalmar linked its various autonomous work groups by investing in its vertical information system, which is used to manage the quality improvement process. Information regarding the level of quality attainment is entered in a data base, which can be accessed by operational groups and by a quality control group. It is the responsibility of the operators to correct their defects without direct assistance from supervisors.

The purpose of creating lateral relations within an organization is to reduce the demand for information transmission. Many organizations get bogged down by information overload and as a result produce untimely feedback and decisions. Increasing lateral relations reduces the demand on the structure by empowering people to make decisions at the lowest possible levels in the system. While there are many ways to increase lateral relations, Figure 13-5 lists several of the most common.

COUPLING

The relationships among work groups extend beyond the degree to which they share information and participate in joint decisions. A work team's potential for autonomy also depends on the framework used in the interface of technology and workflows. Technological requirements for the production of some products or services may cause groups to have relatively high interdependence. Such groups are tightly coupled. Generally tightly coupled groups show a sequential dependency, where one group cannot undertake its work until a prior group in the sequence completes its function. In fast-food restaurants, for example, order takers are tightly coupled with food preparers; they cannot fill orders until the food preparers complete their tasks. In some restaurants this results in a line of customers waiting for their orders to arrive.

The creation of close linkages between groups reduces the autonomy available to either group, thus making it more difficult for such groups to control their own internal processes. Over the years since the STS perspective was

1. Create boundary spanning roles to link groups together.
2. Make decisions for temporary, organization-wide issues by using ad hoc groups comprised of members from a variety of locales within the system.
3. Encourage direct contact between any parties that have a common concern over a shared problem.
4. Develop interfunctional teams to address broad issues.

FIGURE 13-5 ***Methods to Increase Lateral Relations***

Source: Adapted from J. Galbraith, *Organization Design* (Reading, Mass.: Addison-Wesley, 1977), pp. 112–13.

first developed by members of the Tavistock Institute, a number of basic design principles have emerged. Figure 13-6 on p. 330 summarizes the key features of these principles.

One of the most radical experiments in the use of sociotechnical design during the 1970s took place at the General Foods' Gaines Dog Food plant in Topeka, Kansas. The minicase explores the effects of this approach on productivity at the plant.

Gaines Dog Food

In 1971 the Gaines Dog Food Division of General Foods opened a new plant in Topeka, Kansas, with the goal of improving worker attitudes. General Foods managers Lyman Ketchum and Edward Dulworth embarked on a radical plan to stimulate worker participation. Self-managing work teams of seven to fourteen members were created. Team members were trained to perform their usual tasks as well as others, ranging from maintenance to quality control. Work assignments were planned to balance mental activities with physical labor. Foremen were replaced by team leaders, who acted as facilitators to assist people in getting things done. Team leaders were hired prior to the plant's opening to orient them to the new approach and train them regarding new procedures. Although procedures existed, groups were given the opportunity to make their own rules and decisions whenever possible. This was a nonunion plant.

Status differences within the plant were minimized. All operators had a single job classification and received pay increases based on the number of jobs they could learn and perform with proficiency. The plant was designed to encourage interaction and informal gatherings between team members. Workers were able to move freely throughout the plant and fraternize with other workers. Parking spaces were available to anyone; none were reserved for managers.

The primary product of the Topeka plant was dry dog food. Most of the routine tasks in the factory were automated. Team members received reports with the current information necessary to support decision-making within the group. When members were absent, the teams retained the right to make job reassignments to cover the vacancies.

In general, the performance at the Topeka plant was superior to that of comparable plants. Product quality improved in the first several years by over 90 percent. Overhead costs were reduced more than 30 percent. Initially morale was good, and both absenteeism and turnover were relatively low. However, even as early as 1972 participation and openness were reported to have declined. Despite the declining mood in the plant, productivity remained impressive. Over the next four years, however, the quality of life in the Topeka plant declined from the lofty levels of the first two years. Group cohesiveness and trust declined, apparently as a result of fewer team meetings, inadequate indoctrination of new members, the loss of original members, and continued pressure for high productivity and conformance to group norms. In general, it was difficult for both team members and leaders to make the adjustments to the new style of democratic management. The openness and ambiguity in the system caused some members to feel uncomfortable. Rivalries and conflict began to develop between teams, and no problem-solving systems were in place to address these issues.

From 1971 to 1976, overall performance was evaluated as being "very productive" and the Topeka plant remained "a superior place to work" (Walton, 1978). In late 1986, however, Gaines's parent company was acquired by Quaker Oats at a cost of $225 million. The operating profit margin in this unit dropped from 8 percent in 1986 to 6 percent in 1988. In June 1989, Quaker Oats wrote off nearly $61 million in relation to its pet foods division. Lack of sales growth in the dog food market has prompted Quaker to consolidate a number of its pet operations.

Source: R. E. Walton, "The Topeka Story: Teaching an Old Dog Food New Tricks," *The Wharton Magazine* (Spring 1978), pp. 38–46; R. Walton, "The Plant that Runs on Individual Initiative," *Management Review* (July 1972), pp. 20–25; L. Yorks and D. Whitsett, "Hawthorne, Topeka, and the Issue of Science Versus Advocacy in Organizational Behavior," *Academy of Management Review* (1985), pp. 21–30; L. Therrien, "Quaker Oats Pet Peeve," *Business Week* (July 31, 1989), pp. 32–33.

The recent decline of the Gaines Dog Food Division cannot be directly attributed to the work system at Topeka. Changing market and environmental dynamics changed the pet food industry substantially from its characteristics when the program started more than two decades ago. The Gaines Topeka experiment demonstrated that sociotechnical principles can be quite effective in balancing task demands with social needs. However, by reducing structure to a minimum in an organization without a dominant high-performing culture already intact, the Pandora's box of social conflict was opened. Sociotechnical design was never intended to act as a method for resolving social conflict.

WHOLE ORGANIZATIONAL SYSTEMS

The **whole organization system** provides the context for the primary work system. This organization system must be well matched to the primary work system for the entire system to match. On the most fundamental level the organization system is the conduit between the primary work system and the environment. To be responsive to its environment, an organization system must follow the principle developed by Ashby: it must be at least as complex. It also must have more resources available than will be necessary to support the response. This concept has referred to variously as redundancy, slack resources (Galbraith, 1977), and strategic reserve (Obloj and Joynt, 1986). Reserve capacity—whether in terms of variety (Ashby, 1956) or resources (Emery and Trist, 1964)—may serve to protect the organization from overextending itself and losing its balance in a dynamic environment.

PROVIDING FOR RESERVE CAPACITY

Two basic types of redundancy may be recreated in organizations: redundancy of parts and redundancy of functions. The scientific management approach relied primarily on redundancy of parts, which occurs when the components

1. *Human values.* There is a need in organizations to balance technical goals with human needs, specifically the needs of the people who will perform the work. Whatever is important to them must be paramount above the values or beliefs of those people conducting the design.
2. *Minimal critical specification.* When possible, avoid disempowering people by confining them with unnecessary rules or procedures. Supply the minimal directives necessary to ensure that the job gets done properly and leave the rest to the judgment and creativity of the jobholder.
3. *Multifunctional perspective.* Provide sufficient cross-training so that people are capable of performing many jobs. People who are multiskilled are better able to control themselves and require less external supervision.
4. *Social support.* Provide behavioral reinforcement for all desired performance. The views and philosophies expressed by management must support the activities of team members.
5. *Incompletion.* The design process never ends. The implementation of a plan creates new issues, which create the need for new designs or adjustments to old designs.
6. *Boundaries.* People who perform related tasks should be located in close proximity to each other.
7. *Compatibility.* The way work is structured must reflect the work's basic functions and goals. For example, if people are to feel satisfaction in performing their work, challenge, autonomy, and feedback must be designed into the work itself.
8. *Information flow.* Information must be transmitted such that it is available to the people performing the tasks when they need it. For example, information about the characteristics of a beer (sugar content, alcohol content, and so on) must be readily available to the brewer so the ingredient mix can be altered. Waiting too long will cause the beer to be unacceptable.
9. *The sociotechnical criterion.* Variations from what is planned or expected must be controlled as close to their origin as possible. For example, quality measures cannot solely be taken from the final product; components need to be tested for quality to identify defects before they are incorporated in the finished product.

FIGURE 13-6 ***Design Principles for STS***

Source: Based on A. B. Cherns, "The Principles of Sociotechnical Design," *Human Relations,* 1976, *29,* no. 8, pp. 783–92.

of a product are reduced to the simplest and least expensive form possible. For example, when the parts of an automobile can be stamped out on a machine press in high volume, costs are reduced and any single part can be replaced by an identical one in case the original is defective. This allows the system a wide margin to cover any possible defects and still remain cost efficient. While this technique may be useful for attaining efficiency, it creates a mechanistic type of system. Bureaucratic organizations often take this approach with people as well. They reduce jobs to simple tasks, making any one jobholder expendable. The prevailing thought of the time held that this reduced the risks for the organization. However, the negative effects on productivity—alienation and poor morale created by such tactics—were largely ignored. No investment is made in building the capacity of workers and the success of the system depends on the ability of the major decision maker to know which parts require redundancy.

Redundancy of functions is a strategy for building reserve capacity by investing in the capabilities of people within the organization. Redundancy of functions may be created by cross-training employees, which gives them multiple skills and the ability to perform a number of different jobs. By creating redundancy of functions the system also creates flexibility and a greater potential to creatively respond to pressure for change. Creating this type of redundancy is a major platform for designing a STS. Great emphasis was placed on creating redundancy of functions at Volvo. This approach not only builds organizational flexibility but also enriches work by increasing variety and challenge. Thus creating redundancy of functions is also a sound human resources policy. The profound impact that such redesign can have on productivity is shown in the feature below.

Digital Equipment Corporation

Digital Equipment Corporation (DEC) is one of the world's premier manufacturers of computers, computer networks, and peripheral equipment. Its most well known product is the VAX line of computers, including the MiniVAX and MicroVAX. In 1981, in the northern Connecticut town of Enfield, DEC built its first facility based on the principles of sociotechnical design. The Enfield plant employed approximately 200 people, who work together manufacturing printed circuit board modules. DEC's purpose for experimenting with the sociotechnical approach was to increase its international competitiveness. This contrasts with the sociotechnical experiments of the early 1970s, which were largely initiated to respond to issues of poor quality, worker alienation, and turnover. The basic philosophy that governed the plant was to create a "high quality human environment, valuing business and people goals as interdependent and of equal importance" (Proctor, 1986).

Work teams established their own working hours, and there were no security guards. The setting was given a more natural appearance by the presence of green plants at each work station. No walls separated equipment, and equipment was positioned to create as much interaction between people as possible. Each employee had a desk with a telephone and a computer, which was part of a network. People communicated with each other primarily by electronic mail. A strong cultural norm emphasized sharing and learning. Finally, the company provided exercise facilities and classes on aerobics and stress management.

The organization structure in the Enfield plant was innovative. It placed customers and team members at the top of the chart. There were only three levels in the hierarchy, with managers playing a primarily supportive role. There were two technical support functions in the organization, product and process support (see Figure 13-7).

Within the plant were three business units, each containing anywhere from one to four teams of twelve to eighteen members. Each member learned how to build and test the printed circuit module that his or her group specialized in. Eventually, each member was to progress to the point where he or she could perform support activities such as scheduling, personnel selection, and training.

Enfield's plant manager compares the organization of the plant to a volleyball game. "Everyone rotates around the court and can play any position. Wherever the ball goes over the net, there are people who can respond. There are no leaders, the leadership changes as people come forward as they're needed" (Solomon, 1985).

Each operating team was responsible for its own human resources management, budgeting, materials planning, quality assurance and control, maintenance, and problem solving. Technical support teams provided help with implementing and testing new technologies. Administrative support teams offered assistance in relation to compensation, benefits, purchasing, financial reporting, and cash management. Finally, the management teams played one of the most critical roles. The managers at DEC-Enfield were given extensive training on how to avoid overmanaging. As one manager stated, "My job involves developing, training, and coaching, as opposed to calling the shots. I try to develop teams so that they have the skill and knowledge to make the right decisions." The ten managers within the plant acted as coaches and team builders for the manufacturing teams. This required a conscious effort not to fall into the traditional role of the command-oriented manager. Managers also worked on long-range planning, conflict resolution, and resource allocation.

DEC invested substantially in the training of its Enfield employees. Every team member received skill training and social training designed to facilitate the assimilation by employees of the organization's cultural norms, values, and expectations. Each new employee was assigned to a mentor and underwent a week-long intensive training process. Skill training was rigorous as well, placing considerable demands on people's abilities. Upon joining the organization, a new hire participated in a three-month program integrating both technical and social skills. Before becoming an operating member of a team, an employee had to become certified as competent to build a complete module. The floor certification process was administered by current members of the team, who had been floor certified themselves. Once floor certified, team members were eligible to learn basic support functions to help the team to operate more effectively. People were linked to specific support activities based on their personal interest and on the needs of the team. Training for support activities was accomplished through a computer-based learning system. The reward system at DEC-Enfield was based on recognizing people for the number of skills and functions they can master.

In the first three years of operation, the overall performance at DEC's Enfield facility was outstanding. Product manufacturing time was reduced by 40 percent below traditional comparable facilities. The standard fixed asset turnover was six times per year compared to 1 to 2 times per year at traditional plants. Productivity, measured by units manufactured, was double that of comparable plants with half the workforce and half as much space. Finally, scrap rates were reduced by 50 percent compared with a traditional facility, and overhead was cut by 40 percent. In 1991, however, the Enfield plant was closed and moved to an existing DEC manufacturing facility in nearby Springfield, Massachusetts, as a cost-cutting measure.

Source: D. Proctor, "A Sociotechnical Work Design System at Digital Enfield: Utilizing Untapped Resources," *National Productivity Review* (Summer 1986); B. Solomon, "A Plant That Proves That Team Management Works," *American Management Association* (1985).

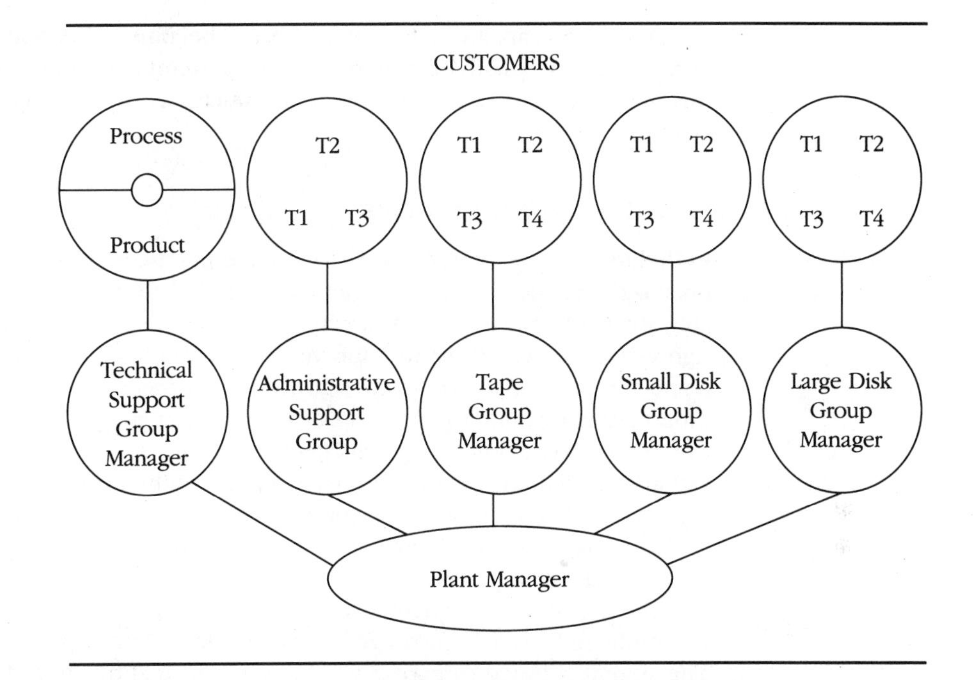

FIGURE 13-7 ***Organization Chart of DEC-Enfield***

Source: B. Proctor, "A Sociotechnical Work Design System at Digital Enfield: Utilizing Untapped Resources," *National Productivity Review* (Summer 1986), pp. 262–70.

The experiments at Volvo, Gaines, and Digital have shown that many benefits result from employing sociotechnical design strategies. The creation of self-managing teams makes sense not only from a systemic perspective, but also in terms of costs and customers. Dent (1990) has identified at least three ways that self-managing teams produce benefits:

> 1. *Resources are conserved.* Solving problems at the lowest possible level reduces costs and increases responsiveness and innovation.
> 2. *Employees who control their own environment and make their own decisions are more motivated and accountable.* This means higher productivity and less management time wasted on futile efforts toward "artificially" motivating employees and fostering customer service.
> 3. *Decisions are made closer to customers,* which means lower costs from fewer bad decisions and higher revenues from more satisfied customers. (p. 34)

The sociotechnical systems approach effectively blends hard and soft systems approaches. The limitations of STS are its static perspective and its lack of a comprehensive cybernetic dimension. Pava (1986) has noted that STS excels in organizing linear, sequential systems; however, it provides few answers in nonlinear, concurrent systems. As he concludes, "The absence of clear beginnings and ends in nonlinear work, along with saturated interdependence, makes the use of conventional STS analysis virtually impossible" (p. 205).

Despite this apparent limitation, STS can become richer and more capable of responding to the dynamism of complex systems. One strategy for attaining this goal is to increase the cybernetic dimension by creating self-designing systems.

SELF-DESIGNING SYSTEMS

By linking sociotechnical systems with the practices of action learning and organization learning, the cybernetic content of STS can be increased measurably. By involving employees directly in continuous experimentation to redesign work, they can learn to improve the way the work is designed as well as refine the process of work redesign itself. Mohrman and Cummings (1989) propose that by building their capacity for self-design, workers gain the ability to translate basic organizing principles into forms that mesh with the organization's technology, people, and environment. When the cybernetic component is added, self-designing becomes a continuous cycle of experimentation and learning, which continues to add to both the design and work capacities of organization members.

The transition to participative work design is not simple; centuries of domination by reductionist paradigms cause this logic to escape many executives. The system is tuned to expect STS to be performed by consultants who provide one-shot solutions. Once such a system is implemented, however, it forces the unfreezing of old mental models. Workers and managers can jointly experience the outcomes of their experiments in redesign (Weisbord, 1987). Weisbord goes on to argue that with minimal guidance workers can produce redesigns that are 80 to 90 percent as effective as those achieved by professional consultants. Additionally, since workers are always onsite, they can continually fine-tune their experiments and learn in the process. A number of such innovative experiments have been conducted in Norway, where thousands of managers have engaged in such action-research redesign work (Elden, 1983).

~ THE MACROSOCIAL LEVEL OF SYSTEMS

The macrosocial level of systems represents an early attempt by Trist to account for the large systems that exist beyond the normal boundaries of any single organization. Trist's identification of the macrosocial environment is extremely important for its pioneering recognition that organizations are significantly related to larger sociotechnical systems. However, when this model was developed, in the late 1970s, there was relatively little consciousness of global systems.

The **macrosocial system** is composed of the collective social and technical systems that exist beyond individual organizations. The technical macrosystem is comprised of all the technologies available to organizations. Technology such as computers and robots—even nonphysical technologies, such as methods for managing people—all have the potential to provide competitive advantages to those organizations that master them. The implementation of innovative technologies has the power to reshape organizations' form and function. The thrust of the concept of the technical macrosystem is that people will be dramatically affected by technology in organizations. The effects of the technological macrosystem are determined by which forms of technology the organization incorporates into its system and how it applies them.

The social macrosystem is comprised of three basic levels: the industry, the community, and networks. Industry-level systems include much more than competitive practices, normally associated with industries. The focus of this concept is on the collaborative efforts of the members of an industry to manage the industry itself.

SUMMARY

STS is an integrative framework for balancing the social and technical dimensions of systems. While efforts to improve productivity in the sociotechnical domain represent a conscientious effort to integrate the two forces, there are still many areas where improvement is needed. STS uses design and engineering modes of thinking, and soft systems ideas; it needs to expand its implementation of cybernetic thinking. Trist (1981) himself notes that as originally designed by the Tavistock group there was a much greater integration of the social and technical dimensions than is evident today.

The process elements of STS have largely been ignored in favor of the design dimensions. The result is that STS has often been applied in a relatively static and linear approach that has severe limitations (Cummings, 1986; Pava, 1986). However, the addition of participative and action-research methods are renewing interest in STS; these enhancements offer a means to transform it into a relatively more integrative framework for managing and for helping managers to think about their work in novel ways. The next chapter looks at an even more integrative approach to managing, the system dynamics approach.

KEY TERMS AND CONCEPTS

sociotechnical systems
joint optimization
technological imperative
primary work systems
autonomous work groups
self-regulating groups
slack resources
self-contained tasks
whole organization systems
redundancy of functions
macrosocial systems

QUESTIONS FOR DISCUSSION

1. Should work designs attempt to meet people's needs or just attempt to avoid creating negative psychological states?
2. Self-designing systems offer great promise as a way to empower workers and improve performance. Do workers, in general, prefer

to have their work designed by someone else or to participate in the process themselves?

3. Imagine the work done by people in a fast-food restaurant. How could you redesign such a system so that both human needs and task concerns are optimized? Draw a sketch of such a system if possible.
4. A class is a primary work system in which considerable work is accomplished. Redesign a class or seminar from a sociotechnical perspective to optimize human and technical concerns. What type of restrictions do you anticipate from other people—such as professors and deans?
5. Does the sociotechnical systems approach attempt to solve existing problems or just improve a situation? Does it make any difference if it is applied in a new organization or an existing one with well-established problems?

REFERENCES

Ashby, W. R. (1956). *An Introduction to Cybernetics.* London: Chapman and Hall.

Bernstein, P. (1988). "The Learning Curve at Volvo," *Columbia Journal of World Business, 23,* Winter 1988, pp. 87–95.

Cummings, T. (1986). "A Concluding Note: Future Directions of Sociotechnical Theory and Research," *The Journal of Applied Behavioral Science, 22,* no. 3, pp. 355–60.

Dent, H. (1990). "Organizing for the Productivity Leap," *Small Business Reports,* September, pp. 31–44.

Elden, M. (1983). "Client as Consultant: Work Reform Through Participative Research," *National Productivity Review,* Spring, pp. 136–47.

Emery, F., and Trist, E. (1964). "The Causal Texture of Organizational Environments," *Human Relations, 18,* pp. 21–32.

Galbraith, J. (1977). *Organization Design.* Reading, Mass.: Addison-Wesley.

Griffin, R. (1982). *Task Design: An Integrative Approach.* Glenview, Ill.: Scott, Foresman.

Hackman, J. R., and Oldham, G. R. (1975). "Development of the Job Diagnostic Survey," *Journal of Applied Psychology," 60,* pp. 159–70.

Hackman, J. R., and Oldham, G. R. (1980). *Work Redesign.* Reading, Mass.: Addison-Wesley.

Herbst, P. G. (1975). "The Product of Work Is People," in L. Davis and A. Cherns (eds.), *The Quality of Working Life.* New York: Free Press.

Mohrman, S., and Cummings, T. (1989). *Self-Designing Organizations.* Reading, Mass.: Addison-Wesley.

Morgan, G. (1986). *Images of Organization.* Newbury Park, Calif.: Sage Publications.

Obloj, K., and Joynt, P. (1986). "Strategic Reserve," working paper, Norwegian School of Management.

Pava, C. (1986). "Redesigning Sociotechnical Systems Design: Concepts and Methods for the 1990's," *The Journal of Applied Behavioral Science, 22,* no. 3, pp. 201–21.

Trist, E. (1976). "A Concept of Organizational Ecology," *Bulletin of National Labor Institute,* New Delhi, 12, pp. 483–96.

Trist, E. (1981). "The Evolution of Socio-technical Systems: A Conceptual Framework and an Action Research Program," Ontario Ministry of Labor.

Weisbord, M. (1987). *Productive Workplaces.* San Francisco: Jossey-Bass.

ADDITIONAL READINGS

Cherns, A. B. (1976). "The Principles of Sociotechnical Design," *Human Relations, 29,* no. 8, pp. 783–92.

Proctor, D. (1986). "A Sociotechnical Work Design System at Digital Enfield: Utilizing Untapped Resources," *National Productivity Review,* Summer, p. 262.

Solomon, B. (1985). "Plant That Proves That Team Management Works," *American Management Association.*

Yorks, L., and Whitsett, D. (1985). "Hawthorne, Topeka, and the Issue of Science Versus Advocacy in Organizational Behavior," *Academy of Management Review, 10,* no. 1, pp. 21–30.

14 System Dynamics

Structures of which we are unaware hold us prisoner. Conversely, learning to see the structures within which we operate begins a process of freeing ourselves from previously unseen forces and ultimately mastering the ability to work within them and change them.

Peter Senge (1990)

1. To learn the basic assumptions of the system dynamics perspective about the relationship between systemic structure and behavior.
2. To recognize how the system dynamics perspective provides a unique framework for interpreting behavior in organizations.
3. To understand the integrative nature of system dynamics.
4. To know the basic techniques of causal-loop diagramming and stock-and-flow diagramming.
5. To comprehend the role of modeling and simulation in the system dynamics perspective.
6. To appreciate the relationship among systems thinking, organizational learning, and system modeling.

SECTION I
FUNDAMENTALS OF SYSTEM DYNAMICS

INTRODUCTION

Integrative systems approaches focus on the need to combine hard and soft systems thinking and cybernetics into a single perspective. The approach that achieves the fullest integration is system dynamics. System dynamics originated in principles established in engineering and control theory. In its early development, it was criticized as being exclusively a closed-systems approach based solely on hard systems thinking and cybernetics. In practice, the effectiveness of system dynamics largely depended on the ability of specialized consultants to build computer models that accurately reflected other people's perceptions of how a system worked.

With the advent of high-speed desktop computers, user-friendly software, (I-Think, Dynamo, STELLA II) and a shift in emphasis toward using model-based insights as a stimulus for organizational learning, system dynamics represents a relatively integrative approach to managing. Through the application of modeling and computer simulation, system dynamics can be used as a creative educational strategy that challenges managers to question their own assumptions about the underlying cause-and-effect relationships that govern their organization. This chapter will provide an overview of the development of the system dynamics approach.

In **system dynamics,** computer modeling and simulation are employed to help managers identify and understand the dynamic patterns of behavior of a system. These patterns are likely to result from the interaction of the various feedback loops within the system. Moreover, these feedback-driven interactions cause changes in the stocks and flows present within a system. **Stocks** are accumulations that build and decline over time; they may be accumulations of anything, but in organizations they are typically accumulations of inventory, cash flow, employees, and so on. **Flows** are the inflows that contribute to the accumulations and the outflows that drain the level of the accumulations. Each flow either fills or drains the accumulation at a certain speed, the **rate.** The rate of inflow or outflow determines the **level** of the accumulation. A level is the amount of accumulation at any single point in time.

To visualize how a dynamic system might behave according to these principles, imagine the following situation. The Stale Ale Brewery uses a single 2,000-liter vat to brew its ale. Because the demand for Stale Ale is so great, the brewery operates at full capacity twenty-four hours per day. Various flows of spring water, hops, barley malt, corn meal, and yeast are added to the mix. The brewmeister varies the rate of flow of each ingredient based on whether he is making a light ale or a regular brew. The level of mix in the vat fluctuates based on the rate of inflow of ingredients and the outflow to the bottling process. The rate at which the ale is bottled depends on the reliability of the machinery that fills the bottles and the loaders who place the bottles on the conveyer belt (see Figure 14-1).

Since the machines at Stale Ale are old, they break down frequently and require quick adjustments by the loaders. Consequently, the levels in the brewing vat fluctuate based on the rate of outflow to the bottling process and the brewmeister's recipe on any given day. A student from Tech University studied the fluctuations in the vat levels over time and found that they followed a specific pattern (see Figure 14-2). In system dynamics, the focus of effort is often directed toward defining the dynamic patterns within a system and then attempting to determine their cause. By recognizing the pattern and determining its cause, the systemic structure may be changed in order to change the behavior of the system. In system dynamics, modeling and simulation are generally used tools to gain insight into specifically how the system should be changed. Computer simulations enable managers to experiment with different alternative problem solving scenarios. By observing the possible consequences to each alternative over an extended time period, managers can develop a better understanding of how the system works.

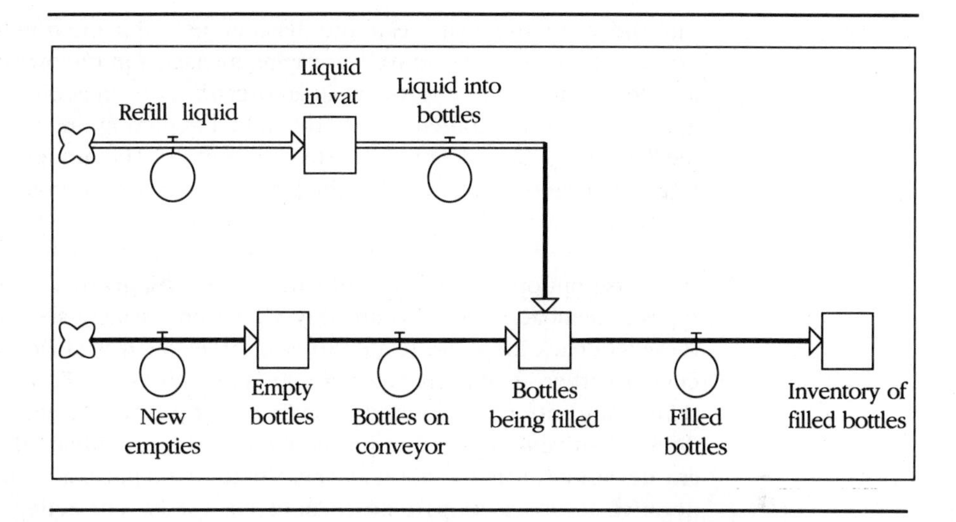

FIGURE 14-1 ***A Brewery as a Stock Accumulation System***

Source: B. Richmond, "Systems Thinking: A Critical Set of Critical Thinking Skills for the 90's and Beyond," Lyme, N.H., High Performance Systems, 1990, p. 12.

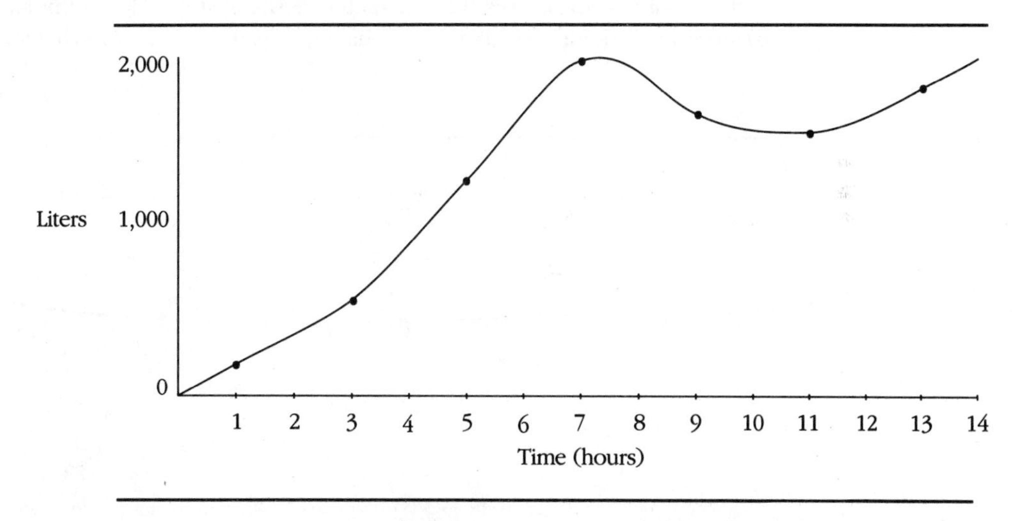

FIGURE 14-2 ***Stock Fluctuation over a Fourteen-Hour Period***

THE PROCESS OF SYSTEM DYNAMICS

System dynamics takes an integrative perspective to developing systems thinking capabilities through the use of computer simulation as a learning tool. System dynamics uses systems thinking as a conceptual tool for gaining insight into the structures that create the dynamic behavior often found in complex systems. It does this by actively engaging managers in the process of modeling a system—such as a department in an organization, an ecological problem, or an urban housing problem—with the model depicting the interactions among feedback loops that define the system's structure. The model-building process forces managers to consciously think about their assumptions of how the system works.

The system dynamics approach encourages model builders to express their assumptions in the form of **causal-loop diagrams,** which identify the types of feedback loops that are at work within a system and then depict how these feedback loops are related to each other. Such feedback loops were discussed earlier, in the chapter on cybernetic thinking. Feedback loops, self-reinforcing patterns of causality, can be either positive or negative (see Figure 14-3). Positive loops tend to promote change by amplifying the magnitude of any deviations from the norm and accelerating the pace of change. Conversely, negative feedback loops tend to limit change by correcting any deviations from the norm and returning a system to its prior state of balance.

System dynamics is a problem solving methodology and a conceptual tool for learning how systems work. System dynamics adopts the perspective that the changes in the performance or behavior of a system are the consequence of the feedback structures that characterize the system. Most importantly, the assumption is made that dynamic behavior is produced by internal structures.

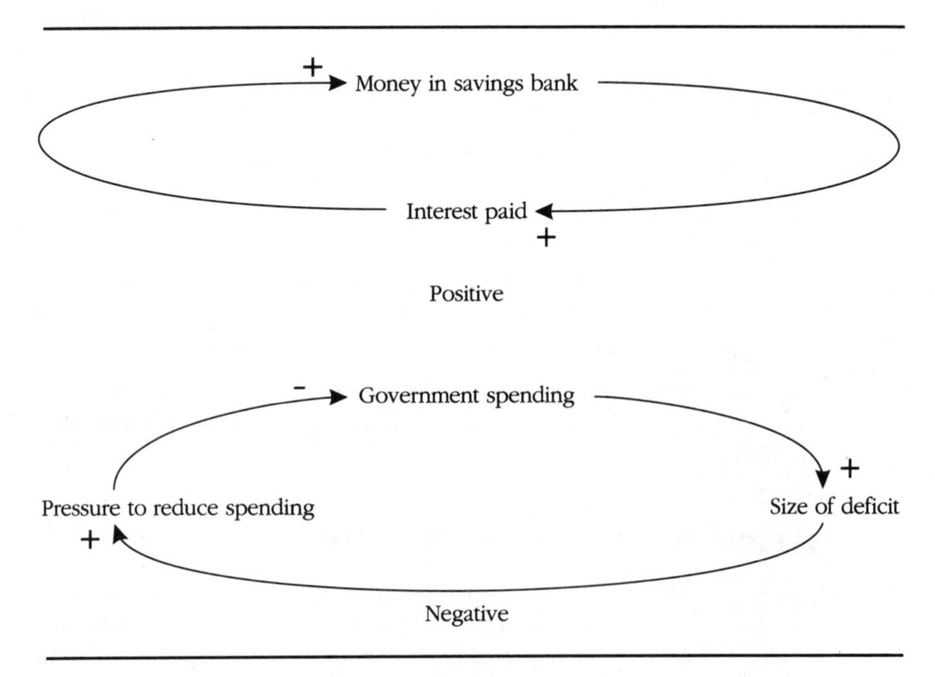

FIGURE 14-3 ***Positive and Negative Feedback Loops***

Consequently, problem solving efforts should have an internal focus (Richardson and Pugh, 1981). In organizations, behavior is assumed to originate in corporate structures and policies. From a managerial perspective, system dynamics argues that the central role of managers is to regulate behavior by recognizing its origins and then establishing policies to govern it (Lyneis, 1980).

Once the model has been constructed, it is used to simulate how the system would behave over extended periods of time. This simulation is based on the causal-loop diagrams designed by the manager. After viewing the results of the simulation, the manager is asked to evaluate whether the system actually functions in this manner. If it does not, he or she must question whether the initial assumptions accurately represented how that system actually works. After viewing the simulation, managers frequently begin to perceive long-term and systemic patterns of interaction that had not previously been evident to them. These patterns are often obscured because time delays exist between cause and effect in most systems. **Time delays** are lags between cause and effect created by slowdowns in information availability. They induce fluctuations in the system's behavior pattern. The simulation helps users recognize these patterns.

The simulation is useful because consciousness of the linkages between the elements of a system and of its feedback loop structure can provide insights that support the process of redesigning the system. Efforts to create new organizational structures often focus on exploring ways that the patterns of information flow may be altered. Modeling and simulation permit a semblance of observation of the patterns of a system's dynamics in real time. This generally provides users with deeper appreciation of the effects of the many interconnections between elements of the system. After reflecting on why these patterns have emerged, managers can redesign the system and witness another cycle of simulation. Ultimately, these cycles are reiterated for as long as possible to promote the greatest possible learning effect in a limited amount of time. The goal of this reiterative learning cycle is to help people modify their mental models to account for the complex relations between structures, feedback loops, and patterns of behavior that would not be apparent otherwise. This is an important contribution that a computer makes to this process. Forrester (1961) has argued that the cognitive limitations of the human mind prevent people from understanding how complex dynamic systems operate. Through these model-based insights, however, managers can change the way that they think to more accurately reflect the dynamic dimensions of systems. As a result of this shift of mind, managers will attach less significance to individual, short-term effects and be more inclined to seek systemic explanations (Richmond, 1990).

An innovative system dynamics application is known as **management flight simulators** (Diehl, 1992; Kim, 1990). In management flight simulators, managers are immersed in a simulated gaming environment that recreates the major characteristics of their work setting (see Figure 14-4). They are given the opportunity to experiment with alternative decisions and note, through

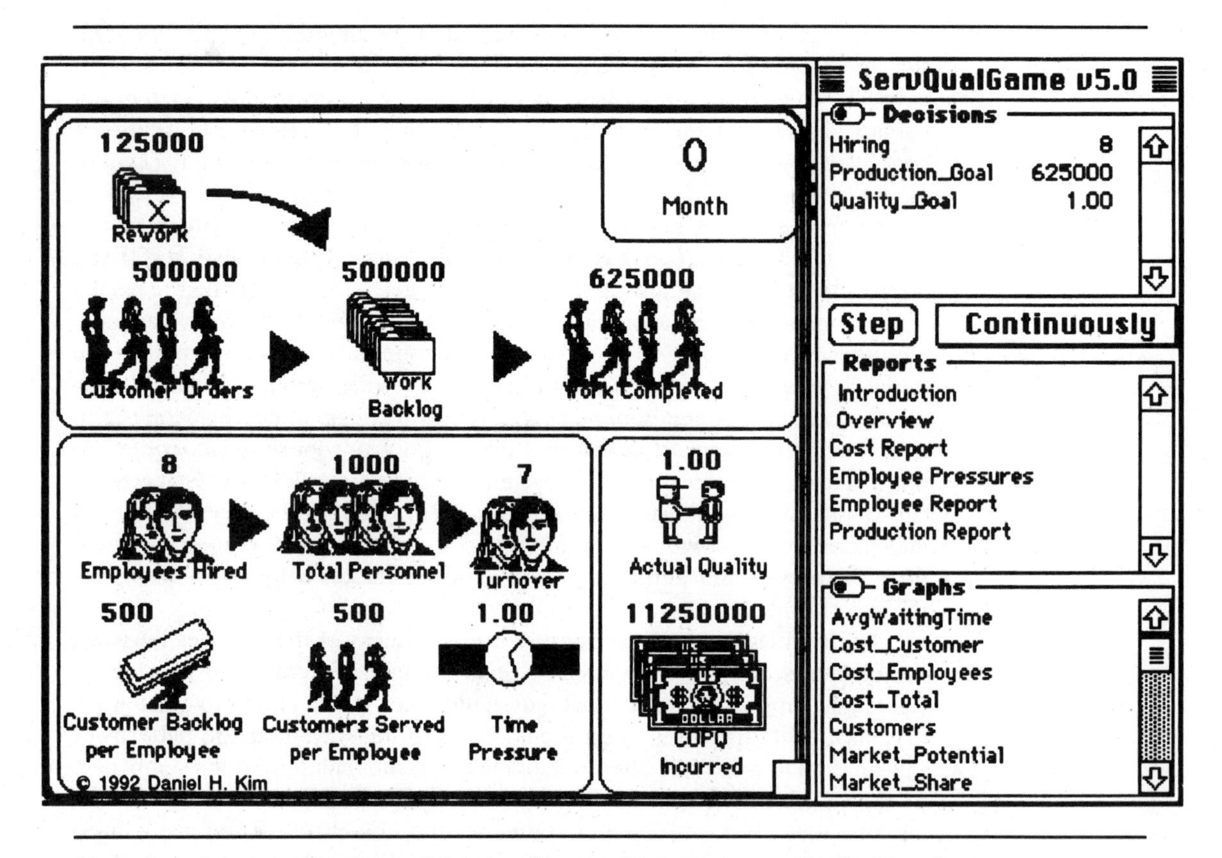

FIGURE 14-4 ***Command Screen from a Management Flight Simulator***

Source: Developed by Daniel Kim, 1991, System Dynamics Group, MIT.

the simulation, the consequences of those alternatives. The outcomes produced by the game players often cannot be accounted for in their current mental models, which stimulates the questioning of basic assumptions. The term *flight simulators* is based on the sophisticated computerized models used to train pilots; these programs can be designed for any situation. They have already been created for managing a major passenger airline, an insurance company claims unit, service quality, oil-tanker prices, a real estate market, and maintenance decisions in a chemical factory.

~ THE INTEGRATIVE VIEW

System dynamics includes all three core forms of basic systems thinking. The hard systems approach forms the basis for much of the model design work. The system is viewed as a set of discrete relations that can be known and quantified. Churchman's (1971) discussion of the issues and challenges of systems design and inquiring systems are especially relevant here (see Chapter 4).

Cybernetics is clearly visible in the design of the framework of feedback loops. The model itself outlines the various cybernetic feedback loops that govern the patterns of interaction in the system. Furthermore, the process of model building, simulation, reflection, questioning assumptions, and redesigning the model is reiterative—and thus cybernetic (see Figure 14-5).

Finally, the learning process of achieving a more objective view of the system by continuously challenging old assumptions is based on soft systems thinking. As Ackoff, Checkland, and others had advocated earlier, system dynamics allows for a dialectic process of inquiry that gradually helps to clarify a more objective view of the system. In learning laboratories, managers are placed in a risk-free environment where they can playfully experiment with their ideas and question their own deeply held beliefs and paradigms. In the political reality of organizations, the process of shifting to new mental models can be extremely awkward because managers have not had the opportunity to fully test new ideas and debunk old ones. Learning laboratories provide that opportunity, allowing managers to exhaust the range of their thinking and develop greater certainty about their own beliefs.

The various strands of management theory that underlie system dynamics are listed in Figure 14-6.

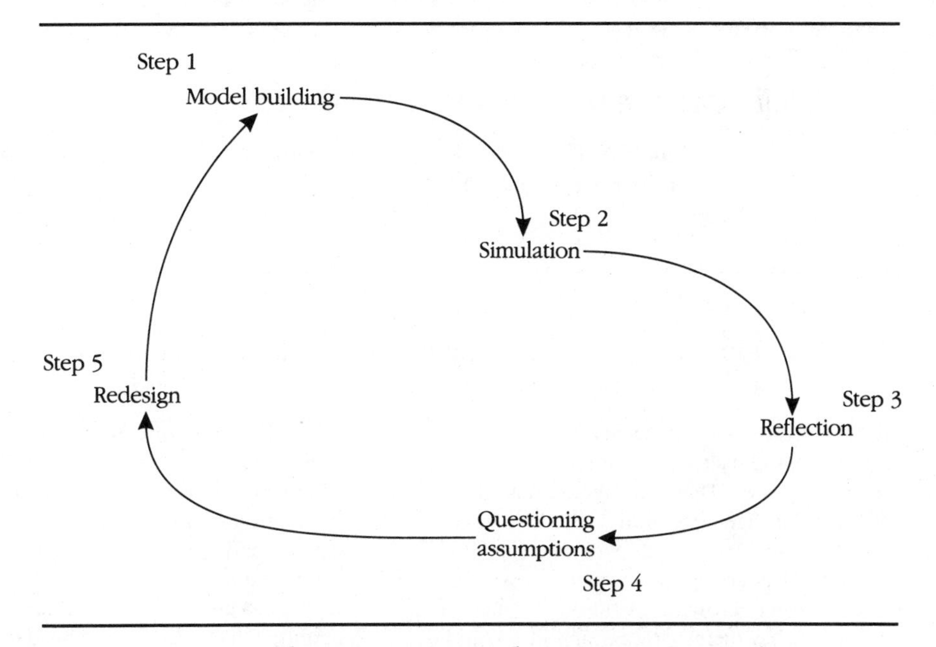

FIGURE 14-5 ***The Reiterative Cycle in System Dynamics***

ORGANIZATION THEORY AND BEHAVIORAL DECISION THEORY
- Organization structure
- Information availability and quality
- Bounded rationality of agents
- Decision making heuristics of agents

FEEDBACK CONTROL THEORY
- Principles of systems
- Structure of decisions
- Link between structure and behavior
- Worldview

DIGITAL SIMULATION
- Accurate determination of behavior from structural and parametric assumptions
- Tools for interactive exploration of microworlds—management flight sumulators

PROCESS CONSULTATION—TOOLS AND PROCESSES FOR
- Eliciting and mapping mental models of clients
- Challenging mental models
- Improving mental models
- Developing systems thinking

FIGURE 14-6 ***Elements of System Dynamics***
Source: J. Sterman, System Dynamics Group, MIT, unpublished.

THE EVOLUTION OF SYSTEM DYNAMICS

Much of the initial development work on system dynamics can be traced to Jay Forrester (1961).

SYSTEMS PROFILE
Jay W. Forrester

Jay W. Forrester, Germeshausen Professor, Emeritus, at the Massachusetts Institute of Technology, directed the System Dynamics Program at MIT's Sloan School of Management until 1989. The field of system dynamics has developed since 1956 under Forrester's leadership to evaluate how alternative policies affect growth, stability, fluctuation, and changing behavior in corporations, cities, and countries.

Forrester was director of the MIT Digital Computer Laboratory from 1946 to 1951 and was responsible for the design and construction of Whirlwind I, one of the first high-speed computers. Forrester invented—and holds the basic patent for—random-access, coincident-current magnetic storage, for many years the standard memory device for digital computers. He was also the head of the Digital Computer Division of MIT's Lincoln Laboratory from

1952 to 1956, where he guided the planning and technical design of the U.S. Air Force SAGE (Semi-Automatic Ground Environment) system for the continental air defense, the most extensive early application of digital computer technology.

In 1956, Forrester became professor of management at the Sloan School of Management. There he applied his background in computer sciences and engineering to the development of computer modeling and analysis of social systems, leading to the creation of the field now known as system dynamics. Forrester's first major book in system dynamics, *Industrial Dynamics,* was published in 1961. He was also an instrumental member of the Club of Rome, an international group of scientists and social scientists whose purpose was to "foster an understanding of the varied but interdependent components—economic, political, natural, and social—that make up the global system in which we all live" (Meadows et al., 1972). In his fourth book, *World Dynamics* (1971), he applied the principles of system dynamics to the study of interactions on a global scale. In 1986, Professor Forrester was honored by Thomas Watson Jr. of IBM, who endowed the Jay W. Forrester Chair in Computer Studies at MIT, and in 1989 he received the National Medal of Technology from President Bush.

The original rationale for system dynamics was that human cognitive abilities are simply inadequate for piercing the fog created by the complexity of some systems but that computers, because they could account for a vast number of possible interactions, were a powerful adjunct to thinking. Consider that a system with ten variables is capable of a minimum of 3,628,800 possible interactions. Work by Simon (1982) on managerial decision making suggests that people have a relatively limited capacity for processing information. Simon (1957) introduced the concept of **bounded rationality** to describe these innate limitations. According to this principle, "it is impossible for the behavior of a single, isolated individual to reach any high degree of rationality. The number of alternatives he must explore is too great, the information he would need to evaluate them so vast that even an approximation to objective rationality is hard to conceive. Individual choice takes place in an environment of 'givens'—premises that are accepted by the subject as basis for his choice; and behavior is adaptive only within the limits set by these 'givens' (Simon, 1957, p. 79).

SECTION 2
BASIC ASSUMPTIONS OF SYSTEM DYNAMICS

The system dynamics perspective makes several assumptions that differentiate it from other nonintegrative systems philosophies. Several of these assumptions have been clearly articulated by Senge (1990) and Forrester (1968).

STRUCTURE INFLUENCES BEHAVIOR

The structuralist perspective has roots in engineering theory and in the behavioral sciences, especially sociology. Perrow (1970) has assessed the value of the structuralist perspective as follows: "Since this approach focuses upon structures, it must deal with the whole rather than with parts of particular processes; it is forced to ask how structures differ" (p. 28). In system dynamics, structure does not refer to organizational hierarchy, but to the systemic structure of feedback loops. A system's structure is regarded as the most powerful factor determining how a system will behave over time. The various feedback loops alone have great power over behavior. When they interact synergistically with other feedback loops, the effect on systems behavior can be even more forceful, though it may be subtle. For example, sales of a new product generate both revenues and word-of-mouth advertising that benefit the seller. The increased revenues enable the firm to engage in a more vigorous advertising campaign, which increases sales further, as well as increasing word-of-mouth advertising even further. These two positive feedback loops are often responsible for creating the steep sales curve that frequently accompanies the introduction of a new product; personal computers and VCRs showed such a dynamic. Figure 14-7 illustrates the powerful reinforcing effects of these loops.

Four basic components of structure occur in systems (Forrester, 1968):

1. The boundaries of the system
2. The network of feedback loops within the system
3. The levels and rates of information flows
4. Goals and goal-oriented actions

In organizations the elements of structure are even more specific because systemic structure can be expressed in many ways. If managers understand the structure of a system, they can determine which are the most dominant and influential feedback loops. Consequently, they can direct their efforts at changing the system to those locales where it will have the greatest impact relative to the magnitude of the intervention. Senge (1990) has labeled such areas of strategic importance for change as **leverage points.** Leverage points are valuable because a relatively small effort can yield a relatively large change.

Systemic structure often diminishes the importance of any single individual or group in determining the long-term behavior of an organization. For example, in the late 1980s, many airlines within the U.S. airline industry went bankrupt or were purchased by competitors prior to becoming insolvent. Although some measure of this systemwide behavior can be attributed to lack of management effectiveness, much of it results from the effects of the systemic structure of the industry over the long term. For instance, we can examine the effects of pricing on competitiveness in the industry. An airline's pricing strategy depends on its plane capacity and its ability to fill capacity. Airlines that fly with less than full capacity have difficulty keeping their rates competitive and have fewer funds to dedicate to advertising. This causes profits to fall, which diminishes their ability to price competitively, which causes sales to fall

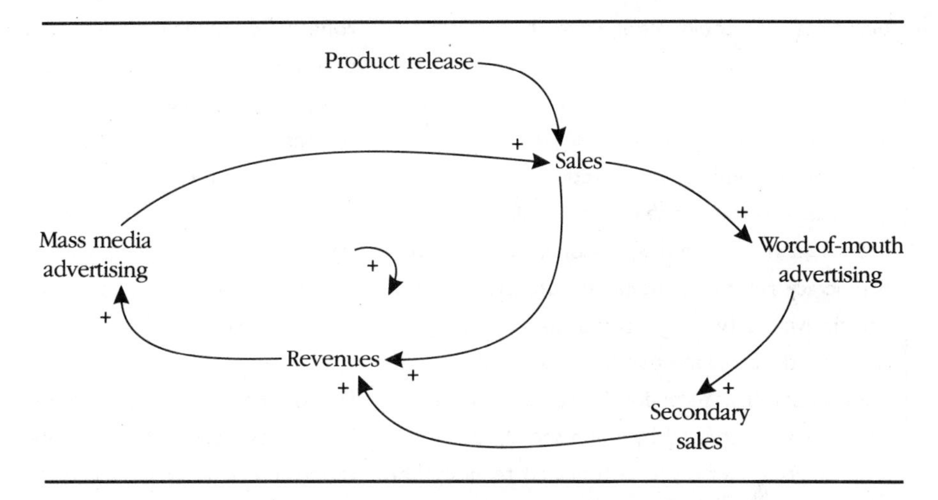

FIGURE 14-7 ***Causal Loop for New Product Introduction***

even further. Consequently, individual personalities and abilities become secondary to the effects of a system's structure in determining performance.

Seen from another side, the framework of a system will, over the long term, neutralize any deviations in work performance resulting from individual initiative. Senge (1990) has noted that different individuals within the same system tend to produce qualitatively similar results. This idea finds expression in the axiom, "If you put a good person in a bad system, the system will always win out." The far-reaching effects of systemic structure on behavior can be seen in the behavior of citizens in various countries throughout the world. Certain forms of government tend to repress citizens and reduce variety, while others promote greater variety. The influence of the disintegration of the systemic structure of the former Soviet republics is evident in the chaotic behavior witnessed in this region in 1992. The global systems spotlight examines future possibilities from a system dynamics perspective.

GLOBAL SYSTEMS SPOTLIGHT

The New Russia: A Dynamic System

The transformation of the former Soviet Union into the Commonwealth of Independent States in 1992 established Russia as the largest member of the confederation. However, Russia itself is also a collective state composed of many republics with varying ethnic heritages. There are also different political factions vying for power in the new Russia. The following scenarios illustrate a number of the potential outcomes that could result from the dynamics at work within Russia in 1992.

Scenario 1: Confusion Reigns

The conflict among rival factions continues to generate political instability in Russia. The most prominent groups are the pro-communists, the pro-democracy reformers, ethnic nationalists, and Russian chauvinists. No single group has the leverage to reconcile the problems that beset Russia in 1997. To maintain stability, a standoff is reached. A coalition government is created by the assembly of various interest groups who are determined to block the ascendancy of any dominant party. This loose coalition is formed in response to the failure of several ill-fated attempts by individual groups to force various economic reforms. The military, theoretically under central control, splinters to support regional governments by acting as quasimilitia.

Economically, efforts to achieve privatization of former state-owned industries have been minimal, accounting for less than 10 percent of Russia's GNP. Russia remains dependent on aid from Western countries and Japan in order to avoid economic collapse. The business climate is perceived by foreign investors as extremely risky. Inflation, unemployment, and bank failures remain high, while the value of the ruble remains low (150 to 400 rubles to the dollar).

Scenario 2: A Return to Communism

A military-backed coup has restored Russia to its former system of totalitarianism and a centrally planned economy following the failure of democratic reforms and the collapse of the free-market economy. Reprisals have been aimed at the former leaders of the democratic movement and their supporters. Economic sanctions are implemented against the new communist government by the European Community and the United States. An underground resistance movement forms to selectively counter the Russian army in a number of regions. High inflation has been dampened by the effects of harsh wage and price controls. The value of the Russian ruble has stabilized at an exchange rate of 400 rubles to one dollar. All foreign business investors are mandated to participate in partnership with state-owned industries.

Scenario 3: The "Russian" State

The emergence of rampant nationalism within the various ethnic areas has led to the disintegration of effective central control and to the formation of several independent states in the former Russian territory. The most powerful of these is the Moscow-centered Russian Republic. Among the other states are the Muslim region of Turkistan and the Far Eastern Republic. This disintegration has caused massive flows of refugees out of ethnic regions to escape persecution for political beliefs or because of cultural identity. The government is being ruled by a Russian-style democracy that is non-parliamentary and offers no minority rights. There has been a sweeping return to the centrally planned economy, and a strong anticapitalist sentiment has arisen. Debt repayments to foreign investors have been dismissed as unnecessary since all businesses are now state-owned. The new central economic planning committee has set the value of the ruble as equal to one dollar as a way of restoring Russian national pride.

Scenario 4: Chile Revisited

In the early 1970s, Chile, under President Salvador Allende, attempted to combine centralized economic planning with decentralized decision making. In 1995 in Russia, a military coup has overthrown the government after a series of half-hearted economic reforms failed and the standard of living fell through the floor, such that the Russian people were faced with serious hardships. The parliament has been disbanded, political parties have been eliminated, and the country is run by a strong, military-backed president. The vision of the New Reconstructionists is to create a world-class economy. The revised economic system is being transformed by a radical, centrally orchestrated move to privatization. The state owns less than 10 percent of all industry, and it distributes vouchers to citizens for the purchase of goods. All price controls, import tariffs, and limitations on foreign investment have been eliminated. Trade within the Commonwealth of Independent States is conducted at world prices, and the ruble trades at fifty to the dollar.

Scenario 5: European-style Democratic Socialism

By 1996, the Russian democratic movement had settled on a multi-party political system patterned on the system that is common in Scandinavian countries. The democratic process in Russia has moved at a snaillike pace, as every issue becomes the subject of a lengthy parliamentary debate. Industries related to infrastructure, such as road building, remain state-owned. The state still retains ownership of around 40 percent of all industry, while most consumer-products industries have been privatized. Most businesses are stock-owned, with stock being held jointly by investors and workers. Individuals and businesses are heavily taxed to provide support for wide-scale social programs. The state continues to maintain strict regulation over the transfer of real estate. State-directed protectionist policies are being used to encourage the development of specific industrial sectors and to ensure that an attractive environment for foreign investment is continued. The ruble is now partially convertible into other currencies and trades at eighty to the dollar.

Scenario 6: New York on the Volga

The 1996 reelection of Russian President Boris Yeltsin has provided a mandate for government by a two-party system, similar to that in the United States. This has followed two years of relative tranquility and growing prosperity. Russia has become the hot spot for vacationing among affluent Americans, Europeans, and Japanese. A sophisticated stock exchange has been designed, with technical assistance from the United States and Japan, to channel the flow of investment within the country. In the booming Russian economy, there are few restrictions on imports, exports, and price levels. The ruble is a sought-after form of currency and is exchanged at twenty-five to the dollar. Economists have predicted that by the end of the decade Russia's GNP will surpass Germany's, putting Russia third in the world behind the United States and Japan.

The causal loop diagram in Figure 14-8 outlines a number of the primary systemic relationships that are likely to determine the future direction of Russia. After studying this diagram, which scenario appears to you to have the greatest likelihood of occurring in 1996?

Source: Written by Igor A. Portyansky, Ph.D., director of the Center for Independent Analysis, Moscow; visiting scholar in the System Dynamics Group at the Massachusetts Institute of Technology; and vice-president of The Compass Group; and Edward Ward of Lexington, Massachusetts.

STRUCTURAL EFFECTS ARE SUBTLE

Although the hierarchies found in organizations are often notable for their clarity and precision, many other forms of structure—forms that are less obvious and dynamic—shape behavior within a system. For example, policies and procedures operate in a less obvious manner than hierarchy. Often taken for granted, they can exert a strong influence on behavior once they become integrated into people's routines. The resistance to change frequently encountered in organizations often arises because people are programmed to act in accordance with procedures and learn to substitute them for judgment. The tradition of always performing tasks in the same way often reflects the power of policies and procedures to influence people.

Nordstrom Stores has purposely tried to eliminate many forms of procedures in order to stimulate innovation. Nordstrom's reportedly uses a one-line policy manual, and that one directive is to make the customer satisfied (Peters, 1987). The belief is that by avoiding reliance on procedure manuals, employees will continually invent new, creative ways to satisfy the customer.

SYSTEMS ARE NETWORKS OF FEEDBACK

Although it may be difficult to imagine, systems are in a perpetual state of either reinforcing past performances or attempting to deviate from them. The state of the system may change as a result of regulation, control, acceleration, adaptation, or coordination. All of these functions are driven by information flows, in the form of feedback loops, that govern the system's actions. It is relatively easy to visualize simple feedback loops, such as a person regulating the temperature of water while taking a shower. In large organizations, hundreds of such loops may be in operation, controlling the range of possible futures of the system. Over time various feedback loops create momentum within the system such that success breeds success—or, conversely, decline may stimulate failure.

FEEDBACK AMPLIFIES OR ATTENUATES BEHAVIOR

The effects of informational feedback loops may be increased or diminished from their original strength. These processes are known as amplification and attenuation. **Amplification** may both strengthen the effect of the information and increase the variety associated with it (Beer, 1985; Forrester, 1961). According to Beer (1985), variety is "a measure of complexity: the number of possible states of a system" (p. 35). The natural forces in systems that are necessary for growth and expansion rely on variety to stimulate experimentation and randomness in the system. When the effects of feedback are reduced they have been **attenuated** (Beer, 1985). For example, when a customer service representative listens to customer complaints on a daily basis, the impact

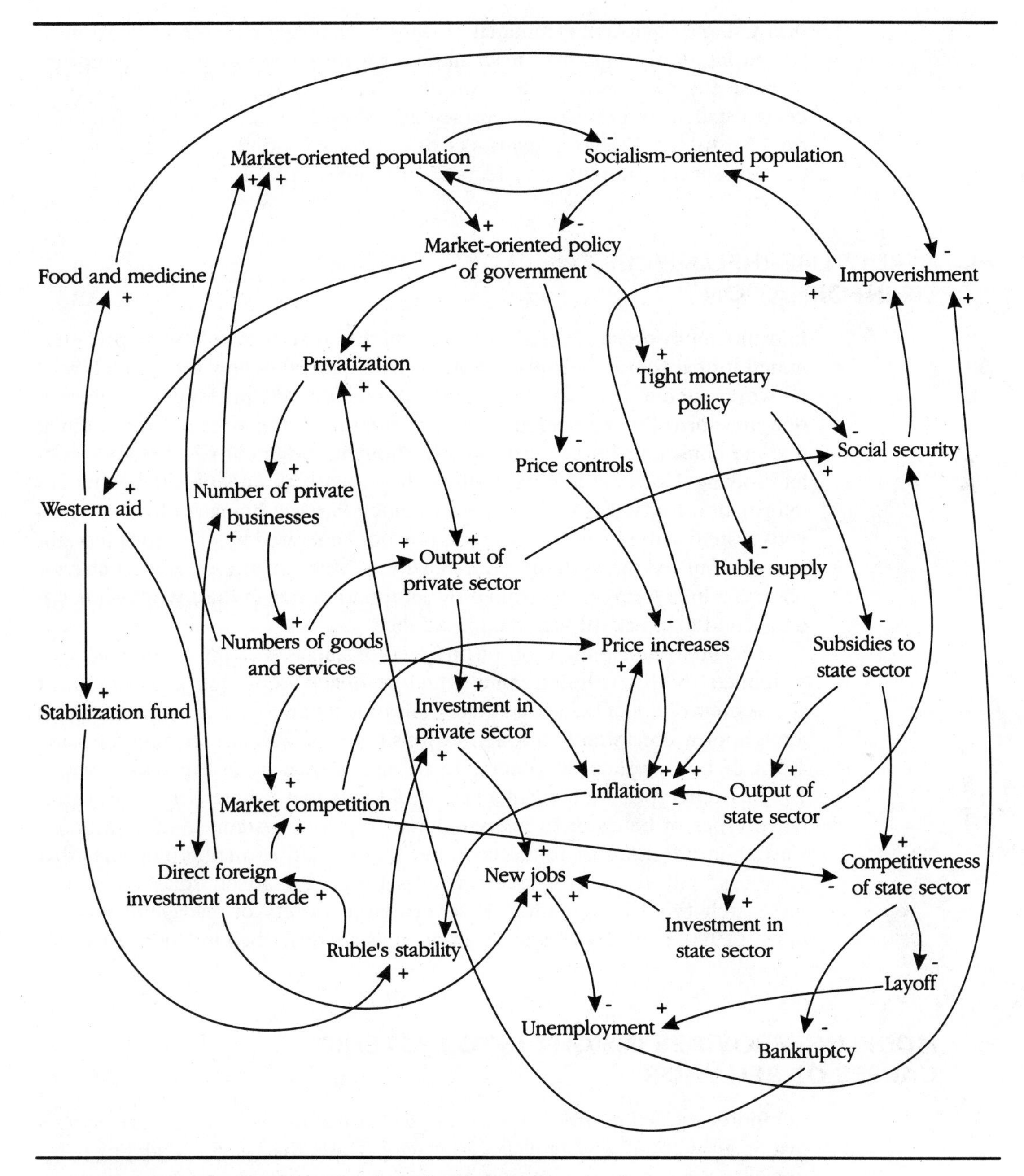

FIGURE 14-8 ***Russian Society in Transition***

Source: Igor Portyansky ©1992.

of any single complaint is minimal. However, a manager who is not accustomed to hearing complaints may react more strongly to any individual complaint because the experience is less routine. Thus, designing a customer service center staffed by permanent representatives tends to attenuate the feedback from customers. Rotating managers through a six-month assignment to this department may amplify the effects of customer feedback.

STRUCTURE INFLUENCES THE FLOW OF INFORMATION

Information in organizations may reach its destination by following predetermined formal routes, or it may create its own pattern of flow through informal networks such as the rumor mill. Paths designed to channel flows of information are normally engineered into the structure of the system. This is much like the conscious human processes of thought, judgment making, and problem solving. However, in organizations there is also an underground system of information transmission that operates much like the human autonomic nervous system—the system of reflexes. It is not generated by conscious thought process, but by the system's built-in wiring. Much of the dynamic behavior observed in systems is controlled by information transmitted within this underground network through feedback loops.

The important dimension of this relationship is that the framework of a systemic structure will determine which feedback loops emerge as dominant and as controllers of behavior and organizational performance. Some feedback loops appear consistently through various types of systems, creating the same types of behavior; these generic forms are known as **archetypes** (Senge, 1990). Archetypes are fundamental causal loops that generate the same qualitative types of behavior in many different types of systems. As Senge notes, "Just as in literature there are common themes and recurring plot lines that get recast with different characters and settings, a relatively small number of these archetypes are common to a very large variety of management situations" (1990, p. 94). Senge's nine archetypes are described in Figure 14-9.

MODELING PROVIDES INSIGHT INTO SYSTEMIC CAUSES OF BEHAVIOR

Computer modeling makes it possible to learn how the various interactions that occur within the system affect behavior within the system. Modeling helps to capture the essence of the way users view the system in question. In doing so, once the fundamental assumptions have been set down, the user is free to concentrate on exploring the patterns of interaction that exist between the elements of the system.

The outcomes of the interaction of various negative and positive feedback loops is virtually impossible to predict through simple observation. The diffi-

1. *Balancing process with delay.* "[Systems,] acting toward a goal, adjust their behavior in response to delayed feedback. If they are not conscious of the delay, they end up taking more corrective action than needed."
2. *Limits to growth.* "A process feeds on itself to produce a period of accelerating growth or expansion. Then the growth begins to slow and eventually comes to a halt, and may even reverse itself and begin an accelerating collapse."
3. *Shifting the burden.* "A short-term 'solution' is used to correct a problem, with seemingly positive immediate results. As this correction is used more and more, more fundamental long-term corrective measures are used less and less. Over time, the capabilities for the fundamental solution may atrophy or become disabled, leading to even greater reliance on the symptomatic solution."
4. *Eroding goals.* "A shifting-the-burden type of structure in which the short-term solution involves letting a long-term fundamental goal decline."
5. *Escalation.* "Two [systems] see their welfare as dependent on relative advantage over the other. Whenever one side gets ahead, the other is more threatened, leading it to act more aggressively to reestablish its advantage, which threatens the first, increasing its aggressiveness, and so on."
6. *Success to the successful.* "Two activities compete for limited support or resources. The more successful one becomes, the more support it gains, thereby starving the other."
7. *Tragedy of the commons.* "Individuals use a commonly available but limited resource solely on the basis of individual need. At first they are rewarded for using it. Eventually, they get diminishing returns, which causes them to intensify their efforts. Eventually, the resource is either significantly depleted, or entirely used up."
8. *Fixes that fail.* "A fix, effective in the short-term, has unforeseen long-term consequences, which may require even more use of the same fix."
9. *Growth and underinvestment.* "Growth approaches a limit which can be eliminated or pushed into the future if the [system] invests in additional 'capacity.' But the investment must be aggressive and sufficiently rapid to forestall reduced growth, or else it will never get made. Oftentimes, key goals or performance standards are lowered to justify underinvestment. When this happens, there is a self-fulfilling prophecy where lower goals lead to lower expectations, which are then borne out by poor performance caused by underinvestment."

FIGURE 14-9 ***Senge's Systems Archetypes***

Source: P. Senge, *The Fifth Discipline: The Art and Practice of the Learning Corporation* (New York: Doubleday, 1990), pp. 272–90.

culties that managers experience as inherent in situations with dynamic feedback loops has been well documented (Dorner, 1989; Kleinmutz, 1985; Sterman, 1989). Senge and Sterman (1991) have examined the effects of feedback loops in the claims settlement process in an insurance company. Their findings suggest that the interaction among the various loops created processes that continually eroded the quality of service and increased settlement costs to the insurance company. The interactive effects of insufficient staffing and lack of skills, experience, motivation, and incentives resulted in overall low performance (see Figure 14-10).

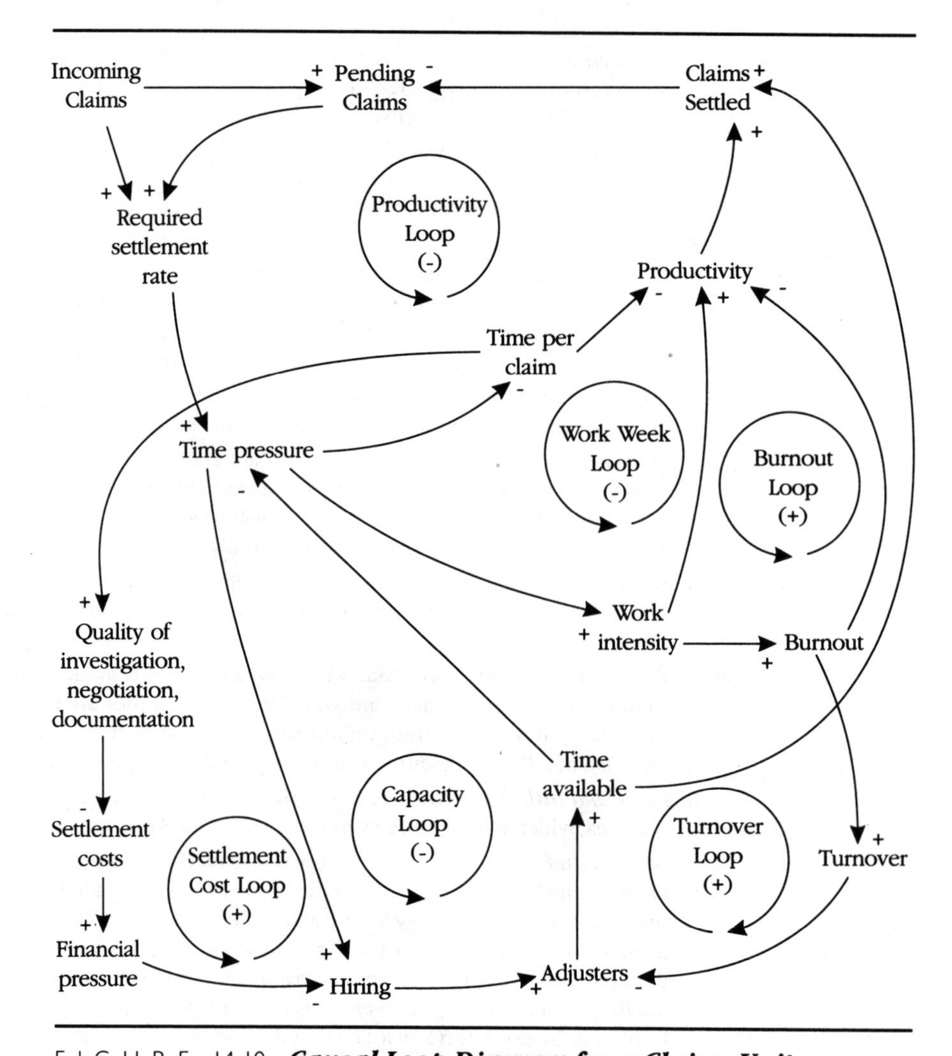

F I G U R E 14-10 ***Causal Loop Diagram for a Claims Unit***

Source: P. Senge and J. Sterman, "Systems Thinking and Organizational Learning: Acting Locally and Thinking Globally in the Organization of the Future," *European Journal of Operational Research* (1990), *59,* pp. 137–50.

INTERNAL FORCES DRIVE PERFORMANCE

A system's capacity to understand its environment and to adjust its internal operation to align it with that environment are critical to its long-run success. Systems overwhelmed by the complexity of the environment may take actions that precipitate harmful environmental responses. Members of a system commonly believe that many failures are the consequence of being at the mercy of the environment. As a result there is a tendency for organizations to attribute outcomes to external forces and feel as though they are victims of unfavorable circumstances. This type of behavior pattern has been labeled by Senge (1990) as the "enemy is out there" syndrome. While there are many threats in the global business environment, managers tend to attribute their failures—or their confusion—to external sources. This pattern of externalizing problems can be a way to rationalize ignoring thorny internal issues.

TIME DELAYS OCCUR BETWEEN CAUSE AND EFFECT

When the behavior pattern of a system is viewed over an extended period of time, fluctuations or oscillations in behavior become apparent. One cause of such oscillation is time delays. Delays between action and reaction cause systems to continue moving in a particular direction until action is taken to change the pattern. Extended delays often trigger overcompensating corrective actions, which induce more fluctuation into the system. For example, think of a driver whose car is not equipped with climate control. Such a person, when driving for an extended period of time, probably will frequently readjust the temperature control on the heater or air conditioner. This is because people tend to overcompensate by raising or lowering the temperature setting too much.

In general, the longer the time between action and reaction, the less likely it is that observers will associate the effect with the cause. Sterman (1989) attributed the poor performance of experimental subjects in a gaming simulation known as The Beer Game to the fact that participants were often confused by the dynamic feedback that was part of the systemic structure of the game and generally tended to misunderstand its significance.

SECTION 3
BASIC ELEMENTS OF SYSTEM DYNAMICS

To this point, the discussion has focused on the idea that a system's pattern of behavior becomes evident over time and primarily results from the interaction of three core factors: the structure of the system, the frequency and duration of time delays in feedback loops, and the extent to which information flows are amplified by the system (see Figure 14-11). As was mentioned earlier, the underlying structure of a system can be expressed in the form of a causal-loop diagram and a stock-and-flow diagram. These conceptual tools can help a manager understand the workings of a system and make a person more aware of his or her own assumptions by requiring that those assumptions be expressed concretely.

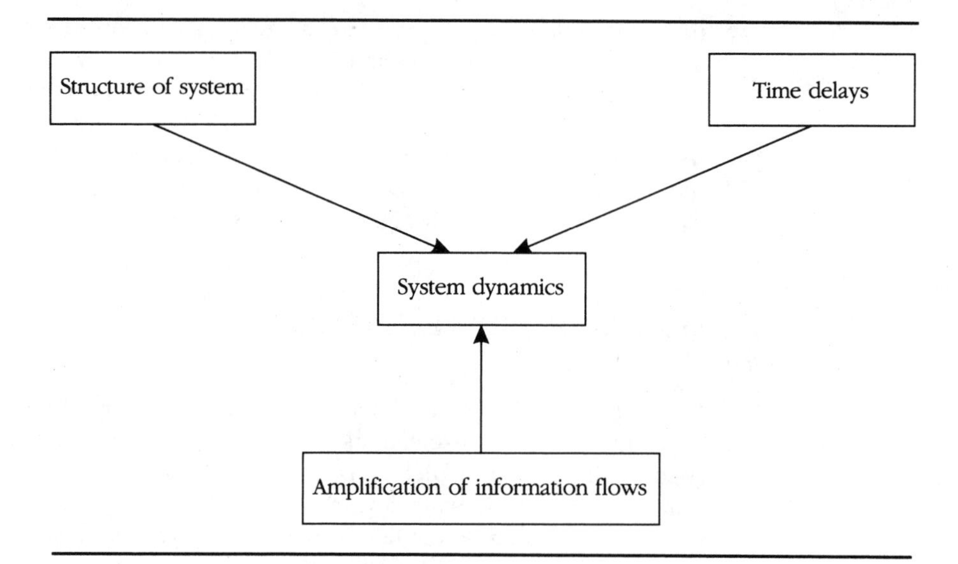

FIGURE 14-11 ***Core Factors of the System Dynamics Perspective***

CAUSAL-LOOP DIAGRAMMING

Causal-loop diagramming helps a manager to graphically express how he or she believes a system works. The construction of such a diagram is an essential first step in developing a working model of how a system operates (Richardson and Pugh, 1981). Causal-loop diagrams show sequences of cause-and-effect relationships to the point where feedback causes a return to the starting point and a loop is formed. Four important dimensions of feedback loops are often represented in such diagrams:

1. The assignment of polarity to causal loops
2. Locating time delays
3. Linkages between loops
4. The assignment of polarity to causal links

Polarity is the designation of whether a cause-and-effect relationship is positive or negative. Polarities can be assigned to individual links of cause and effect and to entire loops. For example, stepping on the accelerator of an automobile causes it to travel faster. An increase in one variable causes an increase in the other, creating a positive feedback link.

A causal loop is formed in a similar fashion. Increased numbers of product defects cause more attention to quality, which in turn causes greater efforts to reduce defects, which reduces defects, which reduces the amount of attention paid to defects. Polarity is assigned to causal loops on the basis of the following principles (Richardson and Pugh, 1981):

- A feedback loop receives a positive designation if it contains an even number of negative causal links.

- A feedback loop receives a negative designation if it contains an odd number of negative causal links.

The polarity of a feedback loop is marked by placing a plus or minus sign in the center of the loop (see the examples in Figure 14-10). Time delays are marked on any causal link where it is appropriate with the mark)). Causal loops are linked to other loops by placing a causal link arrow at the site on the loop where the influence occurs (see Figure 14-10). By using these techniques the systemic structure of an organization may be depicted. A manager may speculate on the effect of alternative strategies and policies by imagining the potential change in the feedback loops. Ideally, this diagram serves as the basis for a computer model that could simulate the effects of a change in policy over various lengths of time. When organizations are managed in ways that fail to account for their systemic nature, the changes made often fail to produce the expected outcome. This behavior is known as *policy resistance.* Managing that appreciates the influence of feedback loops on a system is intended to reduce policy resistance.

STOCK-AND-FLOW DIAGRAMS

Stock-and-flow diagrams depict the systemic effects of various feedback loops on the accumulations and rates of flow within a system. In a stock-and-flow diagram, both inflows and outflows are designated by icons (see Figure 14-12). The horizontal line capping the vertical that stands for a flow pipe represents the valve or regulator that determines the rate of flow into the accumulation. The accumulation, represented by a box, is known as a *stock.* It could be a stock of inventory, money in a savings account, or a person's memory. A stock-and-flow diagram simulated on a computer will actually indicate the changing rates of flow and changing stock levels in a system. If such a diagram were to be observed while a simulation was taking place, fluctuations in the level of the stock would be apparent.

To simulate the action of the system, the stock-and-flow diagram must be coupled with a causal loop diagram, with the result shown in Figure 14-13. By simulating this model, it becomes possible to gain insight into the causes of

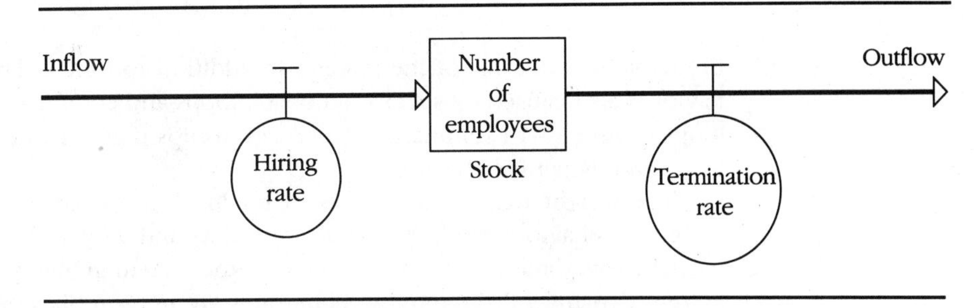

FIGURE 14-12 ***Stock-and-Flow Diagram***

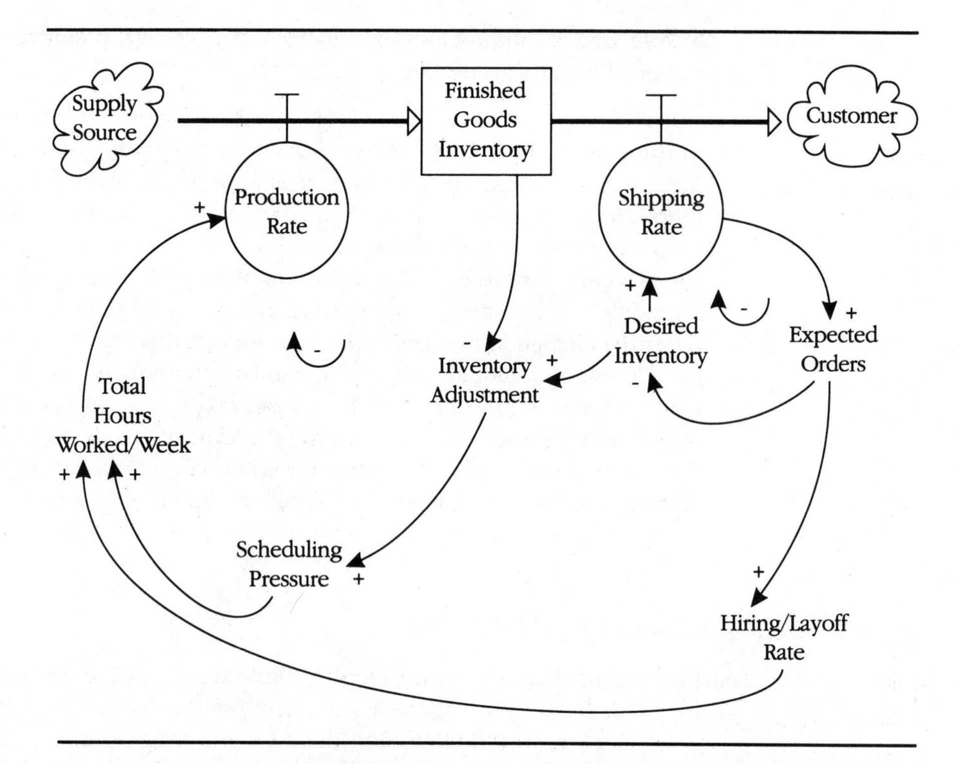

FIGURE 14-13 ***Combined Stock-and-Flow and Causal-Loop Diagram***

TYPE OF FEEDBACK LOOP	DYNAMIC BEHAVIOR
Positive feedback loop	Exponential growth
Negative feedback loop	Exponential decay
Positive loop linked to negative loop	Boom to bust; *or* S-curve; *or* Oscillation

FIGURE 14-14 ***Relationships of Type of Feedback to System Behavior***

the dynamic behavior of the system. In addition to time delays, dynamic behavior is also caused by specific feedback loops and combinations of feedback loops. Figure 14-14 summarizes the relationships that exist between feedback loops and dynamic behavior.

The insight that various types of feedback loops are linked to different systemic behaviors can help managers understand why a system behaves in a particular way and enable them to change the system in highly leveraged ways. System dynamics diagramming techniques are not a replacement for systems thinking but a complement to it. The model-based insights allow managers to enrich their mental models and thereby employ systems thinking more effectively. The minicase describes a situation where systems thinking could have been quite useful.

Coleco Toys

By the early 1980s, Coleco had enjoyed a dramatic rise from a leather works in the Great Depression to a manufacturer of plastic toys such as children's swimming pools in the 1970s. Following the company's rapid rise with the Cabbage Patch Doll in the early 1980s, Coleco's president and CEO Arnold Greenberg was faced with the greatest project of his career. The startling popularity of the Apple personal computer, the IBM PC, Tandy's Radio Shack line, and Atari home video systems had convinced Greenberg that the market was ready for a low-cost computer for sale to the mass market. He envisioned a computer that combined a keyboard, monitor, operating system, and printer all in a single unit and sold for under $600 through discount houses and department and toy stores. The product, named the Adam computer, had the potential to make or break Coleco.

The news of the Adam's upcoming release spread through the industry like wild-fire. Computer magazines reviewed the performance of prototype models and pronounced that the Adam was a true innovation, although a bit noisy. That summer retailers began to place orders to stock the Adam for the Christmas selling season. Unfortunately for Coleco, the FCC held up approval of the Adam for several months with questions about its signal emissions.

The FCC debacle caused difficulties for Coleco. A chain reaction followed that brought Coleco to its knees. Uncertainty over the Adam's actual release date held up production. By the time Coleco had received FCC approval, it was not sure if it should hold up release until the spring season and bypass Christmas. Greenberg boldly announced that Coleco would release the Adam in time for the Christmas season. But by the time Adam units were ready for shipment, many retailers had already canceled or reduced their orders, fearing the Adam would be delayed. The retailers had already filled their shelves and inventories with other products; there was no room for the Adam.

The result was a massive backlog of inventory at Coleco. The company missed the Christmas selling season. In an effort to clear inventory after the holidays, Coleco cut prices. By February, the Adam could be seen on the shelves of drug stores priced at $149. The Adam had lost its luster and was not selling. This was the beginning of the end for Coleco.

MODELING AND SIMULATION

A **model** is a simplified representation of a select dimension of the world; a globe is a model of spatial relations on Earth. British system dynamics theorist R. G. Coyle (1977) has defined a model as "simply a means by which we attempt to represent some aspect of the external world, in order to be able to influence, control, or understand it more effectively" (p. 5). A **simulation** is an operating model that depicts both the configuration of the system and how the system changes over time (Greenblat and Duke, 1975). The use of modeling as a tool for understanding systems and their dynamic patterns has both benefits and limitations. In general, models are only as valid as the insights of the people who created them. Therefore, while modeling may help to clarify one's assumptions about a system, the lessons learned may not be useful in explaining the behavior of other systems. On the other hand, modeling can be an effective and potentially powerful device for translating the abilities of computers into a tool for recognizing the many patterns of interaction within a system.

Sterman (1986) has outlined a six-step process for modeling a dynamic system:

1. *Description of the system.* A summary of information available about the system, and first indication of what may be important to include.
2. *Precise definition of the problem.* A description of the dynamic behavior of interest and the time frame of the process involved.
3. *Formulation of a dynamic hypothesis.* The creation of the feedback loop structure believed to underlie the behavior of the system.
4. *Formulation of a simulation model.* The representation of the hypothesis in equation form.
5. *Analysis of the model.* Comparison of model behavior with the system being studied plus sensitivity analysis and policy analysis.
6. *Communication and implementation.* These two factors should be emphasized from the beginning of the study.

To ensure that the model accurately reflects reality, it is subjected to intense scrutiny by analysts before being used in a simulation. In the system dynamics approach to modeling, continuous efforts are made to refine the model by testing it against the experience of the user group. The goal of this refinement process is to develop increasingly valid representations of the real world. The refinement process is also a learning process for the members of the system involved in building the model. By continually comparing the assumptions used to create the model with their real-world experience they are able to reassess many of the assumptions that are the underpinning for their mental models of reality. The same learning process occurs when users of the simulation compare the outcomes predicted by the computer with what they would normally expect. Model validation should be a continuous learning process that contributes to both the development of the manager's systemic thinking capabilities and to the creation of a learning organization.

SYSTEMS THINKING AND ORGANIZATIONAL LEARNING

Richmond (1990) suggests that the shrinking world created by innovations in transportation and communication has caused various systems that were once loosely related to become more tightly coupled. Increased rates of change in the global economy and increasing tightness of coupling present more threats and opportunities to organizations than ever before. The net effect of these changes is to create issues of concern that seem complex and thus difficult to fully understand. While these changes may intimidate some organizations and cause others to flounder, they need not prevent sophisticated firms from attaining their goals.

Regardless of the specific nature of these global transformations, there is a growing need for managers to think systemically and learn continuously. One of the most enduring strategic advantages that an organization can develop is the ability to understand the forces and interrelationships that create their own patterns. Firms that innovatively find ways to translate these insights into their operations are poised to lead their industries.

A NEW MINDSET

Richmond (1990) has observed, "The systems thinker's forte is interdependencies! Their specialty is understanding the dynamics generated by systems composed of closed-loop relationships" (p. 2). Systems thinking helps people perceive more clearly the dynamic patterns that occur over time. The mastery of systems thinking requires a cognitive shift in a person's mental model of the world. There must be a transition from perceiving isolated single events and causes toward viewing behaviors as extensions of fundamental forces underlying the system. This mental shift is like the difference between viewing the world based on photographs and viewing it as a three-dimensional video.

Kilmann (1984) has suggested that such a perspective will enable managers to visualize organizational dynamics as complex holograms. A *hologram* is a lifelike, three-dimensional image created with lasers; it is able to capture images with much greater depth than a standard picture. The point of this metaphor is that managers should replace their old lenses for viewing the world with new holographic ones. As Kilmann proposes, "To begin with, might it be that organizations are continually applying the same theories and models to manage problems when in fact these no longer capture the way the world has become? The solution is not, therefore, a matter of using the old models better but of adopting an altogether new kind of filter" (pp. 54–55).

ORGANIZATIONAL LEARNING

The concept of organizational learning has yet to be defined precisely, but there is general agreement that learning does occur within institutions or systems (Fiol and Lyles, 1985; Shrivasta, 1983). Generally, **organizational learn-**

ing can be viewed as taking place when changes occur in individual cognitive patterns and are eventually translated into alterations in the collective world view of a system.

Organizational learning is both generative and facilitative. Generative learning is learning how to learn; it is self-reinforcing because it builds learning capacity. Facilitative learning is learning to learn through teaching. In learning organizations people reinforce each other by simultaneously acting as learners and teachers. Facilitative learning can develop rapidly when people become more adept at both learning and teaching. When a system fosters both personal learning and the exchange of insights with others, the overall level of organizational learning is likely to accelerate (Senge, 1990). In such environments the ability to learn from past experience and to engage in continuous learning is enhanced. Ulrich and Lake (1991) argue that significant competitive advantage can be gained by those firms that develop the capabilities of their members. They propose that to build such capabilities "business must adapt to changing customer and strategic needs by establishing internal structures and processes that influence its members to create organization-specific competencies" (p. 75). The core of developing such a capable organization is to develop a shared mindset inside and outside the organization (see Figure 14-15). They state that "mindset represents the patterns people inside and outside the organization use to process, store and retrieve information about the organization" (p. 88). Ulrich and Lake have documented the benefits of building such capacities in such world-class organizations as Marriott and Borg-Warner.

Recent research by Kim (1990) and Senge (1990) suggests that the rate of organizational learning can be accelerated by using the learning laboratories known as Microworlds described earlier in the chapter.

~ SUMMARY

The system dynamics perspective offers a unique, integrated approach to managing that builds on many of the principles developed in the basic systems approaches: hard systems, soft systems, and cybernetics. The hard systems emphasis on structure, closed systems, modeling, and simulation lend a pragmatic value to this approach. Cybernetics are employed as an integrative tool for explaining the mechanism that unites the elements of a system. Finally, the soft systems concepts of organizational learning and systems thinking enhance the value of system dynamics to the practitioner. As currently practiced, system dynamics contributes to systems thinking by acknowledging the indeterminate, open nature of organizations. Organizations are recognized as complex entities that cannot be duplicated through computerized models, but simulations can be used in a continuous effort to integrate theory and reality. Since complex systems generally produce both unintended and counterintuitive consequences, it is doubtful that managers can fully master the system. However, modeling and simulation can help managers learn to expand their thinking styles to break through obsolete paradigms for understanding the world.

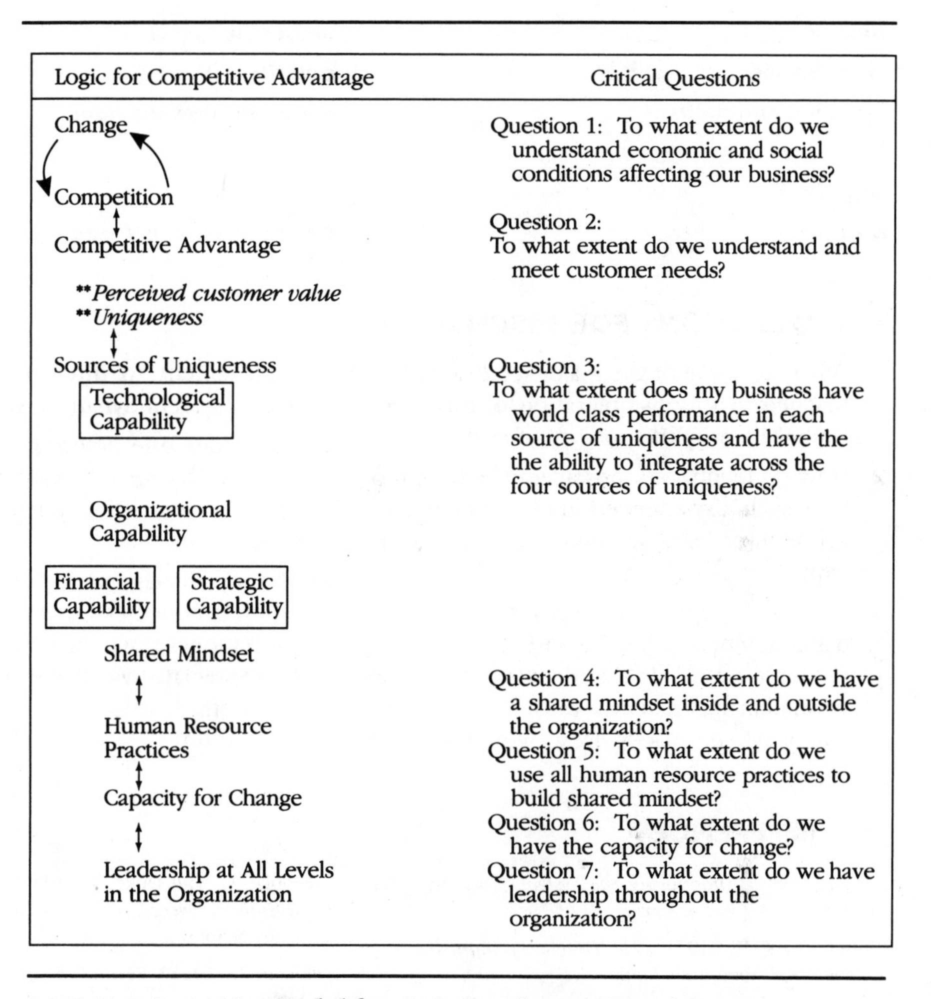

FIGURE 14-15 ***Model for Gaining Competitive Advantage***

Source: D. Ulrich and D. Lake, "Organizational Capability: Creating Competitive Advantage," *Academy of Management Executive* (1991), 5, no. 1, p. 83.

The next chapter examines a number of paradigms that have recently emerged and may contribute to reshaping managerial thinking in the next century. Particular emphasis will be given in that chapter on the dissipative structures model.

~ KEY TERMS AND CONCEPTS

system dynamics
stock
flow
rate
level
causal-loop diagram

time delay
Microworlds
bounded rationality
leverage points
amplification
attenuation
archetype
polarity
stock-and-flow diagram
model
simulation
organizational learning

QUESTIONS FOR DISCUSSION

1. What are some of the major feedback loops that led to the disintegration of the Soviet Union? Identify the polarity of each.
2. How do feedback loops affect the boom-to-bust cycle experienced in most free market economies? Draw a causal loop diagram to explain.
3. At Acme Corporation a group of strategic planning consultants has just been paid $1 million for a report that details how the environment has recently caused most of the company's problems. How would you get management to interpret this information from a system dynamics view?
4. How does one determine whether the dynamic behavior of a system is the result of the structure of the system or of an external factor?
5. The top management team at Acme Corporation is convinced that it does not need organizational learning. How can it be helped to appreciate the value of this managerial concept?

REFERENCES

Beer, S. (1985). *Diagnosing the System.* Chichester, England: John Wiley & Sons.

Churchman, C. W. (1971). *The Design of Inquiring Systems.* New York: Basic Books.

Coyle, R. G. (1977). *Management System Dynamics.* London: John Wiley & Sons.

Diehl, E. W. (1992). "Effects of Feedback Structure on Dynamic Decision Making," unpublished doctoral thesis, System Dynamics Group, MIT, Cambridge, Mass., #D-4290.

Dorner, D. (1989). "Managing a Simple Ecological System," working paper, University of Bamberg, Germany.

Fiol, C. M., and Lyles, M. A. (1985). "Organizational Learning," *Academy of Management Review, 10,* no. 4, pp. 803–13.

Forrester, J. (1961). *Industrial Dynamics.* Cambridge, Mass.: MIT Press.

Forrester, J. (1968). "Industrial Dynamics—After the First Decade," *Management Science,* March, pp. 398–415.

Greenblat, C., and Duke, R. (1975). *Gaming-Simulation: Rationale, Design, and Applications.* New York: John Wiley & Sons.

Kilmann, R. (1984). *Beyond the Quick Fix.* San Francisco: Jossey-Bass.

Kim, D. (1990). "Learning Laboratories: Designing a Reflective Learning Environment," working paper #D4026, System Dynamics Group, Sloan School of Management, MIT, Cambridge, Mass.

Kleinmutz, D. N. (1985). "Cognitive Heuristics and Feedback in a Dynamic Decision Environment," *Management Science, 31,* no. 6, pp. 680–702.

Lyneis, J. (1980). *Corporate Planning and Policy Design.* Cambridge, Mass.: Productivity Press.

Perrow, C. (1970). *Organizational Analysis: A Sociological View.* Belmont, Calif.: Wadsworth Publishing.

Peters, T. (1987). *Thriving on Chaos.* New York: Harper and Row.

Richardson, G., and Pugh, A. (1981). *Introduction to Sys-*

tem Dynamics Modeling with DYNAMO. Cambridge, Mass.: Productivity Press.

Richmond, B. (1990). "Systems Thinking: A Critical Set of Critical Thinking Skills for the 90's and Beyond," Lyme, N.H., High Performance Systems.

Senge, P. (1990). *The Fifth Discipline: The Art and Practice of the Learning Organization.* New York: Doubleday.

Senge, P., and Sterman, J. (1991). "Systems Thinking and Organizational Learning: Acting Locally and Thinking Globally in the Organization of the Future," in T. Kochan and M. Useem (eds.), *Transforming Organizations,* New York: Oxford University Press.

Shrivasta, P. (1983). "A Typology of Organizational Learning Systems," *Journal of Management Studies, 20,* no. 1

Simon, H. (1957). *Administrative Behavior,* 2nd ed. New York: Macmillan.

Simon, H. (1982). *Models of Bounded Rationality.* Cambridge, Mass.: MIT Press.

Sterman, J. D. (1986). "Corporate and Economic Policy Design with Microcomputers: The System Dynamics Approach," working paper #D-3801-1, System Dynamics Group, Sloan School of Management, MIT, Cambridge, Mass.

Sterman, J. D. (1989). "Modeling Managerial Behavior: Misperceptions of Feedback in a Dynamic Decision Making Experiment," *Management Science, 35,* no. 3, pp. 321–39.

Ulrich, D., and Lake, D. (1991). "Organizational Capability: Creating Competitive Advantage," *Academy of Management Executive, 5,* no. 1, pp. 77–91.

~ ADDITIONAL READINGS

Ashby, W. R. (1965). *Introduction to Cybernetics.* London: Chapman and Hall.

Flood, R. L. (1987). "Some Theoretical Considerations of Mathematical Modeling," in *Problems of Constancy and Change.* Proceedings of the Thirty-First Conference of the International Society for General Systems Research, Budapest, *1,* pp. 354–60.

Flood, R. L., and Carson, E. W. (1988). *Dealing with Complexity.* New York and London: Plenum Press.

Forrester, J. (1971). *World Dynamics,* 2nd ed. Cambridge, Mass.: Wright-Allen Press.

Galbraith, J. (1977). *Organization Design.* Reading, Mass.: Addison-Wesley.

Meadows, D. L., Meadows, D. H., Randers, J., and Behrens, W. (1972). *The Limits to Growth.* New York: Universe Books for Potomac Associates.

Roberts, E. (1978). *Managerial Applications of Systems Dynamics.* Cambridge, Mass.: MIT Press.

15 The Future of Systems Thinking: From Equilibrium to Transformation

The organizational world bubbles and seethes. Observed for a lengthy interval, the configuration of organizations within it changes like the patterns of a kaleidoscope.

Herbert Kaufman (1975)

1. To recognize the validity of the dissipative structures paradigm as a concept for viewing the world.
2. To understand the concept of paradigm shift and how it relates to managing.
3. To appreciate chaos as a force for improving the way organizations are managed.
4. To become familiar with transformational management as a way to help managers promote continuous improvement.
5. To recognize some of the implications of self-referencing behavior for the management process, particularly in regard to strategy formulation.
6. To understand the pragmatic value of the process of changing how one thinks about a situation as a prerequisite to innovative management.

SECTION I
THE DISSIPATIVE STRUCTURES MODEL

INTRODUCTION

In the past decade, a number of revelations in the natural sciences have raised questions about the assumptions that dominated scientific thinking in Western culture. Laboratory experiments in physical chemistry performed by Prigogine and Nicolis (1977) have challenged the prevailing Newtonian view of science. Noted futurist Alvin Toffler described Newtonianism as picturing "a world in which every event was determined by initial conditions that were, at least in principle, determined with precision. It was a world in which chance played no part, in which all the pieces came together like cogs in a cosmic machine" (in Prigogine and Stengers, 1984, p. xiii). Prigogine's work has raised substantial doubt concerning this widely accepted premise. Prigogine and Stenger's (1984) comprehensive model explaining the generic dynamics of systems is based on a different world view than the traditional Newtonian principles. This **dissipative structures model** describes a general process in which a system may dramatically alter its basic configuration as a response to intense pressure for change. When sufficient pressure accumulates to drive the system beyond its normal range of equilibrium, the dissipative process is initiated. In a nutshell, Prigogine holds that most systems in the universe are not machinelike, closed systems. Rather, they are open systems that behave dynamically and exchange energy and resources with the environment.

DISSIPATIVE STRUCTURES AND SYSTEM CHANGE

The widely accepted beliefs that stability is the norm in the universe and that change is exceptional run counter to Prigogine's view. He proposes that many systems contain elements that normally exist in a state of continuous fluctuation. Variability in employee productivity levels is an example of such a fluctuation. When sufficient force is generated by a collective series of fluctuations, there may be enough power to cause the existing system to collapse into a state of apparent chaos.

A stable system may be transformed into a chaotic one as the result of either large or small initial fluctuations. Scientists have often marveled at the power of small initial changes to create major effects. The phenomenon is known among meteorologists as the *butterfly effect* from the idea that a butterfly beating its wings in China can create sufficient air turbulence to cause a shift in the weather pattern in New York a month later. Technically known as **sensitive dependence on initial conditions,** this effect states that seemingly small changes in a system can sufficiently destabilize it to dramatically alter its state. Although it is common to look to the environment as the source of such change, these fluctuations may originate within the system as well as without it.

Such equilibrium disturbances can have varying long-term effects on the system. Ultimately, the system may not survive the chaos; it may collapse into bankruptcy or dissolution. Alternatively, the system may establish a new configuration or function in a reformed way that will allow it to be more closely aligned with those forces that are instrumental to its survival. For organizations, these forces include customers, regulators, suppliers, or other elements in the environment. The virtual collapse of the banking system in New England was precipitated by banks that were not aligned with literally all elements of their environment, customers, regulators, the economy, and so on. Out of chaos can come a new, more complex or more sophisticated level of order. This new stability often results from the effects of internal self-organizing processes. Such processes are driven by the basic purposes of an organization and by the values of its leaders.

Maturana and Varela (1980) have termed such self-organizing processes **autopoeisis.** This is the capability of systems to continually recreate their own unique pattern of structure based primarily on internal sources of information. This is a departure from traditional thinking about structure. In this case, organizations take a form governed by their internal rationale rather than on the basis of environmental factors. The result is that a new, more highly evolved structure emerges. This emergent framework will establish a revised pattern of behavior to guide new relationships and processes within the system. Eventually, this rearrangement will alter the basic character of the system. This process of self-organizing may continue indefinitely, as the organization evolves to establish various different points of equilibria over time.

COMPARING THE FUNDAMENTAL WORLD VIEWS

Jantsch (1980) proposed that the perspectives commonly used to explain behavior in organizations are based on one of three distinct world views: the deterministic view, the equilibrium perspective, and dissipative world views. Figure 15-1 compares the main features of these three fundamental perspectives.

The deterministic perspective assumes that the forces operating within the system can be identified and are knowable. These forces are assumed to be governed by universal laws that act uniformly and predictably. Deterministic world views normally take a closed-system view of reality. They are based on the belief that systems behave in a mechanistic fashion, driven by immutable laws of nature. Such systems are envisioned as being built of discrete, well-defined building blocks, similar to those of a machine. The machine metaphor is unmistakably the guiding concept of this perspective.

The equilibrium view suggests that systems are a collection of interactive elements that work together to help the system achieve a beneficial state of coexistence with its environment. Within this approach, all systems are believed to continuously engage in a rational search for the ideal state of being, which is equilibrium. This model accepts that systems are capable of changing, but systems' dynamism is seen as following a set of well-defined forces, operating together to drive the system toward equilibrium. One of the most influential ideas emerging from the equilibrium paradigm is that individual elements of a system are most accurately understood when considered in relation

PERSPECTIVE	TYPE OF CHANGE	KEY ASSUMPTIONS
Deterministic	Gradual	Positivistic, closed
Equilibrium	Oscillation	Phenomenological, open
Dissipative	Pattern breaking	Multi-equilibrium

FIGURE 15-1 ***Three Fundamental World Views***

to their role in the whole system. The equilibrium perspective summarily rejects the reductionist view that systems can be intellectually disassembled without distorting their character.

Most recently, the dissipative structures model has taken the optimistic view that a system may continually evolve through a number of levels of equilibrium, refining its adaptive abilities and improving its chances of survival. This view does not rule out the possibility of attaining equilibrium. Rather, it asserts that a system experiences a number of different points of equilibrium, which each benefit the system in a unique way. This is regarded as an optimistic world view because, as Liefer (1989) notes, "the presence of instabilities . . . does not necessarily lead to chaotic random behavior, but instead offers the opportunity for a new dynamic order, which is able to handle increasing amounts of uncertainty and complexity". As Nonanka (1988) puts it, "Chaos widens the spectrum of options and forces the organization to seek new points of view. For an organization to renew itself, it must keep itself in a non-equilibrium state at all times" (p. 59).

WAYS OF LOOKING AT THE WORLD

All models that describe the processes that guide systems are based on fundamental notions about how things operate in the world. Whether called *paradigms* (Kuhn, 1970), *Weltanschauung* (Churchman, 1971), or *mental models* (Senge, 1990), the notion is the same: a collective set of beliefs that people use to organize their perceptions of events and give meaning to their experience. A paradigm can be viewed as an intricate web of mutually supportive and consistent assumptions that evolve within individuals, as well as within collectives such as organizations or societies. The critical point about paradigms is that they are built on a framework of assumptions; rarely are they based on truths. By their very nature, paradigms are limited tools; they are sometimes useful for explaining how or why things operate as they do in the world, but not always. Unfortunately, paradigms are often mistaken as truth rather than as artificial representations of reality. The result is that people may approach complex systems as they would simple mechanical systems. The illusion of certainty can lead to the oversimplification of problems by using limited linear, static concepts to solve problems when methods with greater powers of explanation are required. It is analogous to trying to forecast the weather by looking at daily weather in a single locale rather than using satellite information and powerful computer models applying to an entire continent.

People in organizations are often completely unaware of the basic assumptions they use to view the world. As a result, they become paradigm prisoners. That is, they are locked into one way of understanding their business that may no longer be aligned with their environment. Not only are they unaware of this lack of alignment, but they are limited in the extent to which they can do anything to change it. The minicase provides an example of the effects.

General Motors

For decades, the Big Three [automakers] of Detroit believed that people bought automobiles on the basis of styling, not for quality or reliability. Judging by the evidence they gathered, the automakers were right. Surveys and buying habits consistently suggested that American consumers cared about styling much more than about quality. These preferences gradually changed, however, as German and Japanese automakers slowly educated American consumers in the benefits of quality and style—and increased their share of the U.S. market from near zero to 38 percent by 1986. According to management consultant Ian Mitroff, these beliefs about styling were part of a pervasive set of assumptions for success at General Motors.

- GM is in the business to make money, not cars.
- Cars are primarily status symbols. Styling is therefore more important than quality.
- The American car market is isolated from the rest of the world.
- Workers do not have an important impact on productivity or product quality.
- Everyone connected with the system has no need for more than a fragmented, compartmentalized understanding of the business.

As Mitroff pointed out, these principles had served the industry well for many years. But the auto industry treated these principles as "a magic formula for success for all time, when all it had found was a particular set of conditions . . . that were good for a limited time."

Source: P. Senge, *The Fifth Discipline: The Art and Practice of the Learning Organization* (New York: Doubleday, 1990), pp. 175–76.

PARADIGM SHIFTS

Throughout history there have been periods of dramatic alteration in basic paradigms. Such transitions are called **paradigm shifts.** A paradigm shift is indicative of a fundamental acceptance of ideas that had been viewed as untenable. While paradigm shifts have been studied in relation to cultures and societies, they also have relevance to other systems, such as organizations, families, and even individuals. Barker (1985) has clearly described the paradigm shift that occurred in the watchmaking industry in the 1970s. While the Swiss watchmakers enjoyed preeminence in making high-quality mechanical watches, they never conceived that watches could be more broadly defined as time-keeping instruments, which could be electronic as well as mechanical. Such fundamental assumptions contributed to the loss of market dominance to Japanese firms such as Seiko.

On the level of the individual, a paradigm shift is known as **metanoia.** Metanoia is often experienced as a flash of illuminating insight that occurs suddenly when one changes the way one experiences the world.

PARADIGMS OF MANAGERIAL THOUGHT

Just as thinking in the sciences has been influenced by dominant paradigms, similar world views have had a major impact on the development of management thought. Many early managerial models were oriented toward maintenance of control and equilibrium. The central managerial task focused on was often to identify the one best way to control the behavior of workers within the larger framework of the production process. The excessive attention paid to the development of external controls eventually created a need to balance the approach with a new emphasis on the emotional and social side of employees. Greater attention was focused on attempting to maintain an equilibrium between the social and technical dimensions of the organization.

More recently, contingency theorists and administrative scientists have emphasized maintaining internal and external equilibrium through the creation of rational mechanisms such as information processing and of structures that are well suited to meet the demands of certain situations. These approaches all operated on the premise that it is of the utmost importance to seek equilibrium in all possible endeavors within the system.

Thompson (1967) detailed how rational techniques such as planning and buffering could be used to dampen the effects of environmental uncertainty and thus preserve the equilibrium of the system. Some organizations would take heroic measures to preserve a stable state because they believed that technological and economic capacities could be optimized under such circumstances. Thompson captured this core belief: "Central to the natural-system approach is the concept of homeostasis or self-stabilization, which spontaneously, or naturally, governs the necessary relationships among parts and activities and thereby keeps the system viable in the face of disturbances stemming from the environment" (p. 7).

Administrative theorists such as Simon (1958) attempted to reduce the level of uncertainty and chaos in systems by creating structures that compensate for the inability of managers to meet the information-processing needs of the system. The system was conceived of as an information-processing network that, if properly designed, would reduce uncertainty by increasing the flow of information to decision makers.

Simon's information-processing concept was expanded upon by organization theorists such as Lawrence and Lorsch (1967), who proposed that the organization must be aligned with its environment. This alignment could be achieved by cultivating a free flow of information between the organization and the environment, resulting in the creation of appropriate structures within the organization.

Emery (1959) and Trist (1981) attempted to balance the previous overemphasis on control and technology. They sought to develop the social and emotional sides of organizations to create a harmony between the task and the people through sociotechnical strategies for work design.

The ultimate purpose of all these models was to promote various management practices that would ensure the attainment of equilibrium. The desire for equilibrium was based on the belief that stability permits the firm to plan and implement plans with a high degree of certainty that conditions will remain sufficiently stable to allow the plans to succeed. Many modern management techniques—such as strategic planning, forecasting, organizational design, and inventory management—have their roots in this rational open-systems paradigm. While this approach continues to be a popular way of looking at systems, it is inadequate for explaining the process of dynamic change that organizations experience. The rationalist view of order, symmetry, and separateness from the environment offers an incomplete representation of the world in which many organizations find themselves operating.

SECTION 2
MECHANISMS OF DISSIPATIVE STRUCTURES

TRANSFORMATION IN SOCIAL SYSTEMS

In contrast to the open systems model, the dissipative structures model offers greater potential to explain the dynamics of change, which are relevant to the process of managing in turbulent times. Although this paradigm has its origins in the sciences, there are many direct applications to the social sciences. However, there has yet to be an equivalent theory created for the social sciences. At this time, the social sciences rely on rough approximations of the scientific theory. Clearly, the dynamics that characterize change in the physical sciences are significantly different from those at play in social environments. While it may be possible for total collapses of structures and complete reconfiguration to occur in physical environments, change of this magnitude is less likely to occur in a social setting. Social systems are more likely to preserve some remnants of their prior structure, incorporating those remains into a new structure. In social environments, the mechanics of change are more apt to exist within a range of limits between evolution and total transformation.

Loye and Eisler (1987) proposed that theories that attempt to explain the basis for such total metamorphosis in the social sciences be known as *transformation theories.* The concept of transformation appears to be more descriptive of the way change takes place in organizations or other social systems. **Transformation** is a significant alteration of the fundamental structure and processes that form the character of the system. It is a reiterative process in which patterns of order have the potential to generate both chaos and order. Transformation may be envisioned as a process that preserves order while giving way to occasional periods of chaos. Transformative processes have the capacity to preserve a degree of order within the system, yet this transformative order is always dynamic and has high potential to generate further change. Similarly, the dynamic periods of change also have an orderliness that encourages stabilization when a new beneficial equilibrium is attained.

SELF-ORGANIZATION

Transformative change is governed by the process of **self-organization,** the ability of systems to reconfigure themselves based on an internal reference of what the system should become. Self-organizing systems have an awareness of their current state and how this differs from a general ideal state. They are able to renew themselves on the basis of information they already possess. Self-organizing systems show **self-referencing behavior.** By comparison, an automobile is not self-referencing; it receives its direction from external sources (the designer and the driver).

Research conducted by Maturana and Varela (1980) indicates that self-organizing systems operate, in some respects, as closed systems. This does not mean that they are isolated from the environment. It does suggest that the system only interacts with the environment on terms consistent with its own internally determined information. The system is open to those interactions that it identifies as being potentially supportive and closed to all others. Although this perspective may run counter to many widely accepted open systems ideas, it offers parallels to a number of psychological perspectives, such as social-learning theory.

The ultimate purpose of self-organization is to help the system renew itself. In living systems, this self-renewal process is an evolutionary mechanism designed to ensure that the system continually becomes more sophisticated. Self-renewing systems have the potential to recreate themselves in ways that preserve their identity but also adjust to change. Natural selection, the mechanism by which evolution occurs, is a prime example of such self-renewing processes for the systems called *species.* Through natural selection, a species becomes better adapted for survival in a changing environment.

AUTOPOEISIS IN ORGANIZATIONS

Self-organizing processes are critically important to the long-term prospects of organizations. Many such processes are handled formally through training and organization development activities. In many firms, however, these activities are often relegated to secondary importance, after the urgency of day-to-day problems.

The earliest widespread interest in the issue of organizational self-renewal can be found in the emergence of the organization development movement in the 1960s. In one of the earliest discussions of the need for self-renewal, Gardner (1965) observed, "What may be most in need of innovation is the corporation itself. Perhaps what every corporation needs is a department of continuous renewal that could view the whole organization as a system in need of continuous innovation." More recently, self-renewal efforts can be identified in the continuous improvement aspects of total quality management programs and in efforts by leaders to create vision statements that draw organization members continuously toward an ideal.

Successful organizations often use such a grand vision to define what kind of organization they should become. When such a common vision is achieved, it serves as the benchmark for all decisions made during periods of transformation. The critical function of vision is clearly evident in the success of companies such as AT&T, Ford, Apple Computer, Canon, Honda, and Komatsu. Figure 15-2 traces the development of the corporate vision at Komatsu over twenty years.

A clear vision not only helps to steer a system through periods of chaos but also generates chaos within the system. A vision statement may totally redefine the alignments between elements of the system and between the system and its environment. Such a vision statement may reinforce certain fluctuations and dampen others. A vision statement has the potential to force members of the system to reconcile differences in the past, present, and future of the system. The various conflicts, contradictions, and paradoxes that emerge from the vision statement represent the opportunity to engage in the process of dialogue and discussion that is the necessary prerequisite for developing a shared vision. While vision statements are often concocted by CEOs, they only become active and enacted when they become shared. Achieving such enactment is not a task that can be managed rationally. Quinn (1988) notes that "paradoxical, uncertain, creative, and idiosyncratic behaviors do not readily lend themselves to the mechanistic assumptions of Western thought" (p. 12). Senge (1990) proposed that such challenges can be met by teams that can continually learn and apply systems thinking to the issues they face.

Regardless of the approach taken to self-renewal, self-organizing processes must be viewed from a systems perspective. The five basic tools—strategy, structure, procedures, culture, and leadership all can be used interactively to support efforts at revitalization. Managers must also have the insight to use these tools in a manner that integrates them with the natural forces of change rather than attempting to dominate or overcome those forces. One of the critical capacities needed to facilitate transformational change is the ability to understand the role of dissipative structures in the self-organization process.

DISSIPATIVE STRUCTURES

A core process guides transformations in organizations alternately from chaos to order to chaos again. This dynamic behavior is driven by a generic framework known as **dissipative structures.** Dissipative structures provide a system with increased freedom to creatively seek new inner arrangements of its elements. The goal of such activity is to increase the sophistication of the system so that it will be increasingly at harmony with highly complex situations. The primary mechanisms of the dissipative structure are **bifurcation points.** Bifurcation points are critical junctures in a system's state at which its equilibrium becomes characterized by "ultra-instability." At this point a system's prevailing equilibrium is at the highest risk of losing support. It is at such bifurcation points that the structures that supported the prevailing equilibrium can literally disintegrate and send the system into a radical process of transformation. This disintegration process may continue until the point where

CORPORATE MISSION: ENCIRCLE CATERPILLAR ("*MARU CAT*")					
CORPORATE CHALLENGE	PROTECT KOMATSU'S HOME MARKET AGAINST CATERPILLAR	REDUCE COSTS WHILE MAINTAINING QUALITY	MAKE KOMATSU AN INTERNATIONAL ENTERPRISE AND BUILD EXPORT MARKETS	RESPOND TO EXTERNAL SHOCKS THAT THREATEN MARKETS	CREATE NEW PRODUCTS AND MARKETS
Programs	**early 1960s** Licensing deals with Cummins Engine, International Harvester, and Bucyrus-Erie to acquire technology and establish benchmarks **1961** Project A (for Ace) to advance the product quality of Komatsu's small- and medium-sized bulldozers above Caterpillar's **1962** Quality circles companywide to provide training for all employees	**1965** C D (cost down) program **1966** Total C D program	**early 1960s** Develop Eastern bloc countries **1967** Komatsu Europe marketing subsidiary established **1970** Komatsu America established **1972** Project B to improve the durability and reliability and to reduce costs of large bulldozers **1972** Project C to improve payloaders **1972** Project D to improve hydraulic excavators **1974** Establish presales and service department to assist newly industrializing countries in construction projects	**1975** V-10 program to reduce costs by 10% while maintaining quality; reduce parts by 20%; rationalize manufacturing system **1977** ¥180 program to budget companywide for 180 yen to the dollar when exchange rate was 240 **1979** Project E to establish teams to redouble cost and quality efforts in response to oil crisis	**late 1970s** Accelerate product development to expand line **1979** Future and Frontiers program to identify new businesses based on society's needs and company's know-how **1981** EPOCHS program to reconcile greater product variety with improved production efficiencies

FIGURE 15-2 ***Komatsu's Vision over Two Decades***

Source: G. Hamel and C. K. Prahalad, "Strategy Intent," *Harvard Business Review* (May–June 1989), p. 68.

new structures form to provide support for a new, more stable level of equilibrium.

The disintegration of the former Soviet Union offers one such example of the process in action. An argument can be made that the kidnapping of Mikhail Gorbachev, and the subsequent rise to power of Boris Yeltsin, represented a major turning point or bifurcation point in the evolution of government in this part of the world. A similar case can be made that the unresolved labor difficulties between Eastern Airlines management and the airline machinists union represented a similar point in Eastern's history. It should be noted that such transformations do not always suggest the failure of a system. Organizations that enjoy rapid growth may go through similar transformation processes in which their existing structures are insufficient to bear the pressures of fast-paced expansion. Toy companies such as Coleco and Worlds of Wonder are prime examples. Although these firms may not be remembered by name, they left a legacy of products. Coleco produced the "Cabbage Patch Doll" and the "Adam" computer, while Worlds of Wonder was known as the producer of "Lazer-Tag."

When a bifurcation point is reached as a result of internal changes occurring within the system, prudent decision making by top management becomes critical. At such points either a decision is made to marshall the system's resources to construct a new pattern of interaction that will lead to more effective operations or the structure of the system disintegrates to seek another level of functioning (see Figure 15-3).

Prigogine and Stengers (1984) captured the essence of the meaning of the

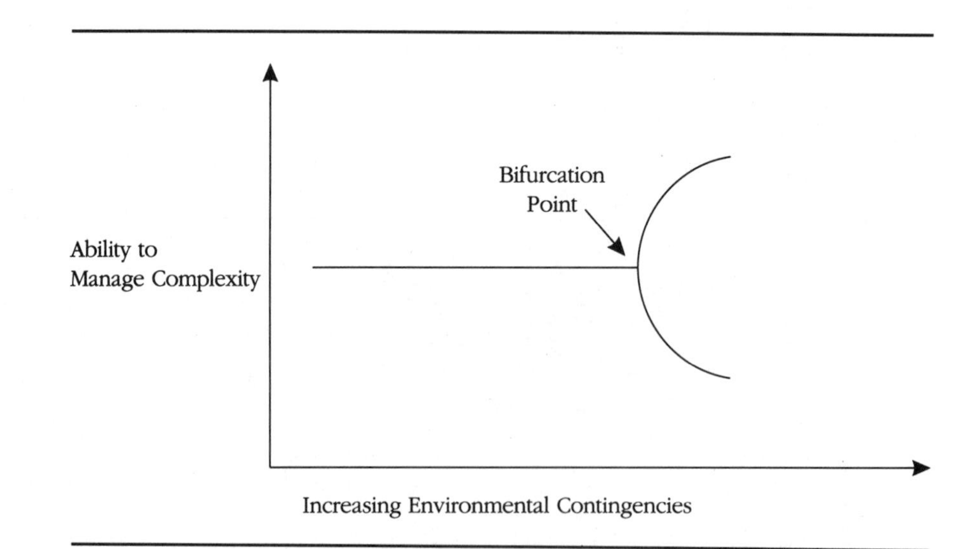

FIGURE 15-3 ***The Bifurcation Process***

Source: R. Liefer, "Understanding Organizational Transformation Using a Dissipative Structure Model," *Human Relations* (1989), *42,* no. 5, p. 905.

bifurcation point thus: "This leads both to hope and threat: hope, since even small fluctuations may grow and change the overall structure. As a result individual activity is not doomed to insignificance. On the other hand, this is also a threat, since in our universe the security of stable, permanent rules seems gone forever" (p. 313). This comment does not mean that systems are naturally unstable; rather, it suggests that systems are capable of change when there is the potential that the change may increase the chances for the survival of the system. The opportunity that the system may ultimately achieve a more appropriate new alignment far from its initial state of equilibrium provides reason for optimism. Such instability suggests that systems possess a perpetual need for adjustment and accommodation in order to move to higher states of functioning.

For example, in order to survive the pressure of massive legal claims against it and the threat of an unwanted takeover bid, Union Carbide became a smaller, more focused organization. On the other hand, as Peter Drucker (1980) has noted, Volkswagen was weaker overall than Chrysler Corporation in 1960 but it reached a decision, global marketing, that strengthened it enormously. A similar example can be seen in the global spotlight on British Leyland Motor Corporation.

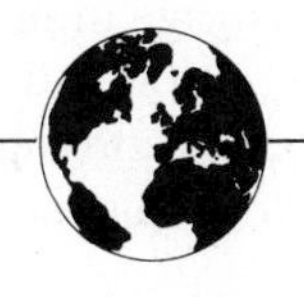

GLOBAL SYSTEMS SPOTLIGHT

British Leyland Motor Corporation

British Leyland Motor Corporation was severely affected by the developments in the international automotive industry in the 1970s, notably those associated with producers in Japan. The 1980s found the company debt ridden and unprofitable. However, BLMC's Rover Group emerged as a profitable private-sector company. The company implemented a strategic vision based on four elements:

1. Long-term commitment to customer-perceived quality in products and services
2. A determination to reevaluate its organizational structure
3. A desire to recognize and integrate various Japanese management techniques by associating with its originators
4. The adoption of a visionary approach to strategic planning.

Rover's structure was revised by implementing a system known as *parallel working*. This system blurred the distinctions between functions and eliminated many barriers. By 1990, these changes resulted in less bureaucracy, quicker adaptation to product needs, higher quality, and greater productivity.

Source: Based on R. Bertodo, "Implementing a Strategic Vision," *Long Range Planning* (October 1990), pp. 22–30.

BASIC ASSUMPTIONS OF THE DISSIPATIVE STRUCTURES MODEL

The dissipative structures model is based on several fundamental assumptions that differentiate it from other perspectives. Figure 15-4 compares its assumptions to those of the open systems (equilibrium) perspective.

DETERMINISTIC CHAOS: ORDER THROUGH FLUCTUATION

Transformation becomes possible because of the purposeful disequilibrium created by microscopic fluctuations within the system. These fluctuations are often random attempts by the system to determine if other alignments with the environment are possible. When an alignment between the system and the environment is created, the fluctuations become magnified through positive feedback loops that connect the system to the environment. If the positive feedback continues, the fluctuations will eventually become amplified to the extent that they destabilize the system, initiating a collapse of the current structure. This dynamic behavior has the potential for transforming the system to a new, more stable equilibrium. This process has been referred to as **deterministic chaos.**

Organizations can engage in many activities to stifle natural change, such as implementing tight controls and instituting mechanistic structures. The dissipative structures effect, according to Gemmill and Smith (1985), becomes possible only under four such conditions:

1. The system must be open to change.
2. The system must be able to break down established system functions.
3. The system must be able to generate new system functions.

FACTORS	EQUILIBRIUM MODEL	DISSIPATIVE STRUCTURES MODEL
Response to change	Dampen fluctuations	Amplify fluctuation
Relation to environment	Alternately open and closed	Open
Amount of information	Low	High
Potential for reorganization	Low	High
Nature of resistance	Unchangeable	Amenable to differences
Environment	Abundance, stability, order	Scarcity, turbulence, chaos
Perception of change	Abnormal	Normal
Order	Via control	Via fluctuation

FIGURE 15-4 ***Comparison of Assumptions of the Equilibrium and Dissipative Structures Models***

4. The system must have inherent stabilities that will facilitate reformulation and restabilization.

Systems that operate as dissipative structures must be capable of experiencing disequilibrium. Whether the source of the pressure that disturbs the equilibrium is internal or external, the system must be sufficiently sensitive to such forces to drive it beyond the limits of its equilibrium. Organizations that employ rationalist techniques such as buffering may prevent the dissipative structures mechanism from ever becoming activated. In the process, however, the system may also grow progressively out of alignment with its environment. This disalignment may ultimately support the forces of entropy, the disintegrative forces, that cause systems to degenerate.

Managers need not serve as passive observers to transformational change. In fact, Nonanka (1988) proposes that management can and should play an active role in facilitating such change. One of the key things that managers can do to foster change is to break existing patterns. These may be patterns of relationships, of communication, of thinking, or of many other behaviors. To remain open to the forces that operate in the dissipative structures mode, a system must be able to purposefully break the patterns that preserve order. Nonanka (1988) proposes that order is reinforced by structures, organizational systems, visions, values, and concepts.

The system must also be able to experiment with new alternative forms as it goes through the transformation process. In nature, this process is often seen in genetic mutations, which are random changes that may produce forms better adapted to an environment, producing evolutionary change. Finally, the system must have the capacity to bring order again by refreezing or reformulating the new configuration. The new alignment must be preserved and established as the preferred standard of stability (see Figure 15-5). These patterns can be changed by reframing or reinterpreting their meaning. For example, when John Sculley replaced Steven Jobs as CEO of Apple Computer, he reinterpreted Apple's vision from a perspective of marketing to individuals to a view of targeting businesses.

Sculley Transforms Apple

Apple Computer prospered in the months following Sculley's arrival. After one and one-half years, however, events internal and external to Apple indicated that Apple was confronted with major problems. First, the computer industry was in the midst of a significant shake-out. Second, problems began to surface with the Macintosh. Third, individuals at Apple began to express concerns about the job that Sculley was doing.

To correct the rapidly deteriorating position of Apple, Sculley made three significant moves. First, he decided that he had to remove Jobs from his position as general manager of the Macintosh division. After a series of confrontations between Jobs and Sculley, Jobs resigned from Apple to begin his own company [NeXT, Inc.].

Second, Sculley reorganized and refocused Apple. He moved away from Apple's decentralized structure because there was duplication in sales, marketing, manufacturing, and product development. As Apple sought to discover its identity, Sculley placed the most creative, passionate, and brilliant leaders in key positions.

Third, with these changes in place, Sculley began to reposition the Macintosh . . . [to appeal to] business people. So as not to compete with IBM on a head-to-head basis, Apple began to "back into" the business market. The most successful of these approaches was with desktop publishing. . . . With desktop publishing and newly developed hardware and software appropriate for business, Sculley restored the shine to Apple.

Source: J. Pierce and J. Newstrom, *The Managers Bookshelf* (New York: Harper & Row, 1990), pp. 131–33.

ORGANIZATIONAL CHANGE AND TRANSFORMATION

During the 1960s it became apparent to many organizations that their ability to survive over the long run depended on their ability to remain aligned with the environment. Accordingly, various strategies designed to promote change were developed. Concepts such as planned or managed change became the guiding principles for relatively mechanical, logical processes of aligning systems with their environment. The goal of such alignment was to restore the system to equilibrium. Practitioners of organization development were told of the importance of creating a detailed picture of a desired end state. However, Quinn (1980) noted that successful firms frequently bypassed the precise rational analytical systems they had devised to guide the change process in favor of a more synthetic approach. The approach, known as logical incrementalism, was a mixture of "various behavioral, power dynamic, and formal analytical approaches" (p. 16). It has since become apparent that logical incrementalism focused on one type of change possible within a system but did not explain the dynamics of the dramatic changes that seemed to have a profound impact on organizations.

These rational approaches to change emphasized internally generated planned change transacted within discrete units of a system, such as a division or strategic business unit. The environment was seen as contributing to the need for the change, but the change was internally controlled through planning schemes and behavioral science techniques. This approach is marked by the assumption that the basic structural forms, patterns of relationship, and processes that have been established will remain largely intact.

The dissipative structures paradigm offers an alternative view, one which encompasses change associated with turbulent conditions. Such turbulence or uncertainty may be characteristic of the environment of the system or of its internal state. In this paradigm, fluctuations in a system become magnified, either by internal disruptions or external events. The focus of this model is on

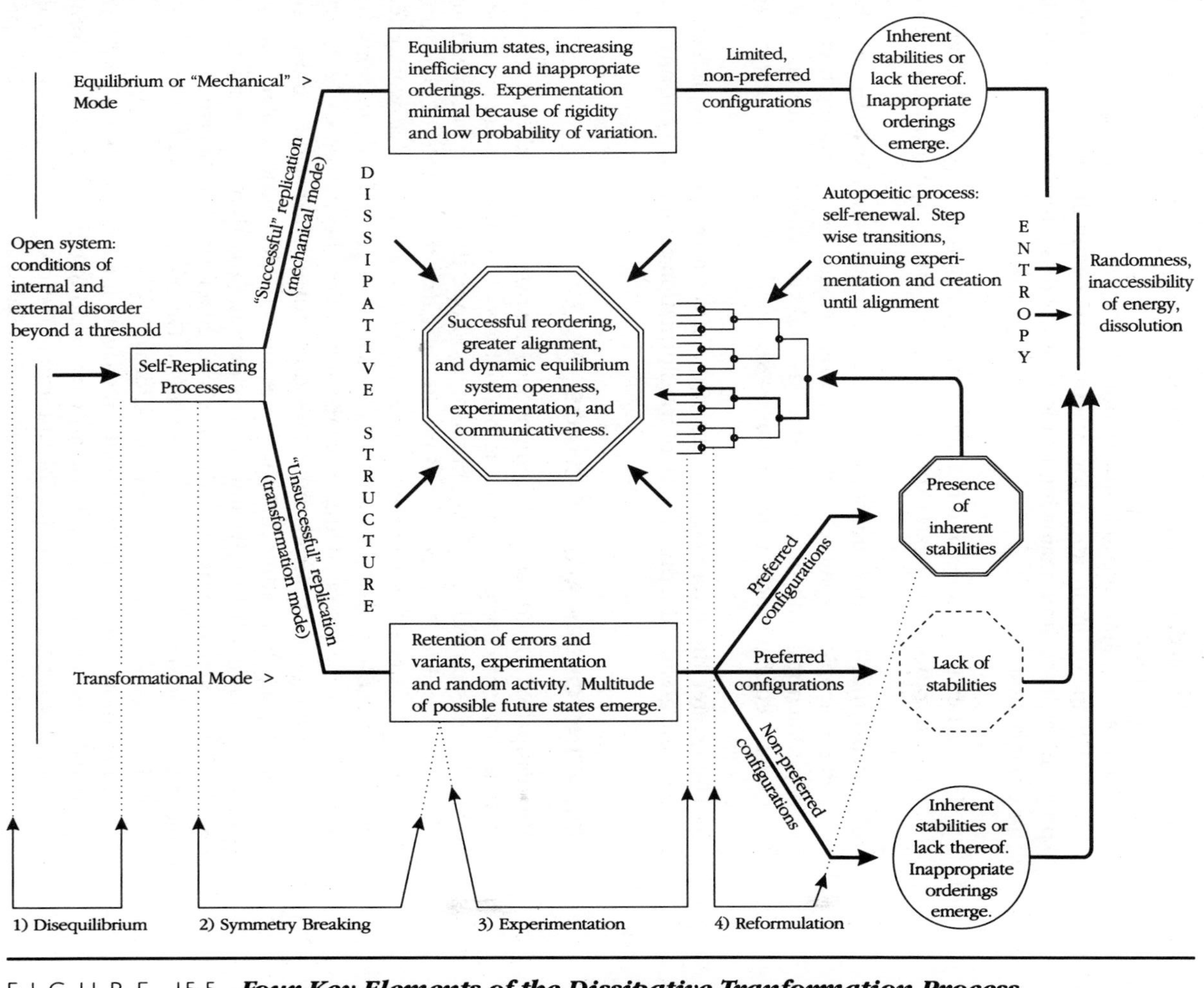

FIGURE 15-5 ***Four Key Elements of the Dissipative Tranformation Process***

Source: G. Gemill and C. Smith, "A Dissipative Structures Model of Organizational Transformation," *Human Relations* (1985), *38*, no. 8, pp. 751–66.

large-scale upheaval that touches many of the core structures and processes, those central to the character of the system. Such change is truly transformational in that it triggers a metamorphosis in the identity of the organization. Not just form is altered, but what the organization is capable of achieving has now been changed. In essence, the new configuration redefines the character of the organization.

Such a profound reconceptualization of the basic processes of systems clearly has major implications for practitioners. Gemmill and Smith (1985) proposed at least three major implications of the dissipative structures model for organizations:

1. The higher the level of internal and external disorder experienced within an organization, the greater the probability of either transformation or entropy.
2. Systems that become transformed, in comparison with systems that become entropic, are more likely to actively engage in symmetry breaking, experimenting with new configurations, and reformulation into an evolutionary configuration that reinforces self-renewing or autopoeitic processes.
3. The greater the level of awareness of the dissipative process within a system, the greater the probability of transformation rather than entropy.

CREATING ORDER AMIDST CHAOS

The dissipative structures model emphasizes the turbulence and total transformation that can emerge despite the best efforts of managers to maintain stability. Yet this model is one of optimism and hope that better organizational forms are possible. Managers need not be relegated to the role of victims or passive observers. They can engage in a number of activities that work effectively with these forces. Research in the field of organizational learning has increasingly demonstrated that organizations have the power to learn and transform themselves purposefully. This does not mean that they must have clear end goals, but rather that they must possess clear visions that continuously interplay with the dynamic forces operating within the organization's world. These visions must have the power to generate wholehearted commitment from the members of the organization. On the other hand, they should never be considered to be more than an idea.

To promote organizational vitality, managers must create an orderly form of chaos. This is accomplished primarily by creating new meanings for events that occur within the system. This interpretative process of reframing is intended to call into question the organization's basic assumptions. Such a process will generate conflict, but in doing so it promotes an ongoing debate or dialectic that is the well spring of energy and innovation in the organization. Nonanka (1988) has proposed that several key activities are necessary to create such beneficial forms of chaos. They are outlined in Figure 15-6.

1. Create chaos by
   Creating a challenging but equivocal vision
   Generating fluctuation by opening the system to market forces, new technologies, and other organizations
   Using leadership and the culture to increase fluctuations
2. Amplify fluctuations by focusing on contradictions
3. Create a dynamic process to resolve conflicts by developing self-organizing teams
4. Transform accumulated information into knowledge to create a learning organization

FIGURE 15-6 ***Marshalling the Forces of Chaos***

Source: Based on I. Nonanka, "Creating Organizational Order out of Chaos," *California Management Review* (Spring 1988), pp. 57–73.

TRANSFORMATIONAL MANAGEMENT

The logic of dissipative structures argues for the presence of many alternative levels of equilibrium within a system. In business practice, this translates into a wide realm of possibilities of how an industry or organization might prosper. The equilibrium level that generates the highest profit for a firm may not be the most suitable for the system. The evolution of the automobile industry over the past thirty years offers such an illustration.

During the 1960s the Big Three automakers, GM, Ford, and Chrysler, earned outstanding profits. Yet at the same time, they virtually ignored customer needs and placed little importance on quality. The industry position was precarious, yet industry leaders were blinded by the existing equilibrium and profitability. In fact, the equilibrium within the industry was relatively stable, yet unsophisticated. Since the demise of many small, local car producers in the 1920s, the Big Three had not been pushed to a more sophisticated equilibrium. (Between 1908 and 1920, the number of auto manufacturers fell from 250 to 125. By the end of the 1920s, the industry had been concentrated down to just 25 firms, with the top seven companies controlling over 90 percent of the market [Lawrence and Dyer, 1983].)

The threat posed by global competition that began in the 1970s caused transformations in the industry and within the various organizations. Ford has become known as a firm innovative in technology and design. Chrysler has become known for its expertise in minivans and for its creative alliances with other car producers, such as Mitsubishi. GM's Cadillac division has won the Malcolm Baldridge Award for quality, and its Saturn division has been recognized for innovative management and for producing vehicles with high value to customers.

All of these dramatic improvements in organizational performance result from radical transformations in these companies. Such transformations have been driven by a variety of leaders and managers who had the vision to break the existing organizational pattern. Implicit in their behavior was the recognition that the company could move toward many possible ways of being. The major restructuring, cultural changes, new strategies, and new world views orchestrated by the leaders of these firms were, in sum, a near total transformation. In this context, dissipative structures and transformative processes become translated into people's gut feelings and visions of a better future for them and the organization.

Although the path to managing transformation is not well understood there are a few key principles for managers to keep in mind. First, organizations need to develop a significant interest in building internal processes that will enhance their capacity for change (Boynton and Victor, 1991). Secondly, the most successful organizations will have the capacity to learn quickly from experience and translate that learning rapidly into action. Learning and knowledge must be tightly linked with action and experimentation. As is evident in the Deming cycle of quality management and the plan-do-check-act cycle, people learn by doing (Senge, 1991). Finally, the processes of creating order and disorder are closely related in organizations. Reframing the way one thinks is the prerequisite to creating order, while breaking patterns and creating a type of controlled disorder are necessary for innovation. Both depend on continuous cycles of experimentation, action, and learning (Nonanka, 1988).

SECTION 3
OTHER NONEQUILIBRIUM PARADIGMS

This chapter has emphasized the importance of the dissipative structures model. We do not intend to advocate this model as the only useful way of understanding radical change. Other nonequilibrium models have shown considerable promise as systemic tools for understanding complex phenomena.

THE POPULATION ECOLOGY VIEW

The population ecology perspective was briefly discussed earlier in relation to organizational strategies. In general, this paradigm takes a naturalistic outlook at the dynamics that occur within industries. In particular, it assumes that the dynamics of an industry are governed by survival of the fittest, the force of natural selection. Much as species and animal populations evolve through mutual interaction and adaptation with their environment, according to this perspective, organizations play according to similar rules. In the population ecology model the traditional distinctions between organization and environment become blurred. Organizations are seen as contributing to the shaping of an environment through their choice of strategies. Organizations that develop superior strategies—those that put their competitors at a disadvantage—are surely contributing to the construction of their environment.

As Hannan and Freeman (1989) have proposed, the population ecology model shifts the focus from individual organizations to the collective effects generated by competition. The practical implications of this perspective are presented in the following feature.

The Natural Manager

Evolutionary biologists argue that periods of stability are punctuated by dramatic events that can overturn ecological orders. A fire, a volcano, or a flood can wipe out a large area. Afterward, species compete to become parts of the new ecological order. Once-dominant species may not be reestablished. Marginal species and mutations—once right only for certain microenvironments—may take the lead.

Savvy executives see the business parallel. New technologies, deregulation, and changes in customer behavior provide the metaphorical equivalent of floods and fires, opening up new competitive landscapes. On such fertile grounds new or transformed industries can grow. For example, the worldwide taste for American-style mass entertainment, plus a proliferation of channels of transmission and methods of display, fuels the growth of global production and distribution ecologies.

In computers, a powerful driver of discontinuous change has been improvements in microprocessors. In the late 1970's, cheap and plentiful microprocessors spawned the computer industry. In the 1980's, more powerful microprocessors enabled small computers to substitute for minicomputers in many applications, deflating the computer industry. Now ultrafast RISC microprocessors are driving another revolution, yielding powerful work-stations so inexpensive that they will be likely to crush much of the personal computer business.

What can executives learn from this? First avoid complacency. Try to anticipate potential cataclysmic changes—technological, regulatory, and customer-based. Play out alternative scenarios of how business ecologies might collapse and be replaced. Anticipate your obsolescence. Consider competing against yourself.

Promote cooperation and invite other companies and customers to become your partners. Ecological competition is a race to richness and diversity. A strong single company can serve as an anchor to an ecosystem, but the most robust business systems have many suppliers and customers evolving in relation to each other.

Source: J. B. Moore, "Taking Cues from Natural Disasters," *The New York Times* (May 19, 1991), p. 11.

Business ecosystems experience dynamic, sometimes dramatic changes that disrupt the stability of a system, yet there are also forces that promote the evolutionary tendencies that help to make such systems more resilient and able to cope with change. This process, known as self-organization, is discussed in the next section.

THE FRANKFURT SCHOOL'S CRITICAL PERSPECTIVE

A number of sociologists in Germany's Frankfurt School have criticized the methods used by social scientists to explain social systems. Habermas (1971) argued that the forms of reason used to explore the physical world are not appropriate for use in the social world. The tools used by physical scientists to discover the laws and principles of science are subject to the test of scrutiny; they either succeed or fail to explain a phenomenon. The criteria for evaluating physical science methods are relatively rigorous and definitive. Social science is a completely different discipline because theory and practice may differ. If practitioners rely on inappropriate hard science explanations for understanding social systems, they are likely to be misled.

Systems theorists such as Jackson (1985) argue for the creation of a new set of tools for the social sciences, tools not based on hard systems assumptions. Checkland's soft systems methodology (SSM) is such a tool. Jackson criticizes approaches such as cybernetics, particularly Beer's work, for being too heavily hard systems-oriented. According to the advocates of this perspective, there is a clear need for the further development of a unique set of fundamental principles to guide intervention into social systems in the future.

THE FUTURE IN PERSPECTIVE

The future of systems thinking lies in understanding how managers can play a more integral role in the transformation process. This is not to suggest that they should control or dominate the process, but rather that they envision how they can apply the natural forces for change in a way that benefits people. The dissipative structures model has the potential to free managers from the responsibility of always being in control to playing a role as active facilitators of those processes that support growth and development in organizations. There are many opportunities that the nonequilibrium paradigm offers to managers to clarify and provide meaning to new forms of management practice. As systems become increasingly defined in global terms, the necessity of managing change in organizations is greater than at any other time in history. This necessitates a shift to more systemic forms of thought as a means for organizational success and survival. In organizations of the future, the old Latin phrase *Non schola sed vita discimus* ("We don't learn for school but for life") will have new meaning.

SUMMARY

How a person manages is largely determined by what he or she believes about how the world operates. Although there are many paradigms that a manager may use to explain why organizations behave as they do, a gradual shift is now taking place in the basic assumptions that form the foundation of management theories. Advances in scientific thinking have made alternative explanations of reality, such as the dissipative structures model, more relevant and accessible to managers. Consequently, many management practices once accepted as fundamental are now being reconsidered. Organizational charts, strategic planning, and centralized management are giving way to organizational learning, total quality management, continuous improvement, and flatter organizational structures. Attempts to forecast the future are being replaced by developing skills and processes that enable faster responses to change. Internal operations once sheltered from the noise and confusion created by environmental change are now opened to such change through flexible manufacturing and high-speed management practices. Finally, order and stability, which were reinforced through rigid structures and tight procedures, are yielding to autonomous work teams and innovative processes focused on meeting customer needs. The trend is away from a mechanical, design-oriented focus toward self-organizing processes. Managers are increasingly aware that the form and focus that an organization takes will likely change, over time, in varying ways. More and more managers accept that the natural fluctuations in organizations and their environment generate profound changes in the world in which they operate.

The discipline of management is itself at a bifurcation point in its evolution. Managers today have more incentive than ever to explore new ways of managing and viewing the world. Transformative, knowledge-creating technologies, such as systems thinking and organizational learning, offer tools that many managers will use to take advantage of future opportunities. Hard systems thinking, soft systems thinking, cybernetic thinking, integrative systems thinking, and the nonequilibrium paradigm together offer managers powerful means of managing that can enable them to transform the organizations of the future to better meet the needs of people.

~ KEY TERMS AND CONCEPTS

dissipative structures model
sensitive dependence on initial conditions
autopoeisis
paradigm shift
metanoia
transformation
self-organization
self-referencing behavior
dissipative structures
bifurcation points
deterministic chaos

~ QUESTIONS FOR DISCUSSION

1. What are the implications of the dissipative structures hypothesis for the way organizations are structured and managed?
2. If organizations are indeed self-referencing, should they ignore their environment or place more emphasis on understanding it?
3. Some social analysts argue that the Western world is experiencing a paradigm shift away from the reductionist-analytical paradigm and toward a more systemic one. Is there any evidence to support this claim?
4. When Lee Iacocca turned Chrysler Corpo-

ration around in the late 1970s, was Chrysler at a bifurcation point, or was it merely experiencing a downturn in its business cycle?

5. Is the pursuit of equilibrium a legitimate goal for a business organization? What are the costs and benefits of an equilibrium position as compared with a nonequilibrium stance?

REFERENCES

Barker, J. (1985). *Discovering the Future: The Business of Paradigms.* Elmo, Minn.: ILI Press.

Boynton, A., and Victor, B. (1991). "Beyond Flexibility: Building and Managing the Dynamically Stable Organization," *California Management Review,* Fall, pp. 53–67.

Churchman, C. W. (1971). *The Design of Inquiring Systems.* New York: Basic Books.

Drucker, P. (1980). *Managing in Turbulent Times.* New York: Harper and Row.

Emery, F. (1959). "Characteristics of Socio-technical Systems," #527 and #529. London: Tavistock.

Gardner, J. W. (1965). *Self Renewal.* New York: Harper and Row.

Gemmill, G., and Smith, C. (1985). "A Dissipative Structures Model of Organization Transformation," *Human Relations, 38,* no. 8, pp. 751–66.

Habermas, J. (1971). *Towards a Rational Society.* London: Heinemann.

Hannan, M., and Freeman, J. (1989). *Organizational Ecology.* Cambridge, Mass.: Harvard University Press.

Jackson, M. (1985). "Social Systems Theory and Practice: The Need for a Critical Approach," *International Journal of General Systems, 10,* pp. 135–51.

Jantsch, E. (1980). *The Self-Organizing Universe.* Oxford, England: Pergamon Press.

Kaufman, H. (1975). "The Natural History of Human Organizations," *Administration and Society, 7,* August, pp. 131–49.

Kuhn, T. (1970). *The Structure of Scientific Revolutions.* Chicago: University of Chicago Press.

Lawrence, P., and Dyer, D. (1983). *Renewing American Industry.* New York: Free Press.

Lawrence, P., and Lorsch, J. (1967). *Organization and Environment.* Homewood, Ill.: Richard D. Irwin.

Liefer, R. (1989). "Understanding Organizational Transformation Using a Dissipative Structure Model," *Human Relations, 42,* no. 10, pp. 899–916.

Loye, D., and Eisler, R. (1987). "Chaos and Transformation: Implications of Non-equilibrium Theory for Social Science and Society," *Behavioral Science, 32,* pp. 53–65.

Maturana, H., and Varela, F. (1980). *Autopoeisis and Cognition: The Realization of the Living.* London: Reidl.

Nonanka, I. (1988). "Creating Organizational Order out of Chaos," *California Management Review,* Spring, pp. 57–73.

Prigogine, I., and Nicolis, G. (1977). *Self-Organization in Non-Equilibrium Systems.* New York: J. Wiley & Sons.

Prigogine, I., and Stengers, I. (1984). *Order out of Chaos.* Toronto: Bantam Books.

Quinn, J. B. (1980). *Strategies for Change: Logical Incrementalism.* Homewood, Ill.: Richard D. Irwin.

Quinn, R. E. (1988). *Beyond Rational Management.* San Francisco: Jossey-Bass.

Senge, P. (1990). *The Fifth Discipline: The Art and Practice of the Learning Organization.* New York: Doubleday.

Senge, P. (1991). "Transforming the Practice of Management," presentation at Systems Thinking in Action conference, November 14, 1991, Cambridge, Mass. Paper #D-4287, System Dynamics Group, MIT, Cambridge, Mass.

Simon, H. (1958). *Organizations.* New York: John Wiley & Sons.

Thompson, J. D. (1967). *Organizations in Action.* New York: McGraw-Hill.

Trist, E. (1981). "The Evolution of Sociotechnical Systems." Toronto, Ontario, Ministry of Labour.

ADDITIONAL READINGS

Goldstein, J. (1988). "A Far from Equilibrium Systems Approach to Resistance to Change," *Organizational Dynamics, 17,* no. 2, pp. 16–26.

Hannan, M., and Freeman, J. (1977). "The Population Ecology Model of Organizations," *American Journal of Sociology,* March, pp. 929–64.

Hofstadter, D. (1979). *Godel, Escher, Bach: An Eternal Golden Braid.* New York: Vintage Books.

Vaill, P. (1989). *Management as a Performing Art.* San Francisco: Jossey-Bass.

APPENDIX

Three Case Histories: The Application of Systemic Managerial Tools

▶ Observation suggests that many of the corporations dominating world markets are managing from an increasingly systemic perspective. The systemic managerial tools—strategy, structure, procedures, culture, and leadership—depicted in the management systems model (MSM) have been used in various configurations by these organizations to achieve and maintain their positions as global leaders. Conceptually, there is no established approach that governs the way that these tools should be integrated. In fact, the complexities of integrating them are massive, both in theory and practice. This is largely due to the dynamism of organizational systems and to the interdependency of these tools. This complexity represents one of the greatest potential challenges to corporate leaders of the next century.

One step toward clarifying the web of systemic relationships that managers and leaders may face in the future is to examine the experience of several leading organizations. Toward that end, this appendix contains cases describing three successful global organizations in detail: General Electric (GE), International Business Machines (IBM), and Toyota. These cases have been written with special emphasis on one of the systemic tools from the MSM. General Electric's focus is on the core tool, leadership. The case looks at the role CEO Jack Welch has played in revitalizing a formerly lethargic GE and transforming the company into a pace-setting firm in a wide variety of global industries. The IBM case pays special attention to culture, looking at the role of IBM's culture as the organization attempts to transform itself into a company that can continue its former dominance into the next century. Finally, the Toyota case is oriented toward strategy. Toyota has gradually positioned itself to be a global leader in automobile production by placing a premium on innovation in all dimensions of its business. The case addresses Toyota's efforts to map out a truly global competitive strategy for the twenty-first century.

Although each case maintains a particular focus, each also introduces information pertaining to the other tools of the MSM. As a result, it becomes possible to visualize the systemic effects that have emerged over time in these firms. It is especially important to recognize that these tools have both synergistic as well as potential disabling effects on the system. One example of such synergy is found at IBM in the important link between IBM's corporate culture and its marketing strategy of providing superior customer service. The culture thrives on extraordinary feats of achievement, which strengthen the culture. The stronger the culture, the more acknowledgment, status, and value placed on the tradition of providing the best service.

Since the subtleties of the MSM should become evident when reading these cases, it is a useful time to reconsider the implications of the MSM model. Two fundamental sets of polar forces operate simultaneously in the MSM, with both sets related directly to the ongoing issues of creating change versus promoting stability:

- Relative openness and closedness of the system
- Conflicts between technical and social tradeoffs

Organizations experience continuous tension over the relative degree that the system and its various subsystems should express openness or closedness to the environment. Openness is necessary for expansion and to freely receive the many benefits that can be derived from the environment. However, openness also tends to be destabilizing. Over time, the environment is likely to change in ways that are not always beneficial for the organization. On the other hand, relatively closed systems are generally more stable, yet less able to take advantage of external opportunities. In practice, this dynamic tension is often played out in decision-making processes concerning the selection of key strategic alternatives. Individual perceptual differences regarding the meaning or significance of specific environmental changes also cause this tension level to build. This tension can also be aggravated by personal values regarding the relative importance of risk taking and rewards. Some decision makers are driven primarily by the attraction of rewards; others focus on avoiding risk. Over the long run, however, regardless of how it is achieved, organizations must have the capability of being both open and closed to the environment. This capability is necessary to facilitate the achievement of strategic initiatives as well as to maintain a wise balance in relation to the environment.

Another tension-filled drama that takes place within organizations stems from the conflict between the polar forces of the social and technical subsystems. As was indicated earlier, the technical subsystem includes technology, but also subsumes other task-oriented tools and resources such as procedures, structure, and material. The social subsystem refers to the organization's people and their relationships. This can take the form of small group interactions and a broader culture. The tension between the social and technical dimensions is evident in many managerial theories ranging from leadership (the Ohio State studies and Fiedler's work) to motivation theories (Herzberg's dual-factor model) to various perspectives on structure (Galbraith, Emery, and Trist).

The source of this tension is that work can be engineered to be accomplished in highly efficient ways, which are stable, programmable, and relatively simple, but people are complex and have individual as well as group social and psychological requirements that must be addressed. The social system of an organization is capable of generating great leaps forward in innovation and providing the creative spark that can drive the system to new, higher levels of equilibrium. On the other hand, a well-conceived technical system can significantly contribute to stability and efficiency. Again, as with the issue of openness and closedness, the forces of change and stability remain a source of tension.

The job designs, structures, and procedures that are most economical are rarely sufficient to meet the social needs of organization members. Strategies designed to resolve this conflict, such as the sociotechnical systems approach, attempt to balance the two competing forces by designing work to address both. The tension that results from the interaction of these opposing forces need not be destructive. Many firms have used this dynamic tension creatively to energize their organizations. GE, IBM, and Toyota have all, at some time in their history, integrated these two forces.

In general, strategy and culture are traditionally used to open a system, while structure and procedures are used to close it. In some cases, however, both strategy and culture have been used to close the system. For example, divestiture strategies of selling off assets are a way of detaching from a given environment. However, as a strategy divestment is an exception to the normal course of business. Organizations may also develop cultures that are internally-focused and resistant to change; this is particularly evident in older, more bureaucratic organizations. There are many examples of corporations that were once vital, dominant firms that suffered from internal decay and gradually faded into oblivion. Among these organizations are U.S. Steel, W. T. Grant, and, more recently, Bank of New England.

Procedures and structure are instrumental systemic tools in that they are used primarily to carry out prior executive decisions. They cannot initiate direction or change the purpose of an organization. They are primarily limited to use within the system. The primary orientation of structure and procedures is toward controlling the technical system of the organization. For example, manufacturing operations in most organizations are largely governed by programmed automation, tight procedures, and a structure designed around the technical core.

However, it is possible to have a system in which culture is used to regulate the technical system; for that to work, the culture must be extremely dense and rigorous. At Quad-Graphics, for instance, the culture places high demands on people to align their behavior with the norms of the organization. In effect, the culture becomes a substitute for procedures. In fact, several of the systemic tools can function as substitutes for each other. Culture can also function as a substitute for structure. For example, an organization can employ an extremely loose structure when a thick, dense culture is in place. This is visible in many R&D programs, such as those at 3M. Conversely, in a new organization, where culture has not yet developed, leadership, structure, and procedures are generally relied on to fill the void.

The leadership function is at the core of the MSM. It is the executive function, not in the sense of its being an exclusively top-management role but in the sense that it is oriented to the processes of deciding and executing. The leadership of an organization has the extremely complex task of reconciling the dynamic polar forces with the various configurations that can exist among the four other systemic tools. The systemic tools, the organization, and the environment have dynamic, interdependent relationships. This set of relations simultaneously confronts leaders with complexity and many opportunities to be creative in how they envision a desired future state for the system. Ultimately, the tension that exists between where the organization currently rests and its desired future state will drive the leadership transformation process. This is where effective leadership has the greatest impact. The ability to depict a compelling future and initiate change toward that direction is a challenging, yet essential, task in the next century's global economy.

The cases that follow, GE, IBM, and Toyota, are examples that portray leading companies that have used the various systemic tools to gain competitive advantage. At the simplest level, these cases are intended to illustrate how systemic tools can be applied. Additionally, they also reflect many of the contradictions and paradoxes that exist in systems that are trying to resolve the dynamic tension of openness-closedness and social and technical issues.

▲

General Electric Corporation

Krzysztof Obloj, Ph.D.
Steven Cavaleri, Ph.D.
Luke Giroux

INTRODUCTION

If one American corporation epitomized U.S. industry for the last hundred years, it is General Electric Company. GE's roots can be found in the Edison Electric Light Company of the late 1800s and thus in Thomas Edison, who, over the course of his eighty-four years, received more than one thousand patents.

The Edison Electric Light Company was originally formed to support Edison's idea for developing an incandescent light bulb. The concept of forming a company to support research basically set the stage for a creative organization that continued to be innovative and successful. This innovation was maintained in the 1910s and blossomed in the late 1920s and early 1930s as most modern home appliances were introduced by General Electric during this period. World War II brought GE what it brought many U.S. manufacturers: a temporary change from manufacturing peacetime products to wartime production. After the war, new technology created an eye-catching variety of products to satisfy repressed demand. This technology carried General Electric, along with other dominant American companies, up to and through the prosperous 1950s and 1960s. Unfortunately, however, this prosperity led to increased size and, eventually, the development of the sluggish bureaucracies that were prominent in the 1970s.

These bureaucracies, with their tight planning and control, were ideal for the more stable and predictable environments of the 1950s and 1960s. However, the 1970s marked the beginning of increased Japanese competition in the United States and international markets, and domestic corporations either had to sink with bureaucracy or swim with expediency. The 1980s turned out to be a decade of great change at General Electric.

With the appointment of Jack Welch as CEO and chairman of the board in early 1981, the company began a revitalization effort not seen in the United States ever before—or since. Welch, with his management experience and vision, turned GE from an overweight, bureaucratic corporation into a lean, efficient, flexible organization capable of reacting effectively to rapidly changing domestic and global markets.

GE is involved in these domestic and global markets through thirteen different major business entities. They are aerospace, aircraft engines, broadcasting (NBC), communication and services, electrical distribution and control, financial services, industrial and power systems, lighting, major appliances, medical systems, motors, plastics, and transportation systems. To go into the evolution of all thirteen businesses is far beyond the scope of this case. We will examine a few examples that delineate GE's evolution over the past several decades from researcher and innovator to world-renowned business leader.

GE LIGHTING

It was over one hundred and ten years ago, in October of 1878, that a company was formed to support Thomas Alva Edison's inventive genius, which had transformed a tiny carbonized sewing thread sealed in a glass tube into a practical product—the incandescent light bulb. This invention launched a new electrical era that transformed the lives of people throughout the world. In 1892, the merger of the Edison General Electric Company with Thomson-Houston Electric Company marked the official formation of General Electric Company.

Edison turned his attention to other endeavors, but the aggressive innovators at newly formed GE began a century-long quest to improve and perfect the light bulb. Over the course of the last hundred years, the principles of lighting have basically remained the same, but lamp and lighting improvements have continued, involving everything from using better materials for the filament to enhancing bulb size and shape. GE's laboratories achieved several major technological breakthroughs, including the photoflash lamp (1930), the fluorescent lamp (1938), and the sealed-beam automotive headlamp (1940).

Edison's invention and these breakthroughs set the stage for what today is known as GE Lighting. Building on its legacy of bringing light into the world, GE Lighting is constantly improving the efficiency, quality, and utility of its products. To further improve its leadership in the United States while addressing its commitment to build world market share, GE acquired a majority interest in Tungsram Lighting Company of Hungary in early 1990. The $150 million transaction, one of the largest postwar investments in Hungary by a U.S. company, allows GE and Tungsram to merge their European distribution and sales operations, significantly strengthening GE Lighting's position in Europe. Tungsram will manufacture and sell products ranging from household lamps and energy-saving fluorescent products to high-technology discharge lamps.

The company's global strategy is evident not only with GE Lighting but also with every other GE business entity (except for NBC, for which a world market is not relevant). The evolution of philosophy illustrated by GE Lighting applies equally well to almost every other GE business. For example, the development of a steam turbine in 1903 led to the establishment of GE Industrial and Power Systems, which has evolved into being a present-day world leader in the design and manufacture of steam and gas-turbine generators. Similarly, the invention of the x-ray tube in 1913 laid the groundwork for GE Medical Systems, which is today a world leader in medical diagnostic imaging technology.

THE GE BUREAUCRACY

Along with the evolution of GE businesses, one could also trace the development of a strong system of planning and management controls. This bureaucracy, fine-tuned in the 1950s and 1960s and still a model for many major companies, was a strength at General Electric, making the company a leader in the development and application of management theory.

Eliminating that corporate bureaucracy was one of the first strategies that Jack Welch put into effect. He "regards bureaucracy as evil because it destroys productivity by distracting attention from useful work" ("Inside the Mind of Jack Welch," *Fortune,* March 27, 1989, p. 19). His ultimate goal is to change corporate structure so as to leave as few layers as possible between top management and the workers who deal directly with customers, research, and production. Welch states: "We like to think of trying to become a big company and a small company simultaneously" (M. Potts, "GE's Management Mission," *The Washington Post,* May 22, 1988, pp. 1–4). He wants managers to spread themselves thin, forcing them to hand off trivial decisions to subordinates so they can concentrate on more important issues. One such issue is attaining Welch's business strategy: to become first or second in every market GE serves or risk being sold or closed.

To achieve this goal, some critics say, Welch has been ruthless. During the 1980s, he sold more than 200 businesses, closed 73 plants and facilities, and eliminated more than 130,000 jobs through divestiture, attrition, dismissal, and contract buyouts. However, in spite of these severe means used to streamline the corporation, GE appears to support his vision. So often is this simple concept of rank repeated around GE that people express it as a single, seven-syllable word: "number-one-an'-number-two," a sure sign that the corporate culture runs thick. In fact, Welch has been successful in this goal (see Figure 1). Concurrent with this strategy, Welch wants his line managers to think and act more like entrepreneurs than bureaucrats, and he has given them the responsibility, authority, and incentives to do so. GE now has leaders (not managers) in each of the thirteen businesses, who figuratively own those businesses. When Welch first arrived as CEO, he had to sell this idea of ownership—a commitment to relentless personal interaction and immediate sharing of information—to 2,000 or so top executives. That process took eight years and a lot of hiring and firing, and it still leaves 99 percent of the organization needing to be sold on his vision.

His current visionary strategy, first implemented in 1989, is one he is attempting to sell to every member of the organization, just below 300,000 at last count. It is a strategy called *Work-Out.* For a gigantic company like GE to be successful in the 1990s, Welch believes, it must move with the agility and flexibility of a small company. According to Welch, Work-Out is absolutely fundamental to GE if it is to become this kind of company. Welch knows that GE has to apply the same relentless passion to Work-Out that it did in selling the strategy of being number one or number two globally.

The basic objective of Work-Out is to remove the backbone of bureauc-

	IN THE U.S.	AND IN THE WORLD
Aircraft engines	**First**	**First**
Broadcasting (NBC)	**First**	**Not applicable**
Circuit breakers	**First** tied with Square D and Westinghouse	**First** tied with Merlin Gerin, Siemens, Westinghouse
Defense electronics	**Second** behind GM's Hughes Electronics	**Second** behind GM's Hughes Electronics
Electric motors	**First**	**First**
Engineering plastics	**First**	**First**
Factory automation	**Second** behind Allen-Bradley	**Third** behind Siemens and Allen-Bradley
Industrial and power systems turbines, meters, drive systems, power transmission controls	**First**	**First**
Lighting	**First**	**Second** behind Philips
Locomotives	**First**	**First** tied with GM's Electro-Motive
Major appliances	**First**	**Second** behind Whirlpool tied with Electrolux
Medical diagnostic imaging	**First**	**First**

FIGURE I ***How a Dozen GE Businesses Rank***

Source : *Fortune,* March 27, 1989, p. 16.

racy: multiple approvals, unnecessary paperwork, excessive reports, routines, and rituals, and to replace them with simpler, more expedient, and more efficient processes of communication. In fact, communication improvement is one of the essential goals of Work-Out: to improve communication to the point where the leader of every business is aware of ideas and concepts of workers in that business at all organizational levels. No two GE businesses approach Work-Out in the same way; a process this intensive cannot be cloned successfully among vastly different enterprises. But the premise of the program is the same.

Ultimately, GE is redefining the relationship between boss and subordinate. The aim is to reach the point where people challenge their bosses every day—challenging them on aspects of their jobs, company procedures, or anything related to the business. Jack Welch and GE are aware that this will not be easy. The norms today in most businesses (not only GE) are to avoid critical issues and to lessen the effects of bad news when approaching the boss. If GE can change this atmosphere—if it can create risk-reward tension and an open exchange of ideas at all levels—it will develop self-confidence throughout the organization. And, in turn, it will create more fulfilling and rewarding jobs, improving the quality of work life dramatically.

MANAGERS AND LEADERS

To accomplish this task, GE needs to place an inordinate amount of faith in leadership, the middle and upper-level managers who are the backbone of any organization as immense as GE. The faith must be placed in their ability to *lead,* not to manage. What makes a manager a manager is being in control. This control becomes a problem when managers repress subordinates' creativity and limit subordinates' abilities. Welch insists on using the term *leaders* instead of *managers,* claiming that if employees are called *managers,* they will start managing things, getting in the way, and working on trivial things that they should entrust to their subordinates. The leader, on the other hand, takes available resources, both human and financial, and creates a vision. More importantly, a leader should have the ability to articulate this vision to employees—to motivate them to focus their intellectual energy toward succeeding in a competitive world. Jack Welch is one of those leaders, and he is one of the best in the business at selling his ideas to his executives. He has had tremendous success since 1989 selling the idea of Work-Out to his upper-level leaders. But what about the lower-level managers? Herein lies a potential problem.

It seems to be no problem for upper-level executives at GE, who work no more than two layers below Welch and who have his direct support, to become forceful advocates of Work-Out and criticize the status quo (a major component of Work-Out). They can ask simple questions such as, "can we improve what we're doing or how we're doing it?" But a problem might exist with lower-level management, who could interpret a subordinate's criticism of the status quo as an assault on their ability or a threat to their position. How does a corporation as large as GE build in a confidence factor, which is crucial to the success of this new strategy, throughout all levels of the organization? The jury still appears to be out on the success of Work-Out. Implemented in the beginning of 1989, the new strategy is expected by executives to take at least five (but probably closer to ten) years before it can make a difference and change the internal workings of the organization.

COMMUNICATION AND MOTIVATION

A potential problem exists for GE in the core of the company. One can see the problem of leading a multimillion dollar corporation in which the personnel are intimidated by their superiors—afraid to be bearers of bad news. This dilemma exists, beginning with CEO Welch and his direct subordinates and continuing to the lowest superior-subordinate relationship. Jack Welch has a reputation for being tough, aggressive, impatient, and intimidating; he is both admired and feared. As a leader, he is a strong believer in taking responsibility for one's own actions. The leaders of each of GE's business entities possess a substantial budget for research and development as well as for proper implementation of new programs, products, or ideas. The unwritten policy states that if one of these leaders is willing to take a risk for any of the above, GE is usually willing to support them both financially and procedurally. If the venture fails, however, Welch insists that those directly responsible for the failure take the consequences. A former GE manager claims that Welch has terminated executives not so much for failure of an effort but for placing blame on those not directly responsible for that failure. The problem is, Welch has also terminated people for their delivering bad news as well.

Senior executives are apprehensive whenever they have to bring bad news to the leader who is striving to open communication throughout the organization, and a similar lack of faith exists between a foreman and the worker on the shop floor—probably even more so. A senior executive, when approaching the CEO, would not feel belittled by another executive officer. But the worker on the shop floor is more likely to feel anxiety about his inferior position when he has to discuss with the plant manager a problem concerning his job. Such situations create a number of barriers between a worker and his superior, and thus hamper the open communication process that the organization is attempting to implement.

One might speculate that the vast number of changes made recently would affect employee motivation. Herein lies yet another problem. It has yet to be established that it is possible to motivate people—to get them positively involved with their corporation—on the basis of making them better people and their jobs more rewarding and fulfilling without additional assistance from a promotion or at least a very basic raise or bonus. Motivation is an important aspect of an organization's productivity and growth needs. When considering an organization's needs, one must also consider the needs of the individual and see that these needs are satisfied. One must examine what motivates employees—at all levels of the organization—to excel. If an organization such as GE reduces the number of organizational layers from more than ten to only four or five, what types of motivational tactics are needed given that the reward of a promotion is no longer available? Is the prospect of job enrichment enough to motivate employees to excel? If so, can it be used repetitively, the way that climbing the steps of an organizational ladder can?

So far, the answer to these questions appears to be yes. GE requires certain types of individuals—people with an inner drive to excel at any task they set out to accomplish; a willingness to be stimulated by and to get involved with a position or job; a willingness to experience change within a position and to make changes that can make things happen; and a dedication to GE, which is number one in their lives while all else is secondary. These people are movers and shakers. If employees do not possess these traits when hired, the corporate culture is thick enough to change those who want to get involved and to do away with those who do not. GE seems to have the ability (in part through aggressive recruiting practices) to obtain employees who can use job enrichment as a motivator to perform well, who can expand employment horizons without moving from one business to the next. As soon as employees find success in a certain work atmosphere or team, GE tends to keep them in place. The object is to strengthen the organization's position—by having people work together and become proficient in a particular area or business.

FUTURE PROSPECTS

It appears as though the next couple of years should give decisive answers to many questions regarding GE's continued success. The future may bring some failures or blemishes to GE, as the company has experienced a few in the past.

> Welch has blown some big ones. He lost over $120 million trying to sell factory automation equipment to manufacturers that were unprepared to embrace his vision of the future. He wrote off more than that on two early acquisitions that turned out to be also-rans at producing computer chips and computer-aided design equipment. With the purchase of Kidder he unwittingly bought trouble in the form of an inside trader named Marty Siegal, who enmeshed the firm in his crimes. (*Fortune,* March 27, 1989, p. 15)

There was also the scandal of supposedly selling defective parts for nuclear reactors to a number of utility companies. Finally, Welch was so hypnotized by the glories of automation that he wasted vast quantities of money, time, and talent on the most modern locomotive plant in the world—at a time when there was no market for locomotives. But for breakeven orders from China, the plant would have been closed in no time.

Given GE's entire record of more than 100 years and Welch's record in the last ten or so years, one would have to conclude that there have been exponentially more successes than failures, however. Just before the arrival of Jack Welch, success appeared to be slipping away as the aging ominous bureaucracy became overweight and sluggish in its reactions to the environment. GE entered the 1980s with a widely diverse and marginally successful set of businesses and major product lines—as many as 350. Under Welch, GE followed the strategy that diversity could only be a decisive advantage if a business was a world leader in its particular market. This appears to be a successful game plan, and so far GE seems to be one of the few serious players. The company began the 1980s with a total market value of $12 billion, ranking eleventh among American companies. It left the decade ranked second, with a total market value of $58 billion—the largest increase of any company in the United States. As a result of many changes and improvements, GE has become an integrated, diversified company held together by shared management practices, which can apparently prepare it for the years to come.

2 International Business Machines

Steven A. Cavaleri, Ph.D.
Krzysztof Obloj, Ph.D.
Daniel Bach

INTRODUCTION

International Business Machines (IBM) has long been a name synonymous with industry dominance. By 1992, the certainty of this assertion was questionable, although IBM remained as the fifth largest industrial corporation in the world and the most profitable. Amidst this seeming prosperity, IBM has witnessed the across-the-board erosion of its market share as a result of intensified competition. Even the once-safe mainframe product line is faced with decay in market growth due to inroads from PCs and higher-quality Japanese competition.

In the early 1980s, IBM rode the crest of a wave of positive publicity through a number of books, such as Peters and Waterman's *In Search of Excellence* (1982). These widely read books identified IBM as exemplary of the best of American industry in categories such as customer service and corporate culture. At the time, it would have been unthinkable to consider a decline in IBM's overall performance. Yet an article in *Fortune* described IBM's unfortunate 1980s thus:

> To understand fully what a disaster IBM has been, and just how blind its own management was to the depth of its problems, step back to a moment in late 1986. IBM was more than a year past a boom period and struggling. Revenue growth was miserable, earnings growth was nonexistent, and IBM's stock, then $125 a share, had lost nearly $24 million in market value from a peak of $99 billion just seven months earlier.... It is now 4½ years later. The stock was recently just below $100, which means another $18 billion in market capitalization has been shredded into megabits. IBM's total revenues have dragged, rising over the past five years at an average annual rate of only 6.6% against 13.4% for the data processing industry as a whole. In unhappy concert, the company's worldwide market share has dropped from 30% to 21%. Each percentage point lost represents $3 billion in annual sales. (*Fortune,* July 15, 1991, p. 41)

Despite the fact that IBM spends more on R&D than any four of its competitors combined, it maintains control over a dying market segment ("Refashioning IBM," *The Economist,* November 17, 1990). It holds about 50 percent of the world market for mainframes, yet the market is not growing at the rate needed to support the company—the mainframe market grew only 5 percent from 1984 to 1989, while the market for minicomputers grew 13 percent and that for personal computers grew an incredible 74 percent during the same time period (*New York Times,* December 10, 1989). However, IBM holds only 15 percent and a little over 10 percent in these two market segments ("Refashioning IBM").

In the more recent race for the lead in the workstation market, IBM entered late. Although growth was only about 2 percent in this market over the past five years, the prospects for the future are promising. IBM currently controls 1.8 percent of the workstation market, but to participate in this market it has been forced to use the operating systems of other companies, such as AT&T's Unix ("IBM is Finally Saying in Unix We Trust," *Business Week,* February 12, 1990).

In an industry racked by fast-paced change, where competitors know what to expect, IBM's CEO John Akers in 1986 laid out a game plan to reverse the company's fortunes. Among the major components of this grand plan are cost-cutting measures, strategic alliances with innovative competitors (Apple and Microsoft), listening more closely to customers, expanding software offerings, and creating a revolution in the company's culture. This cultural revolution, begun in 1989, is built on a systematic drive to push responsibility down through the ranks of the hierarchy. In an industrial giant with 387,000 employees in over 130 countries, however, cultural change may take the form of slow evolution, rather than revolution. Akers's efforts to decentralize IBM's decision making have met with considerable resistance in some quarters, while achieving success in others. IBM's technical managers have surprised competitors by their announcement that they will produce heavy-duty four megabit chips in high volume.

Despite such successes, Akers has been unable to conceal his disappointment with the overall pace of change. His comments to the *New York Times* in May 1991 brought pressure on him to cool his rhetoric. Akers stated, "too many people are standing around the water cooler waiting to be told what to do. . . . Everyone is too comfortable, the tension level is too low" (p. D1). By 1992, IBM had made it clear that it would only reward high-performing employees and would plan to lay off up to 20,000 employees. The news sent panic throughout the company, which had virtually guaranteed permanent employment in the past. The company that was the model for Peters and Waterman's famous search for "excellence" now saw employees give the press anonymous comments like "I feel betrayed. If IBM isn't growing, is it the fault of the employees or of the senior management?" (*Business Week,* December 16, 1991, p. 115).

IBM often has been credited for having a thick, enduring corporate culture built on the values of its founders, the Watson family. Yet, some industry critics have noted that IBM's drive for cultural change must be seen against the backdrop of the company's position as a globally recognized bastion of conservatism. Many IBM employees are reputed to continue the tradition of wearing dark suits and white shirts. One IBM manager noted, "There's no rule about white shirts, but whenever I wear a striped shirt people ask me if I'm going to the beach" (*Fortune,* August 14, 1989, p. 34).

Akers's supporters argue that, in order to understand IBM's current predicament, one must first appreciate the fact that the computer industry is more turbulent than ever before. The industry is ruled by several major competitors, such as Hewlett-Packard, Digital Equipment, Fujitsu, Compaq, Apple, and Sun. The remaining 50,000 competitors are small companies bloodthirsty for market share. Second, Akers's backers say, his predecessors Frank Cary and John Opel left him an inefficient company that self-confidently rested on its laurels while its position was slowly deteriorating. The culture of such a company is incongruous; at times it appears aggressive with clear values and at other times it looks passive and confused. The prospects for IBM's return to its preeminent position within the computer industry seemingly reside on its ability to solve the riddle of its own cultural paradox.

IBM'S HISTORY

IBM began as a conglomerate of three companies brought together by Charles Flint in 1911. The company's original name was CTR, or Computing Tabulating Recording Company. The company consisted of three unrelated divisions with a wide variety of products ranging from coffee grinders to time clocks to a tabulating machine used by the U.S. Census Bureau. Today, IBM's more than 387,000 employees operate within the structure of fifty-nine major subsidiaries. IBM has become so profitable, with sales of $62.7 billion, that its bottom line is five times that of its nearest competitor and its net income is $3.76 billion (see Figure 1).

IBM designs, manufactures, and sells products for the information processing and telecommunications industries. These products have become industry standards for improving productivity in business, government, science, space exploration, defense, education, and medicine. They range from typewriters and copiers to telecommunications equipment, data processing machines, office systems, industrial workstations, and educational and scientific testing materials. IBM is most well known for its mainframe computers, but it also has successfully competed in the markets for workstations, personal computers, and laptop computers. IBM has engaged in a long-running joint venture with software leader Microsoft to supply programs for use on its machines. More recently (1991), IBM entered into a bold new strategic alliance with Apple Computer. The Apple agreement was designed to couple both firms' resources in the development of new hybrid computers that would utilize the next generation of computer operating systems.

IBM'S LEADERS

THOMAS WATSON SR.

IBM's beginning, when it was still CTR, was not such that one would predict that it would someday be a world leader in anything. In fact things were so bad for CTR that it hired a sales manager who had been convicted, sentenced,

	1990	1989	1988
Net Sales	$69.018	$62.710	$59.681
Cost of Goods	30.723	27.701	25.648
Gross Profit	38.295	35.009	34.033
R&D Expenditures	6.554	6.827	5.925
Selling, General, and Administrative Expenses	20.709	21.289	19.362
Income Before Depreciation and Amortization	11.032	6.893	8.746
Depreciation and Amortization	NA	NA	NA
Nonoperating Income	0.495	0.728	0.996
Interest Expense	1.324	0.976	0.709
Income Before Taxes	10.203	6.645	9.033
Provision for Income Taxes	4.183	2.887	3.542
Other Income	NA	NA	NA
Net Income	$ 6.020	$ 3.758	$ 5.491

FIVE-YEAR FINANCIAL SUMMARY

DATE	SALES (IN BILLIONS)	NET INCOME (IN BILLIONS)	EARNINGS PER SHARE
1989	$62.710	$3.758	6.47
1988	59.681	5.491	9.80
1987	55.256	5.258	8.72
1986	52.160	4.789	7.81
1985	50.718	6.555	10.67

FIGURE 1 ***Income Statements for IBM (in billions)***
Source: IBM Annual Reports.

and fined for violations of the Sherman Antitrust Act. This person was none other than Thomas J. Watson. When Watson was hired, he was offered a salary of $25,000 a year plus 5 percent of any profits, a provision that later made Watson the highest paid executive in the country. Within ten years of becoming general manager, Watson turned the company around, tripling sales volume. By this time, Watson had control of the company and decided to change the name to IBM. In 1932, the main products of the company were tabulating machines, typewriters, and punch card equipment. Its single largest customer was the U.S. Bureau of the Census.

Watson ran a tight ship, with strict regulations on dress that included the length of hair of the sales representatives. Drinking alcohol during work hours was prohibited for all employees, including sales representatives. Alcohol was not to be used by company representatives as a sales incentive on company time or company property. In return for compliance with this policy, Watson expressed his respect for his employees by instituting a no-layoffs policy that lasted many years.

As IBM continued to grow rapidly, concerns over fair competition emerged. By 1932, IBM's share of the punch card, typewriter, and tabulating machine markets was so great that a suit was filed by the U.S. Justice Department for violation of the Sherman Antitrust Act. The case was settled when IBM signed a consent decree in which it did not admit guilt but promised to change its selling strategies. For the next fifty years, IBM existed within the continuous shadow of the threat of antitrust action. The influence of the Justice Department was so pervasive that some IBM employees joked that the company should erect a statute in memory of Sherman outside its corporate headquarters.

THOMAS WATSON JR.

Thomas Watson Jr. replaced his father as president of IBM in 1952. Watson Jr. envisioned the company as being primarily a computer manufacturer. By 1968, IBM introduced the 7000 series mainframe computer, the first designed to operate on the basis of transistors rather than vacuum tubes. By 1961, 71 percent of all general-purpose computers were sold by IBM, and by 1964, it controlled approximately 76 percent of the world computer market.

That year, 1964, was a banner year for IBM, as it introduced the 360 series mainframe computer. This computer made all other computers obsolete. Ultimately, the 360 model became the bellwether of the IBM product line for many years to come. Customers' response to this innovative system was outstanding, securing IBM leadership in the computer market. From 1965 to 1970, IBM's sales rose from $3.5 billion to $7.5 billion, primarily due to the success of the 360 series.

T. VINCENT LEARSON

T. Vincent Learson held the position of CEO for only eighteen months. During his tenure, there were no major changes in product line or growth that were directly due to his influence.

FRANK CARY

The seven years during which Frank Cary was CEO saw IBM become increasingly cautious due to wariness of possible antitrust action. The standard procedure of reducing the price of old equipment for clearance was avoided, and a profit margin of 15 to 20 percent above that of its competitors was maintained. Significantly, the company avoided new products in new fields, such as microcomputers, so as not to give the appearance of violating antitrust statutes or attempting to monopolize new markets. Critics argue that this conservative strategy eventually led to IBM's decline in market share. Its overall share of the world market plummeted from 60 percent in 1967 to 40 percent in 1980.

JOHN OPEL

John Opel, IBM's fifth CEO, was very knowledgeable about manufacturing and had a reputation as a brilliant analyst. For the company to maintain market share and high profitability into the eighties, he believed, it would have to reduce production costs, improve technologically, and diversify its marketing strategy. To do this, the company invested $10 billion in plant and equipment. IBM also experimented with selling PCs to consumers through retail stores. Targeting small businesses and individuals, IBM attempted to use its power base to head off rivals such as Compaq and Apple at the retail level.

Opel also designed a decentralized product development system, trying to encourage specific units of the company to become more entrepreneurial. Independent business units (IBUs) and special business units (SBUs) were given greater autonomy and more freedom to innovate to encourage them to grow like new businesses. In doing so, they could flourish without the constraints of bureaucratic decision making that had come to characterize IBM. This major departure from IBM's traditional processes created considerable debate within the company.

This new system was used in the development of the IBM personal computer (PC), the success of which was far greater than ever expected. IBM had projected PC sales to be around 350,000 units, but the product created a spark with consumers and sales rocketed to more than 800,000 units. Unfortunately, IBM was unable to meet the demand for PCs, which opened the door for new competitors to enter the market.

JOHN AKERS

John Akers became CEO in 1985 just as IBM's dominance of the computer industry began to wane. For many years prior to this, IBM had been able to sell its equipment largely on the reputation it had acquired from the sale of its mainframes. In the PC market, however, individual consumers were not as influenced by the IBM name. "It's not an all blue world anymore," remarked Edward E. Lucente, vice president in charge of U.S. marketing ("Big Changes at Big Blue," *Business Week,* February 15, 1988, p. 94).

~ CORPORATE CULTURE

IBM is world-renowned for having a strong culture based on a bedrock of beliefs first laid down by Thomas Watson Sr. The culture of IBM is replete with legends of outstanding customer service and other outstanding feats. The Watson system of values was most simply expressed by three fundamental precepts:

1. Respect for the individual. Respect for the dignity and the rights of each person in the organization.
2. Customer service. To give the best customer service of any company in the world.

3. Excellence. A belief that all products and services should be of the highest quality.

Many people are proud of the fact that IBM is their employer. Competition for IBM jobs is intense, and those who are selected often feel as though they are members of an elite group. Most of their pride is derived from the fact that IBM is the leading company in its field, with sales five times that of its nearest competitor.

The employee's feeling of being special is not solely attributable to IBM's history of achievements. In fact, this pervasive self-confidence may be equally rooted in the core beliefs established by the Watsons, which were institutionalized over the years. Success came from the people that worked at IBM. These attitudes became manifested in the policies set forth by the company's leadership. The primary precept that set the tone for all business dealings within IBM was the senior Watson's insistence on showing respect for the individual. One way he expressed this belief was through his no-layoffs policy, which was adhered to even in the worst of times for IBM. Even when IBM's nearest competitors, all of which have much smaller staffs, trimmed their work force, IBM had not resorted to cutting staff through layoffs. Even when IBM sold divisions (for example, the sale of Rolm), it always sought to ensure that the employees would enjoy continuous employment.

Another belief created early in the company's development is the value placed on rewarding high performance. IBM managers know that they are expected to reward an employee who does something extraordinary. The reward could be a bonus or it could be something that demonstrates the recognition to other employees, such as a gold star on the individual's door or membership in the Golden Circle club for salespeople. IBM's management takes this reward policy seriously, as the following example clearly shows:

> A manager of a one hundred person sales branch rented the Meadowlands Stadium for the evening. After work his salesmen ran out onto the stadium field through the players' tunnel. As each emerged, the electronic score board beamed the salesman's name to the assembled crowd. Executives from corporate headquarters, employees from other offices, and family and friends were present, cheering loudly. (Peters and Waterman, *In Search of Excellence*, New York: Harper and Row, 1982).

Many of the core ideals shared by IBM employees during its rise to prominence are expressed in its song (from T. Deal and A. Kennedy, *Corporate Culture*, Reading, Mass.: Addison-Wesley, 1982):

> EVER ONWARD—EVER ONWARD
> That's the spirit that brought us fame!
> We're big, but bigger we will be
> We can't fail for all can see
> That to serve humanity has been our aim
> Our products are now known in every zone
> Our reputation sparkles like a gem
> We've fought our way through and new

Fields we're sure to conquer too
Forever onward IBM.

ORGANIZATION STRUCTURE

The enduring nature of IBM's culture can, in some respects, be traced to the continuity in its structure. During the Watson years, the structure operated efficiently, albeit mechanically. Thomas Watson Jr. saw the military model as being the ideal approach since its classical bureaucratic structure was regarded as the ultimate efficient machine.

Over the years, IBM succeeded in altering the segments of its structure, but rarely did it embark on such major revisions as it began in the late 1980s. In prior years, IBM often relied on making selective structural changes to unleash its innovative powers. Some of its best products came from project teams such as the IBUs used to develop PCs and the model 360 team.

By 1992, the need for major structural change at IBM became more apparent than ever. As a way to kick-start creativity, a number of divisions were set loose to operate as part of a more decentralized structure. However, IBM critics such as Microsoft's Bill Gates believe that this latest plan does not go far enough to actually release the divisions from the home office reins. Gates noted, "They're going in the right direction, but if they don't do it in a strong fashion, I don't see it happening" (*Business Week,* December 16, 1991, p. 116). Some supporters of radical change even called for spinning off divisions as separate companies with their own stock and board of directors. The thrust of the plan to create what was called "the new IBM" was to break the divisions away from IBM's bureaucratic management committee. Historically, many of IBM's smaller, more agile units were held back from competing aggressively against new competitors by the management committee.

THE MODEL 360 DEVELOPMENT INITIATIVE

The development of the 360 model computer was a remarkable illustration of IBM's capability to unleash a corporate-wide effort to respond to a call for action. It is also a reflection of IBM's inaction when the firm was near its peak. The success of the development of the 360 series was largely due to the structure of the development team. Although it was a large project, incorporating more people than is usually associated with a project team, the project started off fluidly, and the reorganizations took place with great frequency. The project had a specific goal, which was to limit the scope of the directions in which the engineers were going. To unify the direction of the team, the junior Watson called for the creation of a single family of computers, all using the same peripheral equipment and all speaking the same language.

IBM's commitment to the project was absolute, requiring an investment of over $5 billion, leading the project chief to describe the commitment as "You bet your company." Despite this financial support, over time the 360 project was difficult to coordinate, often because decisions were made at so many decentralized locations. The project seemed to keep heading in different directions, and the constant communication created a continual interchange of ideas. No one wandered in one direction very long. The many meetings stimulated a continuous influx of new ideas to be tried. Project team members found themselves able to make binding decisions and commitments, while ideas could be implemented (or then dropped) without any formal repercussions. Consequently, little time was wasted waiting for the formal approval to try an idea. Ideas would be implemented, and then the quality of the ideas could be evaluated at a later date.

This apparent indecisiveness violated basic norms that governed decision making at IBM. The situation bothered Watson enough to lead him to ask Vice President Frank Cary to design a system to ensure against a repeat of the problem. Cary complied; years later, when he was chairman, he got rid of the laborious product development structure that he had created for Watson. As Cary put it, "Mr. Watson was right, it [the product development structure] would prevent the repeat of the 360 development turmoil. Unfortunately it would also ensure that we won't ever invent another project like the 360" (Peters and Waterman, 1982, p. 49). Overall, however, the success of the 360 series was outstanding, making it the single best new product development project in IBM history.

STRATEGY

The convergence of IBM's strategy, structure, and culture can be seen in the way it enters new markets. IBM's goal is to deliver the best service in the world. "IBM is customer and market-driven, not technology-driven" notes Buck Rogers, long-time vice president of marketing at IBM (Peters and Waterman, 1982, p. 161). Thus the strategy for entering new markets is based on its reputation for customer service. IBM typically enters markets late with a product that is reliable, and one that it can support with a service guarantee that surpasses those of all other companies in the market.

An example of this strategy in action is the way IBM entered the computer market. IBM chose not to sponsor the ENIAC project, which produced the first computer and later marketed the first commercial computer, the Univac I. Only after IBM saw the potential of the computer did it start to manufacture its own. Two years after the introduction of the Univac I, IBM introduced its own commercial computer; within ten years, over 70 percent of all general-purpose computers were IBM's machines.

A similar strategy was followed with the personal computer. IBM did not see enough potential in the personal computer market, and so it decided to permit other companies to pioneer development. Such companies as Apple, Tandy, and Commodore all entered before IBM. Within two years after the introduction of the PC, however, IBM was selling more units than all other companies in the market (200,000 per year).

The strategy is not flawless, as IBM experienced in the automatic teller machine (ATM) market. IBM entered the ATM market well after several other companies had established themselves in it. When it finally entered the market in 1985, IBM introduced a new, state-of-the-art unit that could read the magnetic code on personal checks and was able to cash checks, with accuracy to the penny, on the spot. However, few banks were interested in the new machine; they wanted ATMs that had higher operating reliability. They wanted a more dependable version of the machines they currently had to replace old machines and to maintain a degree of uniformity in their stock of ATMs. The resulting drop in sales eventually precipitated the creation of a joint venture with Diebold, in which the Diebold president controlled 70 percent of the new company, known as Interbold.

THE FUTURE

In the early 1990s IBM found itself in a compromising position. Growth had fallen to half of what it was in the 1970s. Its stock price had periodically dropped below the benchmark of $100 per share, and stockholders' earnings had fallen sharply. These problems forced Akers to change the overall structure of IBM, with major cuts in personnel. Up to 1991, Akers steadfastly refused to compromise the fundamental beliefs of IBM but by 1992 he had done an about face and initiated a major program to revitalize IBM. Skeptics, however, claimed that it still was not enough to return Big Blue to its past preeminence.

IBM's goals today are far different than could have been recognized in its plans a decade ago. During the early 1980s, IBM consistently enjoyed a 65 percent gross profit from robust sales of mainframes. Backed by an industry growing at an annual rate in the double digits and controlling 70 percent of that market, IBM saw revenues more than triple from 1971 to 1980. Between 1980 and 1985 sales nearly doubled again. However, by late 1985 the tide had turned, and the growth of the mainframe market fell to about 6 percent. With 40 percent of sales and 50 percent of revenues coming from the sale of mainframes, IBM's growth slowed dramatically.

For the 1990s, IBM's main concern is to get closer to the customer through decentralization of its structure. As IBM moves to expand its role in software and to innovate in the design of computers other than mainframes, a major issue plagues Jack Akers. Can the corporate culture deliver in making the transformation? A company that was once the paragon of corporate culture was now being termed an albatross by critics. How can a group of managers raised in the conservative IBM culture suddenly become revolutionaries and change the culture to which they have been so deeply committed?

3 Toyota Motor Corporation

Steven Cavaleri, Ph.D.
Krzysztof Obloj, Ph.D.
Glen Colley

INTRODUCTION

Toyota is the largest automobile manufacturer in Japan and the third largest in the world, after GM and Ford. The company's principal products are cars, trucks, and buses. Toyota also produces prefabricated housing units and forklifts and other industrial vehicles. The products are sold in Japan and around the world through an extensive network of importers, distributors, and dealers. Although the company is best established in the high-quality, low-price segment of the compact car market, it is trying, through its Lexus line, to become a significant competitor in the luxury car market using the same basic strategy of offering higher quality cars at lower prices.

Despite the fact that Toyota ranks first among Japan's automakers, having turned out four million cars and trucks in 1988 with sales of over $54 billion, it often trails Honda in engineering and Nissan in styling. By August 1989, a number of press reports argued that the company was beginning to show signs of decay. One of the main factors in Toyota's slippage was said to be competitive pressure from Japan's second largest automaker, Nissan Motor Company. Internal factors were causing problems as well. Consequently, Toyota undertook a major organizational overhaul resulting in the removal of two layers of middle management—stripping 1,000 executives from their staff—and the reorganization of its product development unit. Another effort involved a "put the customer first" campaign that is challenging the firm's entire approach to making cars.

Toyota's hold on market leadership was being threatened on many fronts: a war for market share, declining corporate morale, and challenges posed to Toyota by its efforts to expand globally. Having created a major foreign production base in the United States, the firm launched a similar effort in the United Kingdom. Would the strategy that worked so well for Toyota in the 1980s in the United States work as well in the 1990s in Europe?

COMPANY BACKGROUND

The company was founded on August 28, 1937, in Japan. Its predecessor, Toyoda Automatic Loom Works, had started in 1926. Because *toyoda* means "abundant rice field," inappropriate for a car maker, the company changed its name when it entered the auto business. In the spring of 1950, the company suffered through a dramatic decline in sales and a lengthy labor strike. Knowing that the company needed help, then-president Eiji Toyoda visited Ford Motor Company's River Rouge plant near Detroit to get some new ideas. Toyoda studied every detail of the Ford plant but returned home convinced that mass production was not the answer for Toyota or Japan. Toyoda and manufacturing expert Taiichi Ono both believed that the way automobiles were traditionally built needed to be changed radically. The two men worked together to develop an innovative manufacturing system known as the "lean production system." This system uses less of all resources to accomplish the same results as a traditional manufacturing system. Ideally it uses half the labor input, half the material, half the development time, and half the manufacturing space of the usual plant. The lean production system is a way of managing that is based on the principles of continuous improvement and organizational learning.

To accommodate the lean production system when reorganizing for the 1990s Toyota consolidated some 240 members of the product planning division into three major groups: small front-wheel-drive models (for example, Tercel); big rear-wheel-drive cars (such as Lexus); and trucks. This reorganization was driven by changing marketing requirements. According to Kazuo Morohoshi, head of Toyota's Tokyo Design Center, "we have learned that universal mass production is not enough. In the 21st century, you personalize things more to make them more reflective of individual needs. The winners will be those who target narrow customer niches most successfully with specific models."

The challenge of developing a strategy to penetrate the market in Western Europe presents Toyota with a fresh set of obstacles. With Peugeot boss Jacques Calvert as their chief spokesman, European automakers continue to wage a protectionist battle to limit Japanese car sales until the next century. French, Italian, and German buyers are more resistant to non-European cars than are Americans. Toyota has set up a design center in Brussels and is building a plant in Derbyshire, England, with a capacity of 200,000 cars annually. Toyota's experts expect that Toyota will have only 3 percent of Europe's 20-million-vehicle market by 2000 versus 2.5 percent now.

THE AUTOMOBILE INDUSTRY

According to analysts, gradual growth of the auto market is expected in the next five to ten years. The need for cars will increase most in countries where motorization has not progressed very much. In Japan, the United States, and Europe, the trend will be toward high-class, luxury cars and cars that give drivers a sense of individualism. The inevitable collision between the next global automobile market downturn and widespread industry overcapacity can be expected to transform the industry. In 1991, the total automobile overcapacity was estimated at over four million units annually in North America alone, and new Far-Eastern owned transplant facilities on the continent suggest little chance for future capacity reduction. Estimates of surplus capacity in Western Europe range from one to two million units.

In response, major producers attempted to lure consumers with a variety of incentives based primarily on pricing strategies and financing options. For example, General Motors has offered credit based on a consumer's home equity position.

GLOBAL MARKETS AND STRATEGIES

The total European car market was 13.4 million units in 1989, compared with 9.9 million in the United States. Predictions are that by the year 2008, the total European market will represent 24 million cars against only 13 million in the U.S. market. By 1991, Toyota held 7.5 percent of the U.S. market.

Toyota has developed a three-pronged strategy for the future. First, it invested $1.5 billion in a production facility in the United Kingdom and a design center in Belgium to turn out low-priced, mid-size cars for the European market. Second, higher priced, luxury models such as the Lexus LS400 will be produced in Japan for global distribution. Third, American plants will produce cars for the U.S. market. The major U.S. plant at Georgetown, Kentucky, has started slowly, but the production rate has gradually risen as employees become more familiar with the Toyota production system. Toyota executives believe it could take up to a decade before Georgetown is operating at full capacity.

Toyota's marketing strategy in Japan has been innovative and turned the traditional marketing strategy upside-down. The company formed close alliances with dealers in Japan, through which it developed the aggressive selling approach. Aggressive selling focuses on getting to know consumers personally and forming a lifetime relationship with them. The effect is that, in Japan, Toyota no longer builds cars for unknown buyers on speculation. Rather it builds to order for customers it knows well. This allows the production system to more effectively incorporate just-in-time inventory practices and coordinate all order and delivery schedules through a cooperative dealer network. Toyota has also developed an extensive data base of information, demographic and otherwise, concerning household purchasing patterns. This helps Toyota get closer to the customer to clarify their buying needs and preferences.

Toyota has invested millions of dollars to achieve a strong presence in the developing markets of East Asia. With its customary combination of long-term thinking, persistence, and flexibility, Toyota has taken minority stakes in countries such as South Korea, which ban Japanese imports. In countries with uncertain political stability and economic prospects, such as the Philippines, Toyota signed on with local partners, cautiously expanding its presence. In Japan, Toyota is also riding a wave of new products. A total of thirteen new car and truck models have been rolled out since 1988, and the company has managed to hold a commanding 43.4 percent of the Japanese market despite gains from a resurgent Nissan.

The automobile market is highly competitive. It has been characterized by rapid technological advances in quality, performance, and appearance. These advances have substantially increased the range of choices or preferences available to customers. The principal competitive factors in this market are the product's quality and reliability, the relation of price to performance, the manufacturer's marketing and distribution capabilities, the quality of service and support, fuel consumption level, display requirements, leverage, and the company's position in the market. As for quality, Toyota enjoys an excellent reputation (see Figure 1).

THE COMPETITION

As of 1989, General Motors produced 7.6 million vehicles compared with Toyota's 4.6 million. This number includes cars, trucks, and vans. GM holds 16 percent of global car and truck sales. In the United States, Toyota ranks third, alongside Chrysler and Honda and trailing Ford. Ford had a market value of $23 billion with sales of $96.2 billion and profits of $3.8 billion in fiscal 1989. During the same period Toyota had global sales of $55.7 billion and earned $2.4 billion.

Beginning in the 1980s, Nissan made a commitment to a policy of internationalization that would alter its locus of production. The strategy called for 40 percent of sales going to the Japanese market, 30 percent to be exported from Japan, and 30 percent to be manufactured and sold abroad. Part of the production abroad strategy included a $735 million assembly plant in the United Kingdom. However, this project met with many obstacles. First was the European recession. Second, European governments were hostile to cars assembled in the United Kingdom with a partial Japanese content. Third, local

MODEL	COUNTRY	PROBLEMS PER 100 CARS SOLD
Toyota Cressida	Japan	63
Mercedes Benz 300 series	Germany	71
Toyota Camry	Japan	72
Toyota Lexus LS400	Japan	74
Mercedes Benz S-Class	Germany	76
Buick LeSabre	United States	82
Nissan Maxima	Japan	89
Nissan Infiniti Q45	Japan	91
Toyota Corolla	Japan	94
Mazda Miata	Japan	99

FIGURE 1 ***J. D. Powers Quality Index (for October 1990)***

content became a major issue; the government of the United Kingdom wanted 60 percent of the automobile's parts to come from local manufacturers initially and 80 percent within two years. Finally, Nissan experienced some financial reversals at home when it lost 2 percentage points of the Japanese market to Toyota.

EUROPEAN COMPETITION

Toyota's three major European competitors, in order by size, are Mercedes Benz, BMW, and Peugeot SA.

Edzard Reuter, head of Germany's Daimler-Benz, wants to bring his $48 billion conglomerate into the high-technology future. Reuter is shaking up Daimler's conservative, self-satisfied culture, insisting that every unit, from washing machines to jet fighters, innovate with microelectronics and new materials. Transforming Daimler is more complex than many corporate makeovers. In Germany, major industrial groups are interlocked in a web of large banks and government ministries that share board members and shareholders. Reuter's first challenge is to prepare Daimler's core business for the difficult times ahead in luxury automobiles. His strategy includes: rehabilitating diesel models, which have suffered from environmental backlash; revamping A-class models; and moving rapidly into smart cars that feature devices that enhance vision and warn about road hazards. In the United States, sales of Mercedes Benz fell 9.6 percent in 1989 to 75,000 units; the overall luxury import market in the United States declined only 5.5 percent. With $12 billion in cash at its disposal and its faithful, long-term shareholders, Daimler could succeed.

Bayerische Motoren Werke AG (BMW) flourishes by selling to people who feel that they are flaunting their affluence less ostentatiously and more intelligently than if they were driving similarly priced automobiles of other manufacturers. BMW's managers prefer talking engineering to talking marketing, but they are not ignorant of the importance of image. The rivalry between BMW and Daimler-Benz AG, the maker of the Mercedes, is intense. Both sell German craftsmanship and engineering, but BMW currently is outselling Mercedes-Benz in unit volume in most of the world's major car markets. Mercedes still is enjoying a lead in revenues, but its lead is narrowing. BMW has gained on its rival by introducing a lot of new, inexpensive models with slender, raking profiles. BMW, once content to bring new models to the market every decade, plans to introduce either a new model or a new engine every year. Unlike Daimler-Benz, BMW has modest diversification plans. In 1992, BMW announced it would build its first plant in the United States.

Peugeot SA is Europe's third largest automobile company. The company earned $2 billion in 1989, not solely because it makes good cars, but also because France has built a thick wall around its automobile market. Japanese imports are limited to 3 percent of France's sales of 2.3 million cars per year. In Spain, Portugal and the United Kingdom, the Japanese market share is also limited. In theory, quotas like those the French and Italians impose are not supposed to survive the European Community's (EC) 1992 program to forge a twelve-nation open market. In practice, the EC is having a hard time figuring out what to do about Japanese cars. This delay has given Peugeot's Calvert and other hard-line protectionists room to maneuver. Calvert wants the EC to maintain a 10 percent quota on Japanese cars at least until 2002, with continuation of tighter limits such as those in France. Western Europe is now the frontline of the global car wars. Aware of what happened in the United States, Europe's domestic producers are preparing to challenge the Japanese.

MANAGEMENT APPROACH

In a typical U.S. corporation, conservatism, incentives based on individual performance, and decentralized decision making prevail throughout the management hierarchy. These trends are particularly evident in rules and policies involving the delegation of authority and assignment of responsibility to subordinates. Managers are usually specialists with narrow job assignments in which functional relationships are of paramount significance. A rigid top-down approach similar to that in a military organization characterizes U.S. management style.

In contrast, Toyota managers relied on interpersonal skills, emphasized collectivism, employed a holistic concern for employees, and acted as facilitators and symbolic leaders in promoting group harmony within the organization. Group reward systems, such as quality control circles, were encouraged, and interdependence in sharing personal ideas, beliefs, and values was stressed. Managerial ranks were based on status within a unified hierarchy, and decisions flowed from the bottom or middle levels of management to top management. This management style reduced the adversarial us and them attitudes prevalent in the American management system. Also, an increase in promotions on a seniority basis, mutual consultation between low-level management and workers, and a narrowing of pay differentials among executives and non-managerial employees occurred, boosting morale and improving efficiency.

Toyota's executives are most comfortable managing things they can see from their office windows. As the company grew, it clustered most of its facilities around Toyota City in Japan's Aichi Prefecture to ensure prompt delivery of materials and equipment. This allowed the company to perfect ***kanban,*** or just-in-time, inventory delivery, especially as hundreds of parts suppliers are located there. In 1989, Toyota wiped out a layer of middle management, mostly by eliminating positions and cutting back the approvals needed for decisions. Employees no longer must address superiors by titles. The Japanese characters on all business cards are now written horizontally instead of vertically, a symbol of the effort to create a less hierarchical environment. All this is radical stuff for Japanese executives who took two years before agreeing with their United States subordinates that the leather interior in the Lexus LS400 should be wrinkled instead of smooth.

Global growth is forcing Toyota to decentralize, albeit methodically. Evidence of the new decentralization can also be seen in the recent successes of Toyota's Calty Design Research Center in Southern California. Besides expanding Calty and building in Brussels, Toyota has just opened a new design center in Tokyo.

Toyota's goal is to continue to develop a more flexible management structure to respond to the ever-increasing changes in the industry. Among those changes are a maturing of the industry, which means Toyota cannot continue to grow as quickly as before, and an aging of the Japanese population. These changes mean more bosses for fewer and fewer employees, an industrywide problem in Japan. Toyota's managing director Iwao Isomura said that "a U.S. company would respond to such top heaviness with layoffs, but in Japan that's not the way. So our challenge is to find a way to use everyone" (*Los Angeles Times,* October 16, 1989). This philosophy ties in directly with one of Toyota's prescriptive values, respect for humanity.

Both the positive and negative aspects of Toyota's strong corporate culture are visible in their products. The negative aspect—a strong hierarchical, somewhat limiting structure—is evident in the length of time Toyota used to take in making a new design decision. This time lapse put the company in the position of being a trend follower rather than a leader. On the positive side, when Toyota does enter the market, it makes quality improvements.

According to the *Advertising Age* report on the 100 leading advertisers for 1987, Toyota ranked thirty-eighth in advertising expenditures, with outlays of $272.9 million. The Japanese practice is to use a long-term advertising campaign, and Toyota is no exception. This strategy is helpful in an increasingly competitive industry, where image projection must be consistent. The Toyota approach has resulted from longer-term business plans, initially small budgets, and less pressure from dealers to rebound in slow times. Advertising with a consistent theme suggests that a company has clear direction for the future.

DISTRIBUTION

The Toyota dealership, of course, differs from its American counterpart. In Japan, there are only three or four demonstrator models on hand. Since most cars are manufactured to order, there is no need for vast parking areas for unsold vehicles. Moreover, there is no battle over the walk-in customer: the sales team is paid on a group commission. The Toyota customer who buys from the showroom will have his own sales representative, someone who will stay in touch with him and whom he can contact in case of problems. But the showroom purchase leaves the sales force with much more time to devote to so-called conquest sales—those motorists who already have a new car—all the while maintaining up-to-date information on customers and holding their loyalty. This is all part of what Toyota calls its aggressive selling strategy.

FINANCIAL POSITION

In 1990 Toyota enjoyed a net profit of 4.7 percent on sales of $64.5 billion. In 1991, sales were up 7.2 percent but net income declined by 2.2 percent on more than 4.5 million vehicle sales worldwide. Toyota is so well endowed financially that it makes more money on financial investments, including lending to other companies, than it does on operations. Jokingly known as the Bank of Toyota, it sits on $22 billion in cash, enough to buy both Ford and Chrysler at current stock prices, with nearly $5 billion to spare. (See the financial statements that follow this case.)

~ THE FUTURE

As of 1992, the prospects for an economic recovery in the United States were marginal and the growth outlook in the U.S. auto industry remained dimmed. At the same time, Europe was buoyed by the emergence of the European Community and the new economic interests of Eastern European countries such as Poland, Czechoslovakia, Hungary, and Russia. The question facing Toyota was whether it could be as successful in exporting the Toyota approach to auto manufacturing to Europe as it had been in the United States given the restrictive trade barriers and nationalistic spirit that was rising in the new Europe.

	1981	1982	1983	1984	1985	1986	1987	1988	1989
Assets:									
Cash and Deposits	$ 404,321	$ 462,866	$ 852,269	$ 2,697,893	$ 2,648,491	$ 3,756,059	$ 6,073,552	$ 8,584,014	$ 8,451,476
Receivables, Net	2,081,396	2,140,732	2,359,739	2,634,911	2,580,658	3,968,918	4,772,539	6,041,657	6,485,965
Market Securities	481,241	1,423,400	2,880,160	1,736,772	2,273,444	2,734,498	2,512,558	1,849,980	4,992,178
Inventories	176,738	206,360	980,286	1,114,178	1,233,270	1,625,113	1,762,224	1,967,813	2,321,015
Other Current Assets	40,913	56,253	746,441	796,816	1,022,435	1,541,594	2,111,154	2,663,311	7,915,827
Total Current Assets	$3,184,609	$4,289,611	$ 7,818,895	$ 8,980,570	$ 9,758,299	$13,626,184	$17,212,027	$21,106,775	$30,166,461
Property, etc., Net	2,283,733	2,415,509	3,567,046	3,524,241	3,566,644	6,302,329	7,765,926	9,122,759	9,906,495
Investments	1,354,339	1,362,488	2,104,307	2,943,697	3,812,235	6,318,221	8,032,864	10,651,419	9,546,213
Other	305,683	317,398	36,555	40,016	48,436	105,413	124,094	99,315	53,648
Total Assets	$7,128,364	$8,385,006	$13,526,811	$15,488,525	$17,185,616	$26,352,149	$33,134,911	$40,980,268	$49,672,817
Liabilities:									
Bank Loans	$ —	$ —	$ —	$ —	$ 578,649	$ 871,502	$ 969,701	$ 1,233,047	$ 2,460,306
Notes and Accounts Payable	1,289,829	1,353,703	2,469,609	2,397,040	1,845,100	2,812,314	3,134,147	3,879,906	4,068,352
Accrued Expense	389,815	416,247	546,235	960,648	1,057,322	1,823,506	1,991,033	2,318,500	2,984,452
Deposits	443,023	403,016	601,336	668,377	798,564	1,135,446	1,588,079	1,755,589	1,027,669
Accrued Taxes	258,336	530,726	932,092	927,843	1,152,056	743,289	989,730	1,957,187	1,317,263
Other Current Liabilities	9,791	67,625	326,727	333,856	350,585	355,740	721,358	857,953	1,743,225
Total Current Liabilities	$2,390,794	$2,771,317	$ 4,876,013	$ 5,287,766	$ 5,782,278	$ 7,741,799	$ 9,394,018	$12,002,182	$13,601,267
Retirement and Severance Benefits	$ 388,877	$ 430,165	$ 607,256	$ 668,037	$ 707,097	$ 1,159,638	$ 1,417,615	$ 1,693,676	$ 1,633,458
Long Term Debt	$ —	—	—	43,005	32,673	32,162	1,503,372	2,484,906	8,468,491
Other Liabilities	896	894	43,185	67,140	—	—	—	—	101,491
Special Reserve	40,368	31,324	60,088	—	67,360	101,604	105,027	98,299	107,157
Common Stock	345,098	394,706	508,000	508,003	509,839	807,864	906,787	1,011,561	1,300,815
Capital Surplus	288,136	626,959	629,450	629,452	601,645	907,937	1,019,113	1,135,712	1,415,160
Legal Reserve	86,275	89,706	130,861	130,863	127,459	201,966	226,697	250,847	250,426
Retained Earnings	3,587,920	4,039,935	6,671,966	8,154,272	9,357,261	15,399,201	18,562,301	22,303,132	22,794,583
Less Treasury Stock	—	—	—	16	—	DR 24	19	47	31
Total Shareholders' Equity	$4,307,429	$5,151,306	$ 7,940,265	$ 9,422,575	$10,596,206	$17,316,944	$20,714,879	$24,701,205	$25,760,953
Total Liabilities	$7,128,364	$8,385,006	$13,526,811	$15,488,525	$17,185,616	$26,352,149	$33,134,911	$40,980,268	$49,672,817
Net Current Assets	$ 793,815	$1,518,294	$ 2,942,882	$ 3,692,804	$ 3,976,021	$ 5,884,385	$7,818,009	$ 9,104,593	$16,565,194

FIGURE 2 ***Toyota Consolidated Balance Sheets, Nine-Year Summary (in thousands)***

Source: Moody's International, 1981–82, p. 2744; 1983–84, p. 2753; 1985–86, p. 2523; 1987–88, p. 2803; 1989, p. 2845.

	1981	1982	1983	1984	1985	1986	1987	1988	1989
Net Sales	$13,750,637	$15,096,252	$22,368,340	$24,827,618	$27,189,759	$40,280,265	$45,410,957	$54,254,121	$55,701,681
Cost of Sales	12,582,704	13,496,610	17,893,634	19,901,059	21,228,721	32,678,628	38,100,019	45,032,332	46,561,974
Selling and General Expenses	618,195	695,668	2,929,828	2,852,243	3,146,676	4,564,827	4,816,431	5,705,053	5,890,123
Operating Income	549,738	903,974	1,544,874	2,074,315	2,814,361	3,036,809	2,494,517	3,516,736	3,249,584
Interest and Dividend Income	274,978	278,156	376,710	403,170	484,452	785,942	849,398	1,036,482	1,646,604
Interest Expense	16,462	16,619	108,218	94,553	74,466	105,710	129,533	167,636	584,236
Other Income, Net	83,946	32,205	100,218	124,727	115,255	199,151	199,357	202,800	32,877
Income Before Special Items or Income Tax	892,200	1,200,716	1,913,584	2,507,659	3,339,602	3,916,193	3,413,739	4,588,382	4,344,829
Special Items	CR 14,574	DR 30,172	19,029	21,008	—	—	—	—	—
Income Tax	386,275	615,295	973,987	1,355,948	1,853,167	1,943,254	1,758,681	2,368,623	2,065,831
Minority Interest	—	—	—	—	—	6,465	—	11,108	6,797
Amort. of Consol. Diff.	—	—	—	—	—	264	(103)	(3,435)	CR 2,494
Equity in Earnings of Unconsolidated Subsidiaries and Affiliates	—	—	—	—	—	127,939	111,397	132,768	129,904
Net Income	$520,499	$555,249	$958,408	$1,238,692	$1,629,743	$2,094,146	$1,773,497	$132,768	$2,404,599
Earnings per Share*	$0.269	$0.277	$0.396	$0.512	$0.641	$0.785	$0.665	$0.848	$0.801
Brought Forward	100,236	118,744	5,842,609	6,671,967	7,794,043	14,120,957	17,284,818	20,516,227	20,599,420
Other Additions	—	—	—	—	—	—	—	—	3,389
Dividends	96,627	105,494	130,324	152,401	173,831	306,988	317,376	370,838	363,017
Legal Reserve	4,109	12,401	26,664	—	2,377	9,617	—	287	18,740
Transl. Adjustment	—	—	—	—	CR 53,609	DR 498,693	(174,646)	(200,307)	CR 189,711
Other Credits or Deductions	—	—	—	—	58,358	2,952	—	15,513	16,061
General Reserve	400,000	411,765	9,609	—	—	—	—	—	—
Directors' Bonuses	1,255	1,255	1,836	2,270	2,284	3,557	3,992	4,487	4,718
Carried Forward	118,744	143,078	6,671,966	8,154,272	9,357,261	15,399,201	18,562,301	22,303,132	22,794,583

*As reported, adjusted for retroactive effect of free distribution of shares.

FIGURE 3 ***Toyota Consolidated Statements of Income, Nine-Year Summary (in thousands)***

Source: Moody's International, 1981–82, p. 2744; 1983–84, p. 2753; 1985–86, p. 2523; 1987–88, p. 2803; 1989, p. 2844.

Index